STORM
WARNING

STORM WARNING

Gambling with the Climate of Our Planet

Lydia Dotto

Doubleday Canada Limited

Canadian Cataloguing in Publication Data

Dotto, Lydia, 1949-
    Storm warning: gambling with the climate of our planet

ISBN 0-385-25782-1

1. Global warming. 2. Climatic changes. I. Title.

QC981.8.G56D677 1999   551.6   C98-932861-9

Jacket photograph courtesy of First Light
Jacket design by Andrew Smith Design
Text design by Heidy Lawrance Associates
Printed and bound in the USA

Published in Canada by
Doubleday Canada Limited
105 Bond Street
Toronto, Ontario
M5B 1Y3

BVG 10   9   8   7   6   5   4   3   2   1

To my mother,

who has survived sixty Alberta winters

and

to my friends, Aranka and Joan,

who patiently endured two years of

listening to me talk about the weather.

CONTENTS

ACKNOWLEDGEMENTS

I owe a particular debt to Henry Hengeveld, Environment Canada's science advisor on climate change, with whom I had many valuable conversations and who was generous enough to read and comment on the draft manuscript of this book.

I'm also very grateful to the many other scientists mentioned in the book who provided me with information in personal interviews, via email or by sending me papers. Their comments and advice were invaluable. However, the responsibility for any errors is mine.

For their help in bringing this book to publication, I would also like to thank Jennifer Barclay, Greg Ioannou, Marjan Farahbaksh, Pamela Murray and John Pearce.

FOREWORD

There are few phenomena in the natural world more spectacular or mysterious than an approaching thunderstorm. The horizon fills with curtains of gray turning black. Bursts of thunder and veins of lightning split the sky. High above is a boiling panorama of clouds, black-shadowed, blue-shadowed, and glistening white. Pale streamers blow from their frosty heights.

Inside the thunderstorm is wild disorder, a convergence of furious forces that can be neither comprehended nor predicted. But one thing is certain. The storm will release its chaotic, non-linear energies on whatever is ahead and below. The temperature will drop. Water will fall in the form of rain or hail. Winds will blow and lightning will strike. People who find themselves in its path should assess the risks and quickly take reasonable precautions.

This is the theme that runs through this honest, enlightening, and sometimes frightening book.

In 1974, Lydia Dotto was assigned by the Toronto *Globe and Mail* to write an article on our undersea work in Canada's high Arctic. It was our fourth expedition and we were conducting a series of dives under the ice of the Northwest Passage to study human and equipment performance and marine biology. Lydia flew north to visit our camp on the ice. She was full of probing questions about our science and technology. Then she asked if she could make a dive. I explained that she would have to wear a bulky air-filled suit, put on heavy lead weights, and drop into a hole we had cut through six feet of ice. A writer has to know everything about her story, she told me, even if it means taking risks. She made two dives in the lethal, near-freezing water and wrote a wonderfully insightful article.

Lydia has taken a considerable risk in writing this book. It is a complex subject laced with controversy. It took two years of intensely hard work to research and write. It meant saying things that most people will not want to hear. But her reporter's instincts told her it was an urgent story about the future of our planet. "The most important question," she says, "is not whether global warming projections will come true but whether we—or more likely our children and grandchildren—will be prepared to deal with the consequences if they do."

For more than 25 years, Lydia has done a remarkable job explaining science to Canadians. The subjects of her numerous articles and books range from sleep to spaceflight and from computers to genetics. She has been covering the unfolding story of climate change for most of those 25 years. In the following pages she takes us on a compelling journey into the ocean of air that protects the planet, and shows us how we are raising its temperature. As well, she reveals the strange human behavior—in science, politics, and diplomacy—that keeps us in a constant state of denial.

It's all here: thoughtful answers to the major questions about global warming, detailed descriptions of extreme weather events, the multiple effects on human health and safety, and the steps we need to take to avoid and adapt to the oncoming storm. Lydia makes us realize that global warming is a threat we must take seriously, and that the most important concept to keep in mind is not *proof* but *risk*.

This is a timely and indispensable book. It compels you to think differently about the forces we are releasing into our overheated sky and how they will affect our future.

Dr. Joseph MacInnis
C.M., M.D., LLD (Hon)

THE STORMS
OF '98

It was dubbed "Beauty and the Beast."

The blanket of ice, several centimeters thick, certainly gave the trees, roads, bridges, and buildings a winter-wonderland quality. But beneath the beautiful exterior lurked a beast without mercy, one that brought much of the northeastern part of North America to its knees for several weeks and strained the social and technological resources of the most advanced industrial society in the world to a shocking degree.

On Monday, January 5, 1998, a huge mass of warm air laden with moisture barreled up from the Gulf of Mexico, causing severe flooding along the U.S. east coast, and slammed into a shallow layer of cold air hunkered down in the Ottawa and St. Lawrence valleys. Unable to displace the dense, cold mass, the warm air climbed on top and, as it cooled, dumped copious amounts of water into the frigid air below.

With temperatures hovering around 0°C—warmer than usual for the time of year—the result was freezing rain and lots of it. Because the cold air mass was shallow, the water didn't have time to make it all the way to snow; instead it became supercooled droplets that froze on contact with every exposed surface, creating layer after layer of a very tough, adhesive glaze. As Environment Canada senior climatologist David Phillips put it, "This is the hardest kind of ice—you can only break it with a hammer."

It would have been bad enough if the storm had simply dumped its moisture and moved on, which is the typical pattern for such weather events in Canada. But the large weather system that caused it persisted, dropping rain in three waves over five days. On Tuesday night, Ottawa's weather forecast was ominous: "A series of disturbances is approaching...bringing episodes of freezing rain, one after the other. By Thursday morning, 15 to 20 millimeters of freezing rain is expected to have fallen. Thursday into Friday remains to be seen—pray for plain rain."

Unfortunately, "plain rain" was not in the cards; Thursday night into Friday brought yet another wave of the frozen stuff. By week's end, the Ottawa and Montreal regions got a total of about 80 hours of rain, far in excess of the 45 to 65 hours they typically get in a whole year. An estimated 70 to 110 millimeters fell in parts of eastern Ontario and southern Quebec—possibly more in some parts of the "black triangle," an area south of Montreal that suffered massive power losses. Ultimately, the region got about two years worth of freezing rain in five days. It was roughly two to three times worse than the worst ice storms of the past, which had dropped about 25 to 35 millimeters of rain.

Several factors influence the severity of ice storms, including the amount of precipitation that falls, the duration of the event, and the extent of the area affected. This storm would have been extreme by any of these measures, but all of them together made it catastrophic. By Saturday, when the rain finally stopped, an enormous region stretching from the middle of Ontario to southern Quebec, east to the Atlantic provinces and south into the New England states had been coated with a deadly glaze measuring up to 9 centimeters thick in the worst-hit locations. This was more than double the thickness of ice deposited in the worst previous ice storms in 1986 and 1961.

Trees snapped under the strain, showering the ground with broken branches and wrist-thick shards of ice, the sound echoing through the streets like volleys of gunshots. Power lines, looped like icy garlands on a Christmas tree, hung to the ground until, burdened beyond bearing, they too snapped and tumbled into the snow, hissing and crackling with energy that had nowhere to go. It didn't take long for the crackling to stop, though, as utility poles and transmission towers began to crack and topple, littering the bleak winter landscape with piles of fractured wood and twisted clumps of metal. Like decapitated behemoths, crippled pylons stretched as far as the eye could see, resembling the abandoned relics of some otherworldly war.

The lights went out. The heat went off. At the height of the storm, more than four million people in four Canadian provinces and four U.S. states were left freezing in the dark as the electrical lifeblood of modern industrial civilization was relentlessly choked off by the ice. "The only person with power around here is Mother Nature and clearly we're no match for her," one Montreal resident told the CBC.

Power to the island of Montreal, the second largest city in Canada, hung by a thread; four of five transmission lines feeding high-voltage bulk electricity from James Bay hundreds of kilometers to the north were cut off. On Thursday night, with ice still accreting on every surface, Montreal came within a hairsbreadth of a total blackout. "It was a very, very close call," Quebec Premier Lucien Bouchard revealed later. "The line was watched …the whole night. We really thought we might lose Montreal." City officials said that Montreal had been on the verge of losing its water supply, as well as heat and electricity, and that firefighters had been prepared to bulldoze buildings if necessary to keep fires from getting out of hand.

By the second week, up to 100,000 people had moved into shelters; thousands more took refuge in hotels or with family, friends, or complete strangers who still had heat. But most people tried to tough it out in their frigid homes, with periodic visits to shelters or friends' homes for a hot meal, a shower, or a chance to get warm for a while. The elderly, distraught at the idea of abandoning their homes, often refused to leave until police were given extraordinary powers to force their evacuation if necessary. Some who left never returned—at least one heart attack was attributed to the stress of moving—but some who stayed met an equally unhappy fate, dying of hypothermia or carbon monoxide poisoning from using poorly ventilated heating devices. In the end, at least 25 deaths were attributed to the storm.

Hospitals and nursing homes struggled to care for their patients and to deal with the influx of storm-related injuries and illnesses, counting on overworked emergency generators to keep chaos at bay. In Canada, more than 14,000 soldiers were deployed, the largest peacetime military mobilization in the country's history. More accustomed to peacekeeping in far-flung war zones, they were glad to help their own people but astonished at the damage nature had wrought. A soldier who'd served in Bosnia commented that Montreal was just like Sarajevo "without the bullets"— although dodging falling ice projectiles couldn't be much of an improvement. U.S. Vice President Al Gore, touring the worst-hit areas of Maine, said the area looked like it had been hit with a neutron bomb.

It was hard to imagine that things could get worse, but then winter returned with a vengeance; in the second week, a cold front moved in, accompanied by heavy snow and high winds. In Quebec and Ontario, temperatures dropped into the double digits below freezing; in the Montreal area, where more than a million people were still without heat, nighttime temperatures plunged to −35°C with the windchill. Then it started snowing. Upstate New York and New England were clobbered with up to 45 centimeters of snow.

Utility companies in the region had an unprecedented crisis on their hands. More than 5000 utility workers were pressed into service, working around the clock in 16-hour shifts, battling freezing temperatures and blowing snow in a desperate struggle to patch together a fragile and badly crippled power grid. Utilities from all over North America sent people and equipment. (One linesman from Hawaii, who commented wryly that he'd been surfing just the day before, had to borrow clothing suitable for −20° temperatures.) In the first few days, when the rain just kept coming, it was a hopeless task. "We're fighting a battle we can't win right now," said one weary worker with an ice-encrusted face after repairing a line for the third time.

In the early stages, many people were unable to accept that technology would not quickly get the upper hand over nature. It took several days for the full import of the crisis to sink in. People who live in this part of the world are no strangers to vicious winter storms; they've seen snow and ice before. But this storm left them shell-shocked. Many who were camped out in freezing homes or shelters for weeks remarked how humbling it was to have been brought to such a state so abruptly. It had shaken their faith in both their own resilience and that of the social order.

It was clear by the end of the first week that the storm would likely become the most costly natural disaster in Canadian history. Early predictions that it would become the first billion-dollar loss for the Canadian insurance industry proved true. The total tab for property and business losses and repairing the power system is expected to exceed $2 billion.

For many, the most disconcerting aspect of the storm was that it took barely three days to cripple a power system that had taken five decades to build—a system that boasted half the installed hydroelectric capacity in North America. Hydro-Québec said its power lines had been built to the highest standard in North America, able to withstand ice up to 4.5 centimeters, nearly four times the national Canadian standard. Yet it was facing daunting losses: an estimated 40% of its distribution network would have to be rebuilt at an estimated cost of about $650 million. More than 30,000 hydro poles worth $3000 each had been downed, as had 130 major transmission towers worth $100,000 each and more than 800 smaller pylons. Hydro spokesperson Robin Philpott tried to put these numbers in perspective. "Never have we lost a tower before," he told the CBC. "This is beyond nature."

Montreal *Gazette* journalist Mark Abley wrote that from November to April, daily life in most of Canada depends on holding nature at bay. We pretend we've conquered nature and nature mostly "plays along, jabbing us occasionally in the ribs. You could even say that Montreal, like every other city in a harsh climate, is built on a foundation of denial." The ice storm cracked that foundation as surely as it did transmission towers and trees. This was perhaps a good thing, for in the immediate aftermath some people obviously started thinking ahead; companies that deal in alternative energy technologies like solar energy reported a jump in consumer interest. But it wasn't long before many people started saying that the technological infrastructure should have shielded them better from the wrath of nature or at least put them back in the game more quickly, regardless of the storm's unprecedented severity. Those who remained without power a month after the storm were clearly bewildered about why the system couldn't bounce back more quickly. "This is not a third-world country," complained one woman.

The ice storm would have been bad enough if it had been an isolated event, but it was only an early salvo in a barrage of bad weather that hammered many places in early 1998, thanks largely to El Niño, a periodic warming of the waters in the eastern tropical Pacific that affects worldwide weather patterns. In early 1998, both coasts of the United States were mercilessly pounded by a relentless series of storms that brought heavy snow and rain, freezing rain, high winds and waves, severe thunderstorms, and tornadoes. The Appalachians were buried in more than 120 centimeters of snow in two successive blizzards; at one point, about 300 cars were stranded on an interstate highway in Tennessee.

Southern states were hit with heavy snow, torrential rains, and high winds. In March, after an unseasonable warm spell caused fruits and vegetables to bud, crops were badly damaged when a vicious cold snap dropped temperatures below freezing. One TV report described the frost-encrusted plants as "cropsicles." Agriculture Secretary Dan Glickman called it "one of the strangest winters in decades."

Following the warmest and wettest January and February on record, the U.S. tornado season got off to an early and aggressive start. In late February, a dozen powerful tornadoes ripped through central Florida in the middle of the night, flattening hundreds of homes and businesses, overturning tractor-trailers, and capsizing boats. One boat ended up on the second story of an apartment building. The intensity of the tornadoes, with wind speeds reaching 330 to 400 kilometers per hour, was unusual for Florida. At least 39 people were killed and more than 260 others were injured. The conditions that created the tornadoes were quite rare; Florida was hit with as many twisters in one night as normally occur in three months in an El Niño year.

It was the most destructive and widespread tornado barrage in the state's history and the deadliest weather event to hit the region since Hurricane

Andrew in 1992. Insured damages were estimated to exceed US$100 million. The devastation taught everyone a lesson about El Niño's two-faced nature; while it had spared Florida a bad hurricane season the previous fall, what it delivered in the winter of 1998 was not a happy alternative.

In March and April, outbreaks of multiple severe tornadoes hit half a dozen eastern states. Powerful twisters cut a swath across the southeast, destroying or damaging thousands of homes and killing more than 100 people. In Alabama, a rare F5 twister nearly a kilometer wide with winds in excess of 400 km/h remained on the ground for 34 kilometers, barely missing the city of Birmingham. "There is really no place to hide from a storm of this strength," a local official told CNN, saying the tornado "picked homes up and wiped them completely off their foundations."

By mid-April, 1998 was already the most deadly tornado season in 14 years; 102 deaths occurred in the first three and a half months, about twice the normal average for January through April. In fact, these deaths exceeded the total *annual* count for any year since 1984. Tornado experts said that, while the number of tornadoes was not unusually high, more were hitting densely populated areas.

On the U.S. west coast, the name El Niño was mud—literally. California was hammered by week after week of storms that dropped record amounts of rain and whipped up ocean waves that ate away at cliffs and bluffs. They shed rocks and soil—and houses—into the ocean like glaciers shed chunks of ice. The heavy rains produced flash floods throughout the state and put 20 square kilometers of vineyards in Napa Valley under water. Rain-sodden hillsides collapsed in mud slides that took hundreds of homes with them; one half-million-dollar home was deliberately demolished to keep it from falling on another one below. Many owners engaged in an agonizing death watch, sometimes waiting weeks while their condemned and vacated homes teetered on the brink before taking the final plunge. One woman told CNN she had thought El Niño was "El Nonsense ...but I guess it's not."

Huge sinkholes cut gaping wounds in roads, highways, and overpasses, abruptly plunging vehicles and their unsuspecting drivers into torrents of mud and water; in San Diego, a 180-meter-long hole bisected an interstate highway on-ramp. Rivers of mud rolled through the streets, sweeping up cars, furniture, people, and everything else in their path. Thousands of cattle died from exhaustion caused by trying to wade through the deep, heavy mud. Meanwhile, higher elevations received huge amounts of snow; up to 90 centimeters fell in the Sierra Nevada mountains. Authorities warned that spring melting of the snowpack could cause flooding well into the summer.

By the end of February, many parts of California had received twice the normal rainfall for the season, and San Francisco and Los Angeles had beaten century-old records. Nearly two-thirds of the state was declared a disaster area and estimates of the damages topped US$500 million.

By April, hard on the heels of a very wet winter, much of the U.S. south was hit with record-breaking heat and drought. Florida experienced an unprecedented heat wave—24 days of temperatures above 35°C. Forest fires burned out of control in Florida, forcing 12,000 people to evacuate. More than 300 homes were burned. Timber losses and firefighting costs are estimated to be more than US$400 million. In Texas crop losses from drought were estimated at more than US$1 billion. By August, southern Texas faced a stunning reversal when torrential rains from tropical storm Charley caused extensive flooding. One town received nearly six times as much rain in one day as it had in the previous eight and a half months.

The U.S. National Climatic Data Center (NCDC) calculated that a 17-day heat wave in Texas, which would normally recur once in a thousand years in a climate that was not warming, would become a one-in-three-year event by the middle of the 21st century if global warming continues at its present rate.

The extreme weather events of recent years have certainly attracted public attention but it's still questionable whether they've prompted people to give serious thought to the implications of global warming. During the 1998 ice storm, the general reaction seemed to be that nature runs amok once in a while, and there's not much we can do about it, but fortunately nature only rarely runs badly amok. Quebec Premier Lucien Bouchard commented that it was not in the government's power to "issue orders to God. What happened never happened before." Although Hydro-Québec would look into new technology to prevent future power outages, "we'll never change the weather," Bouchard said. A news report about the debate over costly measures to bury hydro lines said consumers must decide whether to spend huge amounts to protect against events that "might only happen once a century."

The fact that a storm like this "never happened before" is not the issue; what's important is how often it might happen again. Can we really be sure such events will only happen "once a century"? Was the ice storm just one of those times when nature ran amok or was there more to it? Rather than saying, "we'll never change the weather," we should be asking if we're already changing it—whether these record-breaking extremes are linked to global warming caused by greenhouse gases we're emitting into the atmosphere.

Later in this book, I'll explore evidence that global warming will indeed increase the frequency and severity of extreme weather events.

And this is only one of many potentially serious effects it's likely to have: climate change is expected to affect everything from agriculture, water supplies, and food production to the spread of human diseases and conflict between individuals and countries.

The evidence is especially disquieting given that the nations of the world seem reluctant to tackle seriously the central problem of global warming: reducing greenhouse gases emitted by burning fossil fuels, our primary energy source. Unless we bite the bullet soon, the generations of the 21st century will likely face an even more daunting series of floods, droughts, storms, and heat waves than we've endured recently. The history books may well record that the 1990s were merely a dress rehearsal.

one

WHY WE'RE GAMBLING WITH THE CLIMATE

Imagine for a moment that you've been diagnosed with a serious medical condition. The diagnosis was difficult but the doctors think they've caught it early. The prognosis is uncertain: they can't fully predict what's going to happen, but they do know you face difficult choices because this condition gets worse before it gets better. As the patient whose fate is at stake, you naturally have questions.

Is there really a problem? How sure are the doctors that you have the condition? Admittedly it's difficult to pin down, but most experts believe, from their observations and tests, that it's far more likely you have the condition than not. They warn you it would be dangerous to assume the negative. There are some, however, who argue that the tests are too equivocal and imperfect to provide definitive *proof* that you have the condition. This is true, but the only way to obtain proof is to wait for debilitating illness to develop—and then it will be too late to do anything.

What will happen to me? There are many uncertainties and expert opinion is somewhat divided. Most experts believe the symptoms will be mostly unpleasant, but they are unsure how bad they'll be or when they'll occur. Some believe the condition will utterly destroy your quality of life, perhaps even kill you. There are others, however, who tell you that you'll hardly notice the effects, that there could even be some benefits from having the condition.

What should I do? There is a treatment. Should you start right away or wait for more definitive symptoms? There are up-front costs and there could be unpleasant side effects, but no one knows for sure how bad they'll be. Some predict the side effects of the treatment will be dire and overwhelming—worse than the disease itself. They recommend you wait, saying treatment might not even be needed. Others say the side effects will be negligible or at least manageable and that delaying treatment would be foolishly optimistic. If you don't start right away, you may never beat this thing.

What happens if you decide to postpone treatment, or at least reject the most aggressive treatment, and instead try to live with the condition? Will you be able to adapt? Improving your ability to handle the symptoms is a good idea and you might be able to cope quite well at first—provided you can afford everything required. Not everyone has the resources and, even if you do, there's no guarantee that your condition won't eventually overwhelm your resources. Moreover, there's real concern that if you put off essential treatment, some treatment options available to you now will disappear in the future. In fact, the condition could become so far advanced that no treatment will work and no coping will be possible. The most prudent strategy might be to start treatment while at the same time shoring up your ability to cope with the condition.

In many ways, these are precisely the questions and dilemmas we face in coming to grips with the risks of global climate change or what is more popularly known as global warming, the "condition" in question.

Is global warming really a problem? There's virtually no question that the earth's surface temperature has increased about 0.5°C over the past century, and that it's warmer now than it was during pre-industrial times. Nor is there any doubt that greenhouse gases warm the earth's atmosphere. These gases—primarily carbon dioxide (CO_2), methane, and nitrous oxide—are emitted by both natural and human processes, notably burning fossil fuels. There's no question that industrial processes, energy consumption, land use practices, and other human activities are adding greenhouse gases to the atmosphere. Most climate experts believe there's a very high probability that the warming of the past century has been caused largely by greenhouse gas emissions from human activities. In other words, they say, we've already seen the human "fingerprint" in global warming. There are skeptics, however, who question whether warming has occurred or, if it has, whether it's been primarily caused by human activities.

Computer simulations of future climate change indicate that if we continue emitting greenhouse gases as we are today, the climate will warm between 1 and 3.5°C, with a best estimate of 2°C, by the end of the next century. This is an amount and rate of warming unprecedented in the past 10,000 years and represents a significant climatic change over

a very short period of time. (To compare, there's only a 4 to 5°C difference between the earth's current temperature and the last ice age.)

What will happen as a result of global warming? A rapid climatic change
of this magnitude will undoubtedly have widespread environmental and
socioeconomic consequences, but there are still uncertainties about when
and where these impacts will occur and how bad (or good) they'll be.
There may be some benefits in some regions, such as lower heating costs
in cold climates or longer growing seasons for food crops. However, most
scientists expect the negative impacts to dominate.

One potentially serious consequence of global warming is an expected
increase in the frequency and severity of extreme weather events such as
heavy rains, floods, droughts, storms, and heat waves. There's considerable evidence that some types of extremes (e.g., heavy precipitation) have
already increased in some places. Global warming may also intensify El
Niño. El Niño has been getting stronger and lasting longer than ever
before, coincident with a period that includes most of the warmest years
since global record-keeping began in the mid-1800s.

There's also concern that heating of the oceans and melting of mountain and polar ice will cause a substantial rise in sea level that will threaten
highly populated coastal regions. Disruption of natural ecosystems and
the migration and extinction of plant and animal species are also
expected consequences of global warming. Possible socioeconomic
impacts include disruption of human activities dependent on natural
resources (e.g., forestry, agriculture, fisheries); major changes in freshwater lakes and rivers, with effects on water supplies; migration of pests and
diseases to new territories with effects on human health and food production; and increased loss of lives and property to weather-related disasters, as well as increased costs and psychological stress caused by more
frequent disasters.

What should we do about global warming and when should we do it?
Those who believe that global warming is already well underway and
bound to increase say we must immediately start cutting back greenhouse gas emissions by reducing our use of fossil fuels. Their urgency
stems from the fact that it takes the atmosphere and oceans a long time
to respond to variations in atmospheric concentrations of greenhouse
gases and the resulting temperature changes; any cutbacks we make now
won't greatly slow the warming trend for the better part of a century.
And if we don't start right away, the warming will get worse for a long
time and become increasingly difficult to turn around. There's even a
possibility that the climate may cross a threshold into a new state that
we cannot reverse.

The idea of immediate drastic cutbacks in greenhouse gas emissions makes many people uneasy, however. They fear it could be very costly in the short run and could slow economic growth—and it might be all for nothing if the projections of global warming turn out to be overblown. Others argue that many energy efficiency and conservation measures and the development of new "green" technologies could actually produce significant economic benefits. Thus, even if global warming projections turn out to be wrong, these measures will have been worth doing anyway.

What happens if we postpone cutting greenhouse gases and focus instead on adapting to the climatic consequences? This is a touchy subject among many environmentalists who argue that talk of adaptation is tantamount to giving up the fight to stop global warming and plays into the hands of those trying to forestall reductions in fossil fuel use. Others, however, feel that examining our adaptive capabilities is a matter of facing political reality. They say that while reducing emissions is the best "treatment" for global warming, few nations are prepared to make draconian cuts in their energy use. Some warming—perhaps quite a bit of warming—is all but inevitable over the next century, so we must examine whether we're prepared to deal with what it may bring.

The escalating social and economic costs associated with extreme weather in recent years provides strong evidence that we're not as well equipped to cope with weather-related disasters as we'd like to think. Even technologically advanced industrial societies have had trouble dealing with the barrage of record-breaking storms, floods, droughts, and heat waves, and costs have risen at an alarming rate. Less-developed societies have suffered even more terrible losses and they have fewer resources to cope with disasters. If, as predicted, global warming causes more severe or frequent extremes, the number of weather-related disasters and the associated socioeconomic costs may increase dramatically.

The nations of the world are currently pondering—and fighting over—what to do about global warming. In 1997, in Kyoto, Japan, they met to negotiate an international protocol that would, for the first time, set legally binding targets for reducing greenhouse gas emissions in developed countries. This was the culmination of a protracted political process that began before the adoption of the Framework Convention on Climate Change (FCCC) at the 1992 United Nations Conference on Environment and Development in Rio de Janeiro (the "Earth Summit") and continues today. Fifty-five nations have to ratify the Kyoto agreement before it will take force, and there are powerful groups lobbying against it, notably the multibillion-dollar fossil fuel industry and others in the business, academic, and political communities.

There are also serious divisions between industrialized and developing countries over the extent and timing of greenhouse gas cutbacks each should undertake. Developing countries resist binding targets, saying they must continue growing economically to rise out of poverty and that it's up to industrialized countries to make cuts first, since they've benefited economically from emitting most of the greenhouse gases to date. Developed countries argue that developing countries will soon overtake them as the world's major emitters of greenhouse gases and they must start cutting back if we're to make any progress against global warming.

This political wrangling makes it unlikely that radical emissions cuts will be the aggressive treatment of choice for global warming in the near future. Even if the Kyoto protocol is ratified and implemented, the reductions it mandates fall short of what's needed to halt global warming; at best, they will only slow it down and buy us some time. As a result, it's virtually certain that by the end of the 21st century, atmospheric greenhouse gas concentrations will be at least double the levels in pre-industrial times. Whatever consequences flow from that—however uncertain they may be at present—will be left for future generations to deal with.

A Question of Proof

While reading this book, the reader should remember that one of the great ironies of the global warming controversy is that so much of it has focused on the wrong question—specifically, whether there's *proof* that global warming is happening, that human activities are causing it, and that it will have adverse effects on human society and the environment.

This misdirection of the debate was not accidental. The emphasis on proof was the key strategy employed by global warming "skeptics" in a well-organized (and, so far, largely successful) effort to forestall adoption of policies to cut greenhouse gases. Major players among the skeptics include the fossil fuel industry; conservative politicians; some economists, especially those associated with politically conservative research organizations; a sizable segment of the business community and the business-oriented media; powerful lobby groups representing the fossil fuel and energy industries and oil-producing countries; some labour, agricultural, and consumer groups that fear the impact of energy cutbacks on consumer prices and jobs; and a small group of vocal scientists, only some of whom have credentials as active climate or weather researchers.

On the other side of the debate are environmental groups, environmentally oriented research organizations, segments of the media, and political leaders who champion greenhouse gas cuts. A sizable majority of scientific researchers who are experts in the atmospheric and environmental sciences also believe that global warming is real and that it presents potentially serious threats to the well-being of human society and natural ecosystems. Many of these scientists participate in national and international programs aimed at understanding the nature and extent of the threat and providing advice to policymakers. Some 2000 scientists from around the world contributed to the most far-reaching of these efforts—the United Nations' Intergovernmental Panel on Climate Change (IPCC). The IPCC has produced several reports summarizing the state of scientific knowledge about climate change and its impacts. The latest, a massive three-volume tome released in 1996, concluded for the first time that "the balance of evidence suggests that there is a discernible human influence on global climate"—a statement that provoked fierce attacks from global warming skeptics.

The scientific community has found itself uncomfortably caught in the midst of the political wars. Politicians, the media, and the public look to them for answers, but the complexity of the climate system precludes quick or simple solutions. Though many climate experts believe global warming is happening, no one claims to understand the full dimensions of the problem. Exactly what will happen, where and when, and how bad it will be are subjects of robust debate within the scientific community, and climate researchers acknowledge there are still many uncertainties in projections of how global warming will play out. In fact, they make these disclaimers so often and so doggedly that many nonscientists, especially the media and policymakers, find it exasperating.

The anti–global warming contingent, on the other hand, loves disputes among scientists. They use the talk about uncertainty to bolster their claim that there's no proof that global warming is real. Uncertainty is the weapon they use to stigmatize as "junk science" virtually all research that suggests negative impacts from global warming. Scientists who warn of serious social and environmental consequences, however cautiously they do it, are dismissed as fearmongers using scare tactics to increase their research funding.

For good measure, some skeptics also argue that increasing CO_2 in the atmosphere is really a good thing because it makes plants grow. Like canny lawyers trying cover all the bases, they argue: "My client didn't do it ...and we're all better off if he did." As far as the skeptics are concerned, global warming will always be a couple of horsemen short of an apocalypse.

To the average person, the argument that we should wait for proof before doing anything to cut greenhouse gases probably seems reasonable. After all, we like to think of "innocent until proven guilty" as a fundamental tenet of civilized society. Unfortunately, while this philosophy makes sense in a criminal justice system designed to punish past illegal behavior, it doesn't translate particularly well to protecting the earth's environment from future assaults. Basically, it addresses the wrong question. At issue here is not so much whether we know with certainty that a particular event will happen, but what we stand to lose if it does. In short, what do we *risk* by doing nothing to stop global warming?

This question is rarely addressed in public debate because the overwhelming focus on the question of proof has been brilliantly successful in diverting attention from the risks involved. We have no difficulty dealing with risk in other ways—we take prudent action as a matter of course to reduce our vulnerability to catastrophic loss. We do so not because we have *proof* that catastrophe will strike, but because *we do not want to risk losing everything if it does.* We protect ourselves against such losses even if we believe their probability of happening is low.

There's no *proof* that global warming will have adverse effects, but the *probability* that it will is very high—certainly as high as many risks we routinely protect ourselves against every day. If we're willing to invest in precautionary measures against catastrophic loss for the sake of our health, our families, our property, and our homes and businesses, why is it so difficult to do so for our planet, the life support system that sustains everything else?

The chapters that follow explore in greater detail the nature of this "condition" that confronts us, the side effects it may produce, and the disputes over how best to "treat" it—or whether to treat it at all. This is not the unabridged story of global warming, for no single book can fully embrace a subject that involves everything from chemistry, physics, and biology through economics, sociology, and politics to ethics, religion, and philosophy. This book should be viewed as a snapshot of a moving target. It explores some of the reasons why we're struggling to come to grips with an environmental problem that has potentially devastating environmental and socioeconomic consequences but entails many uncertainties about the exact extent, nature, and timing of those consequences. It's about our difficulty in processing a body of scientific information that is, paradoxically, both unmanageably voluminous and woefully incomplete. It's about how we've allowed ourselves to be lured into procrastination—indeed, a state of near paralysis— by our inability or unwillingness to deal with the realities of scientific uncertainty. It's about why we have in effect chosen, mostly by default, to gamble with the earth's climate.

two

WEATHERING
THE WEATHER

The images have become all too familiar: houses swept away by rising floodwaters, sliding down cliffs in tides of mud, going up in flames from raging brush fires, pummeled into matchsticks by hurricanes and tornadoes. Millions of people retreating from howling windstorms, being rescued from rooftops, treetops, and submerged vehicles, trapped in homes buried in snowdrifts, dying in sweltering heat waves, freezing in the dark. These relentless, ruinous assaults have left many people with the sense that something weird is happening to the weather, that the swings from one extreme to another are becoming more rapid and intense.

Climate scientists are quick to point out that they can't directly link any individual weather event to global warming, but they do say that many of the weather extremes we're experiencing are consistent with what would be expected in a warmer world. For example, it's expected global warming will increase the frequency of heavy precipitation events—not only rain and snow but perhaps also ice storms like the one in 1998.

There's growing evidence that many parts of the world are already experiencing increases in heavy precipitation, particularly during the past two decades—a period coincident with a rapid rise in the earth's surface temperature. El Niño too has increased in frequency and intensity during this period, becoming bigger, hotter, and more long-lasting than before. Scientists who've analyzed this odd behavior concluded it's highly unlikely to be due to natural causes, which raises the question whether global warming is, in effect, putting El Niño up to all the mischief it's causing. If so, intensifying El Niño may be one mechanism by which global warming will influence extreme weather worldwide. (See Chapter Four.)

Even a small sampling of the disasters in recent years gives a sense of what we may be up against in the future.

These events are notable for their intensity, the incredible amount of destruction they caused, and the fact that so many each year broke records decades and even centuries old.

Extreme Weather in Recent Years

Nineteen ninety-five was a year for wild weather all over the world. It was the hottest year in more than a century since record-keeping began in the mid-1800s (although it was later beaten by 1997 and 1998). Following record-breaking cold in April, Moscow experienced its highest June temperatures in a century, leading to outbreaks of malaria and cholera.

Parts of the U.S. and Canada experienced severe flooding. In California records were set for the heaviest rainfall in a 24-hour period and for peak river flows. Millions of people lost electrical power and some 10,000 homes were damaged or destroyed. The fruit and vegetable industry was damaged and virtually the entire state was declared a disaster area. Total losses amounted to about US$3 billion.

Severe flooding also occurred along the Mississippi River just two years after the "flood of the century" in 1993. This event, along with an outbreak of torrential rains, hailstorms, and tornadoes across the south and southwest, cost an estimated US$5 to $6 billion.

Extreme heat waves hit other areas of the U.S. Much of the east sweltered through humidex readings as high as 42 to 49°C in mid-July. In Chicago, about 700 deaths were attributed to the heat, roughly 500 occurring during one heat wave in July.

Asia, Russia, Europe, and Australia were all plagued by extreme heat. Japan experienced its worst heat wave ever. In India and Pakistan, more than 500 people died when heat and humidity pushed the effective temperature as high as 54°C.

Heavy summer rains inundated parts of China; the southern Yangtze River experienced record floods, which displaced three million people and killed 450. Other parts of China were plagued by drought; a 650-kilometer stretch of the Yellow River dried up entirely.

Most of Europe experienced severe drought. Britain endured one of its hottest and driest summers in at least two centuries. Then, in December, temperatures hit a record low.

The eastern U.S. had its second-driest summer on record and severe drought in the southern plains caused an estimated US$4 billion in damage. Many parts of Canada experienced prolonged dry spells as well, with Vancouver and Montreal setting records for days without rain. The hot, dry summer gave Canada its second-worst forest fire season on record. More than 3000 firefighters battled hundreds of fires simultaneously and firefighting costs soared to $250 million for the year. Five of Canada's worst forest fire years have occurred since 1980, and average annual losses over that period have increased 140% over the previous three decades.

June was the month of "hell and high water" in western Canada. While fires raged out of control in northern Alberta and Saskatchewan, torrential rains produced record flooding in B.C. and southern Alberta. The Alberta floods forced 4000 people from their homes and caused an estimated $100 million in damage. At the end of August, thunderstorms with winds up to 100 km/h flattened buildings in southern Saskatchewan and left behind knee-high piles of hailstones. Across Canada, storms caused $200 million in damage during the year and crop damage from hail added another $100 million to the tab.

The summer of 1995 was Canada's third warmest this century; June and July were the hottest in 101 years. Ontario and Quebec had the most uncomfortably humid summer on record; on July 14, the humidex reading in Windsor, Ontario, exceeded 50°C, a Canadian record. The hot, humid summer resulted in higher-than-normal visits to hospital emergency rooms by people with heart problems, asthma, and migraines.

The Atlantic experienced its second-most-active hurricane season since 1871. The season set records for the fewest days without storms and the highest number of days with ongoing hurricanes. From June through October, the storms traveled relentlessly across the Atlantic from Africa, so numerous that several bumped into each other, sometimes spiraling around like dancers in a deadly gavotte. Five hurricanes made landfall, causing more than US$8 billion in damage in the U.S., Mexico, and the Caribbean. Canada's Atlantic coast was clipped with high winds and waves from five of the storms. On average, this normally happens only once a year.

After a record-breaking summer, winter hit Canada early and hard. By November, Ontario and Quebec were suffering from record-cold temperatures and above-average snowfall. Ottawa got three times its normal snowfall in November and twice the average in December. Some cities in central and western Canada had exhausted their entire snow removal budget before the first official day of winter. In Europe, lack of snow in several countries threw World Cup ski competitions into chaos.

For most of the world, wet was the word for 1996. Rainfall records were set in southern Europe and parts of northern Africa and the Middle East;

large areas of Australia, New Zealand, and South America were much wetter than normal. China, Korea, Vietnam, Thailand, Indonesia, India, and Bangladesh all experienced devastating floods. The U.S. northwest set new precipitation records, while several eastern states recorded their highest levels of rain and snow. Canada had its wettest year since record-keeping began in 1948.

However, Great Britain, central Europe, and Russia were very dry; England and Wales experienced the third driest year in more than two centuries.

In the U.S., January and February were characterized by a disconcerting "seesaw" pattern of alternating extremes. The year started with the "Blizzard of '96," which broke a dozen snowfall records throughout the Appalachians and New England. Two subsequent snowstorms brought snow accumulations to record levels in several major cities.

A sudden warming following the blizzard was equally disastrous. Heavy rains and rapid snowmelt caused massive flooding throughout most of the east from the Gulf of Mexico to Michigan. More than 200,000 people were driven from their homes. More than 11,000 homes, 1500 roads and bridges, and 78 parks were seriously damaged or destroyed. Estimated combined losses from the blizzard and the flooding came to about US$3 billion.

In February, the eastern two-thirds of the U.S. was hit again with a severe cold snap, this one setting 400 records for daily minimum temperatures and 15 all-time-low readings. The cold reached far enough south to cause US$50 million in damage to Florida's vegetable crops.

Shortly after, another rapid warming occurred. Temperatures in the northern plains jumped 55°C in a week. Within 18 days, Tulsa, Oklahoma, broke both its record low and its record high for February. The Pacific Northwest was hit with severe flooding in February as a result of heavy rains and rapid snowmelt from unusually warm weather. More than 30,000 people were forced from their homes and many regions were declared disaster areas. All told, the winter of 1995–96 caused an estimated US$3 to $4 billion in damage throughout the U.S.

In Canada, the winter was long and grim. For five months, from November to March, most of the country experienced below-average temperatures; in February, most of western and central Canada was gripped by record-low temperatures of –40 to –50°C. Meanwhile, in Nova Scotia, temperatures climbed as high as 10°C, confusing local amphibians into thinking spring had arrived.

Snow followed the same extreme pattern as temperature: it snowed well before the first day of winter in 1995 and well after the first day of spring in 1996, and there were record amounts in many locations. Overall, November to March were among the coldest and snowiest months ever in Canada.

Heavy precipitation also caused trouble in summer. In July, the Saguenay region of Quebec was hit by a flood that was believed to be the worst it had experienced in 2000 years. A rainy summer had saturated the thin layer of soil in the region, so there was little forgiveness when rain started falling on the afternoon of July 19 and didn't let up for two days. Weather forecasters predicted this storm would drop 50 to 80 millimeters; in reality, 245 millimeters fell within 36 hours. With the ground saturated, the water ran into rivers and lakes, overflowing their banks. Some rivers expanded to 15 times their normal width, and for two days the water flow was 60 to 80% higher than the maximum that would be expected to occur only once in 10,000 years. Water levels in reservoirs rose 1.5 meters. Violent torrents crashed through more than 50 communities, damaging and destroying thousands of homes and 20 bridges. More than 100 landslides occurred, damaging roads and rail lines.

About 15,000 people were evacuated, including hundreds who were airlifted by the military. Five hundred houses were damaged, 3000 were totally destroyed, and 1000 families were permanently relocated. As Canada's first billion-dollar natural disaster, it was the most costly up to that time, but it would retain the title for barely 18 months, until the 1998 ice storm.

In China, the worst flooding in 50 to 150 years hit widespread areas bordering the Yangtze River from June through August. An estimated 20 million people were affected; two million were made homeless and an estimated 2700 died in the floods or landslides. The economic loss was estimated at US$26 billion, making this the most costly natural disaster of 1996.

For the second year in a row, the Atlantic experienced above-normal hurricane activity, making 1995 and 1996 the most active consecutive years on record. Two intense storms—Bertha in July and Fran in September—were the most destructive; they made landfall in the U.S., causing US$270 million and US$5 billion in damages. An even stronger storm, Hortense, hit the Caribbean in September, causing about US$127 million in damage. It also clipped Nova Scotia in mid-September, causing about $3 million in damage. It was the first hurricane to make landfall in Canada since 1975.

Much of the western half of the U.S. and Canada was hammered by wild winter weather in November and December, extending into January 1997. The mix included everything from blizzards and record-low temperatures to thunderstorms and high winds. The Pacific Northwest was hit by successive snow and ice storms that knocked out power to 300,000 people. Snow fell as far south as Texas. Record amounts of snow fell throughout western Canada; by late November, Edmonton had already received half its yearly average in snow.

In late December, the lower mainland of British Columbia was crippled by several days of heavy snow and freezing rain. Economic losses were estimated at $200 million. Victoria—a normally balmy coastal city that possesses neither snowplows nor a snow removal budget—was buried in an unprecedented 85 centimeters of snow. A total of 65 centimeters fell in one 24-hour period, breaking not only its own one-day record but those of more traditionally snowbound cities like Ottawa and Calgary. Cities and towns throughout southern B.C. were left with snowpacks that ranged from 10 to 55% above normal, creating conditions ripe for spring flooding.

The U.S. west coast also experienced heavy snow, rain, and freezing rain in December 1996 and early January 1997, leaving snowdrifts up to three meters deep in places. Unusually warm temperatures then caused rapid melting, which resulted in record flooding; one-in-a-hundred-year floods occurred in 18 locations. It was the third year in a row that the west coast had suffered severe flooding. Three-quarters of California was declared a disaster area. About 1200 square kilometers were submerged, 300,000 people displaced, and some 16,000 homes damaged or destroyed. As many as half a million people in six states were evacuated during the period and total damages were estimated at US$2 to $3 billion.

Across the Atlantic, a two-week deep freeze killed 200 people and stranded 15,000 tourists in Europe in late December and early January. Some European cities registered colder temperatures than those in Greenland. The cold snap extended as far south as Portugal, Spain, and Italy, where a thin sheet of ice formed on the canals of Venice. West of London, the River Thames froze for the first time in 50 years.

In 1996, natural disasters claimed 11,000 lives and caused an estimated US$60 billion in economic losses around the world, an all-time record. The number of events—600—was also a record; all were weather-related except for 80 earthquakes and volcanic eruptions.

The wild winter of late 1996 continued into 1997 without respite. Much of North America was hit with a succession of extreme and record-breaking events. In early January, the U.S. southwest experienced unaccustomed cold, snow, and freezing rain that decorated Arizona cacti with icicles. A huge blizzard engulfed the upper Midwest and north-western plains, forcing the closure of hundreds of kilometers of highways in three states. One man had to dig a tunnel to deliver food and supplies to his parents, whose house was buried to the roof in snow.

The winter's heavy snowfall, combined with heavy spring rains, created extreme flooding conditions throughout North America. A record-breaking April snowstorm dumped up to 90 centimeters of snow in the northeastern U.S. In the West, North Dakota and southern Manitoba were shut down by a vicious April storm with winds up to 85 km/h that dumped 60 to 100 centimeters of snow in areas that had already received more than 300 centimeters, two to three times the normal average for the season. The storm set 24-hour precipitation records in Manitoba and was the worst April storm there since 1876.

This set the stage for the worst spring flooding in more than a century along the Red River, which flows north into Lake Winnipeg. As thousands of people desperately manned sandbag brigades, the river rose inexorably for nearly two months, engulfing towns, cities, and acres of farmland from Fargo, North Dakota, to the southern boundary of Winnipeg. In Grand Forks, North Dakota, the downtown core was simultaneously submerged in icy, sewage-infested water and gutted by fires that could not be fought because flooding had destroyed the town's water-pumping system.

Meanwhile, 2000 square kilometers of farmland in southern Manitoba were covered by a huge lake—dubbed the "Red Sea"—dotted here and there by tiny islands of sandbagged farmhouses and towns. A 24-kilometer dike was desperately thrown up south of Winnipeg to keep the Red Sea from circumventing the city's multimillion-dollar floodway. The flooding caused an estimated US$1 to $2 billion in damages in the U.S. In Canada, insured losses and government compensation for the flooding amounted to an estimated $500 million.

The winter of 1997 just didn't want to give up. In April, mid-continental areas of the U.S. experienced record cold temperatures. In late May, Calgary got 15 to 25 centimeters of snow, while British Columbia basked in record warmth.

Between April and June, scientists started warning that an intense El Niño forming in the eastern Pacific was likely to cause severe weather conditions around the world in late 1997 and early 1998. From June to September, El Niño–enhanced rains caused flash flooding in the western countries of South America. Parts of central Chile received 10 times their normal annual rainfall in four days. In the fall, severe drought hit Australia, Indonesia, and Malaysia. In Indonesia, the monsoons failed completely, creating the worst drought conditions in 50 years. Fires, some deliberately set to clear land, burned out of control, creating unprecedented levels of smoke pollution throughout Southeast Asia, creating an estimated US$1.3 billion in damage.

The Atlantic hurricane season—initially predicted to be the third consecutive above-average year—was cut down to size by El Niño. There were only three hurricanes, and only one made landfall, causing US$100 million in damage. For the first time in more than 30 years, there were no hurricanes in August.

The Pacific saw much more cyclone activity than the Atlantic in 1997, as expected in an El Niño year. Hurricane Linda was the most intense cyclone ever measured in the eastern Pacific, with winds of almost 300 km/h. El Niño expanded the range of Pacific hurricanes in the eastern and north Pacific, moving their influence as far north as southern California and Arizona.

In the western Pacific, typhoons caused widespread devastation throughout Asia in the summer and fall of 1997. In early August, Typhoon Victor hit Hong Kong and southern China, causing about US$230 million in damage, while Typhoon Winnie hammered Taiwan, China, and the Philippines, damaging 65,000 square kilometers of farmland and causing $US2.2 billion in damage. Almost 800,000 people were evacuated from coastal towns and cities. In early November, Typhoon Linda hit Vietnam; an estimated 120,000 homes were damaged or destroyed and 4000 people were killed. In December, super-typhoon Keith hit the island of Guam with winds up to 380 km/h, the highest ever recorded.

In October, Hurricane Pauline caused extensive damage along the west coast of Mexico, including the tourist resort of Acapulco. Ten-meter waves, raging floods, and massive mud slides occurred; boulders the size of cars were swept down hillsides and Acapulco was buried in mud more than a meter deep in places. Vehicles were stacked on top of each other and dozens of bodies were washed up on tourist beaches; there were outbreaks of cholera and dengue fever. An estimated 400 people were killed and 300,000 in poor and middle-class areas were made homeless, although the luxury hotels weathered the storm quite well. Barely a month later, Hurricane Rick delivered a second punch to many areas still struggling to recover from Pauline.

In late 1997 and early 1998, El Niño turned weather patterns upside down. Florida and Mexico were hit with snow and temperatures as low as −19°C; Guadalajara, Mexico, had its first snowfall in 116 years—40 centimeters. Meanwhile, the normally frigid and snowbound Canadian prairies basked in record-setting warmth. Daytime temperatures were in the mid-teens—higher than those in Mexico City—and golfers flocked to courses normally covered in snow.

In Alberta, warm, dry conditions and an uncharacteristic lack of snow cover gave rise to unprecedented winter wildfires. In southern Alberta, a prairie fire swept across 100 square kilometers of grassland, destroying ranches, killing cattle, and leaving many more without winter feed. It was the worst natural disaster to hit the ranching area in 50 years.

In Britain and Ireland, powerful winds up to 145 km/h left tens of thousands without power on Christmas day.

As we've seen, El Niño carried its miserable legacy into the winter of 1998 without missing a beat. Peru was hard hit by El Niño–spawned torrential rains and its worst flooding in 50 years. In Ica, a desert city that normally gets almost no rain, a knee-deep river of muddy water swept through the streets, destroying thousands of homes. High temperatures and a break in the sewer system contributed to the spread of disease. In Cuzco, the main hydroelectric plant was submerged in 60 meters of water. More than 200 people were killed and 250,000 were made homeless.

The effects of El Niño were not all bad, however. Southern Ontario had its warmest winter in 60 years and its warmest February since record-keeping began in 1840. The average temperature for the month was almost 5°C above normal, breaking the previous record set in 1984. The coldest night in February was only –10°C, at a time of year when nighttime temperatures below –20°C are common. It was also far less snowy than usual; for the first time ever, the Niagara region had no snow at all in February, while Toronto received only 0.8 centimeters, well below the previous low of 4.9 centimeters set in 1976 and very much less than the average of 29 centimeters.

Europe was hard hit by winter storms in 1998. In early January, England and France experienced rare tornadoes and wicked gales with winds up to 185 km/h lasting several days. Up to 100,000 homes in England lost power. The roofs of houses were torn off and waves up to nine meters high were whipped up by the wind.

In northwest Australia, record rainfall in early 1998 caused extensive flooding. In one town, aircraft were turned away from an airport when runways were invaded by hundreds of kangaroos escaping floodwaters. Residents also had to watch out for crocodiles that had overflowed their river homes along with the rising waters.

After enjoying an unusually warm and snowless winter, parts of southern Canada and the U.S. Midwest were slapped with vicious snowstorms and record freezing temperatures during March. Heavy snow shut down the Trans-Canada Highway for several days. A cold snap extended all the way from the Canadian prairies to the U.S. southeast, threatening crops that had started budding during the earlier warm weather. The Calgary area was hit with its largest March snowfall in 113 years of record-keeping just a few days before the official start of spring. Southern Ontario, Quebec, and the Maritimes spent the first few days of spring digging knee-deep piles of snow.

In late March and early April, a sudden and dramatic spring thaw caused widespread flooding in parts of eastern Ontario and southern Quebec. Temperatures shot up as high as the mid-20s in places—15 to 20° above normal for the time of year—and rapidly melted ice and snow left behind by the January ice storm and heavy snowstorms in early March. High levels in Lake Ontario, the St. Lawrence River and many secondary rivers threatened thousands of homes throughout the two provinces; in some places, 100-year flood levels were exceeded. Many of those affected were battle-weary victims of the ice storm. "What's next— a drought? Locusts?" grumbled one beleaguered resident.

By spring, flooding was also a serious problem in Florida, Georgia, and Alabama. Florida received about 300% of normal rainfall during what is usually its dry season.

Throughout the summer, extreme flooding devastated many countries, particularly China, Mexico, India, Bangladesh, and Sudan. In India, 5000 villages were submerged. The worst flooding along China's Yangtze River in more than 50 years left more than 3000 dead and millions homeless. More than 9000 medical teams were dispatched to battle widespread disease outbreaks.

Swings between extremes were common in 1998. In March, the NCDC reported that January and February were the warmest and wettest in more than a century of record-keeping in the U.S. In July, they reported that while the northern half of the country continued to get exceptionally wet weather, parts of the south and southeast had experienced the driest April through July in more than a century. Then, in August, some of these regions experienced record rainfalls.

The litany of weather extremes is not confined to just these four recent years. After a period during the 1950s and 1960s that was relatively quiet, the number of extreme weather events started increasing in the mid-1970s, a period during which global temperatures were climbing rapidly. Because the 1982–83 El Niño created havoc, many people took serious notice when scientists predicted the 1997–98 event could be bad. In 1988, a severe drought in North America, which caused more than US$45 billion in economic losses, focused public attention for perhaps the first time on the possible long-term ramifications of global warming. However, that concern faded temporarily when a two-year cooling period occurred following the 1991 explosion of the Pinatubo volcano in the

Philippines, which blasted dust high into the atmosphere.

The events of 1993 again focused attention on the relationship between global warming and extreme weather. North America experienced a series of severe events—a massive spring storm/blizzard, summer drought, and heat waves, as well as extensive flooding along the Mississippi River and autumn wildfires in California—that caused US$20 to $28 billion in damage. Of that, US$15 to $20 billion was attributed to Mississippi flooding, while the March storm/blizzard, which spread record-breaking snow, sleet, tornadoes, low temperatures, high winds and waves, and storm surges across more than a thousand kilometers of the U.S. eastern seaboard from Florida to New England, added about US$6 billion to the tab. The total volume of snow was estimated to be equivalent to 40 days' outflow of the Mississippi at New Orleans. Twenty-five times the size of Hurricane Andrew, it was the largest storm of the 20th century and the most costly non-tropical storm ever in the U.S. Its death toll of 270 was three times that of hurricanes Hugo and Andrew combined.

These events have primed people to consider the links between global warming and weather extremes. "While extreme events have always happened from time to time, how can you explain so many occurring within the space of a few years?" asked Henry Hengeveld, Environment Canada's science adviser on climate change. "How many one in one-hundred-year events can you have in a row before you concede that this can no longer be considered natural?"

The Significance of Extreme Weather: Increasing the Stakes

Whether the extreme weather events of the past two decades have been caused, even in part, by global warming that has already occurred, these events have provided valuable lessons about our vulnerability to weather extremes and their devastating consequences. Comfortable assumptions about our coping ability have been sorely tested—assumptions about the strength of the shield provided by modern technology and the swift and sure functioning of the emergency response capability of a modern industrial society. All of these weather stories comprise a cautionary tale; they force us to examine the implications of a future in which global warming may bring the challenges of extreme weather more often into our lives. They force us to confront four major issues:

• Global warming is not just about warming.
• Extreme weather probably will be the agent through which global warming will affect us most.

- We must ask hard questions about our vulnerability to weather extremes and reexamine assumptions about our ability to cope with them.
- The possible relationship between global warming and weather extremes greatly increases the risks associated with a delay in reducing greenhouse gas emissions.

Global warming is not just about warming. Perhaps nothing has done more to undermine our understanding of the threat posed by climate change than the name by which it is most widely known: global warming.* Unfortunately, this name has engendered a widespread but misguided belief that what we're facing is a slow, steady rise in temperature, something that many people frankly find difficult to view as an unwelcome prospect.

The perception that global warming isn't such a bad thing is prevalent among the very people who are being called on to take the lead in combating it—those who live in the industrialized mid- to northern latitudes of the Northern Hemisphere. Whenever blizzards and deep freezes put in their inevitable appearance, there's much lamentation along the lines of "where's global warming when you need it?" and suggestions that scientific theories about climate change must be so much hot air. David Runnalls of the International Institute for Sustainable Development once gave a talk about global warming during the depths of an Ottawa winter. With a –40°C windchill and the Ottawa canal resembling "the wind tunnel from Siberia, it was a tough sell," he admitted.

The media are particularly fond of jocular dismissals of global warming in the midst of cold spells—which only demonstrates that they are either unable or unwilling to understand the complexities of climate change. These simplistic sentiments are often tapped by politicians and lobbyists who oppose measures to reduce greenhouse gas emissions. "Let us pause, especially those of us who have lived in a northern climate, to give thanks to the greenhouse effect," remarked Preston Manning, leader of Canada's right-wing Reform Party, during a debate in the House of Commons. A representative of the oil-producing province of Alberta, Manning has been a foe of greenhouse gas cuts.

Comments about the benefits of global warming put one in mind of the old adage about being careful what you wish for. If global warming

---

* An Environment Canada poll found that 45% of Canadians used this phrase to describe climate change, followed by 37% who used the term "greenhouse effect." Few people use the term "climate change."

were to manifest itself *only* as a slow, steady rise in temperatures, it might well have positive side effects in cooler climates. If the negative effects occurred slowly enough that we could adapt, we could probably handle them fairly well—at least in industrialized countries. But slow and steady may not be what's in the cards. "Global warming means something more than just earlier springs and slushier winters," says Hengeveld. "It will also likely mean an increase in the number and intensity of unpredictable storms and weather surprises."

Until recently, few people seemed aware of this dark side to climate change. So there's a need to educate the public that one of the most important things about global warming is that *it's not just about warming*. It's about increasingly large swings in the weather, back and forth between hot and cold, wet and dry. Coping with intense and unpredictable is altogether a different matter than dealing with slow and steady.

Extreme weather probably will be the agent through which global warming will affect us most. Most of the public's awareness about global warming has focused on the rise in average global surface temperature that has occurred this century. This has diluted understanding of the true potential impact of climate change. While trends in global averages are useful to scientists as indicators of change occurring in the climate system, they're not directly relevant to our day-to-day functioning. What is relevant, what we *experience* in our daily lives, is not climatic averages but *weather*: relatively short-term and localized phenomena such as a thunderstorm or hailstorm sweeping through within an hour or so, or hot and humid conditions lasting several days, or rock-bottom temperatures for a couple of weeks, or an ice storm that just won't quit. Weather can swing dramatically between extremes; it can be here one minute and gone the next, and a hundred kilometers down the road it can be something altogether different.

Climate is far less of an in-your-face phenomenon, being an aggregate of weather conditions over larger regions and longer periods of time: seasons, years, decades, centuries, millennia. It's often described as "average weather." The climate changes far more slowly than the weather, in ways that are more subtle and harder to detect. To the layperson, this "averaging" of weather tends to blunt the true impact of climate change. It masks the swings between extremes, as well as the significant differences in weather that may occur in different places. If there's a lot more rain than normal in one place and a lot less rain than normal in another, averaging the two sets of figures suggests that everything is "normal." But for people living in those regions, things are not "normal"—they have to contend with very troublesome weather.

Extreme weather events, says Hengeveld, "may become one of the key determinants of whether climate change is dangerous or not."

Unfortunately, scientists cannot predict the effects of global warming on local and regional weather conditions with any precision. This is one of the great dilemmas and frustrations of climate science, because it's precisely these local and regional scales at which the impacts of global warming will be most significantly and directly felt.

We must ask hard questions about our vulnerability to weather extremes and reexamine assumptions about our ability to cope with them. Even though scientists cannot definitively prove that the extreme weather events of recent years were caused by the global warming that has taken place so far, they say that many of these events are consistent with what's expected to happen in a warmer world. "Recent events offer examples of the kind of situations that would be expected to become more frequent," says Thomas Karl of the NCDC.

It's important, therefore, to examine how well we've coped with these events and, more importantly, how well we're likely to cope if they become more severe or frequent. It's one thing for society to deal with a one-in-one-hundred-year event if it actually happens only once every hundred years; it's quite another matter to cope with—and recover from—the chronic disruption that would occur if such events started happening every few years or decades.

The relentless march of billion-dollar weather events during recent years has given us a sense of what might be in store as the planet heats up. For many, it's been a sobering, even shocking, experience. These events don't just cause problems while they're happening; they also knock out infrastructure, which takes time and money to rebuild. They drain social, economic, and psychological resources. They weaken the social safety net, leaving it more vulnerable to the next assault.

As Kevin Trenberth, head of the Climate Analysis Section of the U.S. National Center for Atmospheric Research (NCAR) put it, "Past experience is not as good a guide to the future as we've assumed. People are beginning to realize that we have to go back and question a lot of assumptions. The evidence is coming together quite well that increasing intensity of weather should have a higher profile as part of global warming."

The possible relationship between global warming and weather extremes greatly increases the risks associated with a delay in reducing greenhouse gas emissions. One of the biggest controversies associated with climate change concerns whether we should act immediately to curb greenhouse gases, even though there are still uncertainties about how widespread or serious the effects of global warming will be. The fossil fuel industry and many business and political leaders advocate a wait-and-see approach before taking actions they claim could have crippling economic impacts.

On the other hand, environmental groups and many scientists argue that climate change is one of those slow-building problems that will get worse before it gets better; the longer we delay taking action, the more serious the effects will be and the longer they will last. Even if we don't yet know precisely what will happen, they say, we know enough to reasonably conclude that what we're doing to change the climate could have serious, perhaps even catastrophic, consequences. Indeed, many scientists have characterized what we're doing as an uncontrolled experiment on the global climate system—an experiment that we can't easily shut down if we find we don't like the way it's turning out.

This debate will be dealt with in subsequent chapters. What's important to note here is the role played by extreme weather events. Simply put, they increase the stakes. If we had good reason to expect that global warming would occur only as a slow, steady rise in temperature, the wait-and-see advocates would have a stronger case. But the probability that global warming will increase the frequency and intensity of extreme weather events changes the odds. It greatly increases the *risks* associated with a wait-and-see approach. It means we run a greater chance of losing the gamble inherent in such an approach. If we allow global warming to proceed unchecked and if this in fact brings about more weather extremes as expected, we'll be in a world of trouble that won't go away for a long time.

There's good reason to believe we're facing increasing weather disruptions and disasters. While we don't yet know exactly where and when these events will hit, nor exactly how bad they will be, such ignorance is hardly cause for equanimity.

three

TEMPERATURE RISING:
Is the Earth Getting Warmer?

More than 170 years ago the French scientist Jean-Baptiste Fourier first suggested that the earth's atmosphere functioned like the windows of a greenhouse to keep the planet warm. Although the analogy is not precisely correct, the image stuck and gave rise to the phrase "greenhouse effect," which so many people now use to describe the process by which atmospheric gases trap heat rising from the earth's surface and prevent it from escaping into space.

The greenhouse effect is the ultimate good news/bad news story. The good news is that life on this planet owes its very existence to a natural greenhouse effect. It keeps the earth about 33°C warmer than it would otherwise be—on average, about 15°C instead of –18°C. Even the ice ages were only about 4 to 5°C colder than today's average temperature.

What keeps the temperature within a range that allows abundant life to flourish is the presence of several naturally occurring gases in the atmosphere—water vapor, CO_2, methane, and nitrous oxide. They allow roughly three-quarters of the sun's energy to pass through the upper atmosphere to the lower atmosphere and the earth's surface. Some of this energy is reflected, but most of it is absorbed by the atmosphere and the earth. This causes heating, which sets in motion the air and ocean currents that redistribute energy around the globe and, in the process, cause the day-to-day weather that makes our lives so interesting.

If this was all there was to it, however, the earth would just keep getting hotter. It avoids this by radiating energy back into space. But the
outgoing radiation is at longer wavelengths (infrared radiation or heat)
than the incoming energy and it can't get through the atmosphere as easily; greenhouse gases trap most of the energy radiated at the longer wavelengths and prevent the earth's heat from dissipating rapidly into space.
This is why the planet remains warmer than it would if the atmosphere
didn't contain these gases. Although the actual process is more complicated—for example, clouds both reflect and absorb energy in complex
ways—fundamentally, greenhouse gases act like a blanket on a chilly winter's night, holding heat inside and keeping the cold at bay.

Perhaps the most impressive thing about greenhouse gases is that they
exert such a powerful influence on the climate while making up less than
1% of the volume of the atmosphere, which comprises mostly nitrogen
(78%) and oxygen (21%). It's precisely *because* their natural concentrations are so minuscule that relatively small added doses can potentially
cause significant climatic effects.

If the *natural* greenhouse effect is the good news, the *enhanced* greenhouse effect is the bad news. It refers to the way in which human activities
are boosting the natural greenhouse effect, causing the earth to warm more
than it would from natural processes alone. The greenhouse gases of greatest concern are CO_2, methane, and nitrous oxide (produced by industrial
activities, energy consumption, and land use), plus industrial chemicals
called halocarbons (used as spray can propellants and refrigerants).

It's the role of greenhouse gases *from these human activities* that we're
primarily concerned about when we talk of *global warming*, the *greenhouse effect*, or *climate change*, even though, technically, all are terms that
can be applied to both natural and human-induced effects. Scientists
often add the qualifier *anthropogenic* when discussing climate change
caused by human activities, but most people use these terms to mean
human effects only.

The Framework Convention on Climate Change (FCCC) defines climate change as a change "attributed directly or indirectly to human activity that alters the composition of the global atmosphere and which is in
addition to natural climate variability observed over comparable time
periods." This definition was also adopted when referring to *future* climate change by the Intergovernmental Panel on Climate Change
(IPCC), a United Nations–sponsored organization comprising international climate and policy experts, which prepares periodic reports summarizing the findings of climate change research all over the world.

The terms *global warming* and *climate change* are often used interchangeably but they don't mean precisely the same thing. Technically, global warming refers to an increase in the average temperature of the planet. Climate change is more encompassing, including changes not only in temperature, but precipitation, wind patterns, and other aspects of climate and weather, including changes in the variability of climate and changes in extreme weather events. It seems paradoxical to hear of cold snaps and blizzards being blamed on global *warming*, so it's important to recognize that, in common usage, the term does not refer solely to rising global temperatures, but rather to the complex effects of rising temperatures on the intricate tapestry that is the earth's climate and weather system.

The difference between the concepts of global warming and climate change is more than just semantics. The ubiquitous use of the phrase "global warming" in popular communications has done a considerable disservice to the climate change debate; by focusing attention on rising temperatures, it's masked the true complexity and most significant dangers associated with climate change, which relate more to how fast the climate is changing and the increasing frequency and severity of extreme weather that may be associated with it. Understanding these complex and often counterintuitive consequences of climate change is among the most serious challenges we face in dealing with this problem.

That complexity is apparent when we examine how the climate system works. It involves every component of the earth—not just the atmosphere, but the oceans, land masses, ice and snow, plants and animals, even rocks and soil. And, of course, humans. The intricacy of the links and feedbacks between these natural systems is staggering and, despite centuries of research, many of their secrets remain elusive, which is why predicting future climate change is so difficult.

How greenhouse gases cycle through this maze—where they come from, where they go, and how long they stay there—is key to understanding their effect on climate. There are numerous *sources* that release greenhouse gases into the atmosphere, some natural and some human in origin. Several processes (e.g., chemical reactions or absorption) remove greenhouse gases from the atmosphere into what are known as *sinks* (e.g., the ocean and terrestrial plants and soils). The amount of greenhouse gases in the atmosphere at any given time—their *concentrations*—represents the balance between what's released (*emissions*) and what's removed. The *lifetime* of greenhouse gases refers to how long they remain in the atmosphere before being removed. Different elements of the climate system interact in complex ways, which can create *feedbacks* that may either accelerate or slow down climatic changes.

Among the most significant features of greenhouse gases is that they remain in the atmosphere for a long time—decades to centuries, and even longer in the case of some halocarbons. The atmospheric lifetime of CO_2, the most plentiful greenhouse gas produced by human activities, is roughly 70 to 100 years. For CO_2, the most significant sinks are the oceans, plants, and soil. Plants, which use carbon to build tissues, remove CO_2 from the atmosphere relatively quickly, as do the surface layers of the ocean. Less rapid is the transfer of carbon into soil, deeper layers of the ocean, and ocean sediments. In many ecosystems, the amount of carbon stored in organic "litter" in the soil is far greater than that stored in above-ground plants.

When carbon is locked away from the atmosphere, it's said to be *sequestered.* However, sequestration is not necessarily permanent; when living matter dies and decomposes, or is burned, carbon stored in biological tissues is again released. This is why deforestation is considered a source of greenhouse gases; it contributes to increasing CO_2 concentrations both because it reduces the number of trees sequestering carbon and because burning and decomposition of trees releases sequestered carbon into the atmosphere.

Methane is the second most important greenhouse gas produced by human activities such as agriculture and landfill operations. It has a warming effect 21 times greater than CO_2 but, at present, there's much less of it in the atmosphere and it's removed faster by chemical processes. Nitrous oxide is also less plentiful than CO_2 but has a comparable long lifetime; it's broken down by the sun's energy.

Industrial chemicals called halocarbons are present in even smaller concentrations. Some have been banned because they destroy the ozone layer in the upper atmosphere (the *stratosphere*) which protects life on earth from the sun's damaging ultraviolet radiation. However, some halocarbons are far more potent heat-trappers than the three main greenhouse gases so their climatic effect is significant despite their low concentrations. And because they're removed from the atmosphere so slowly, they're expected to affect the climate for thousands of years.

Their climatic influence is complex, however. As greenhouse gases, their direct effect is warming, but they also *indirectly* cause cooling by destroying ozone in the stratosphere. Ozone is a form of oxygen that's also a greenhouse gas, so when it's destroyed, this reduces the warming *it* causes. Worldwide, stratospheric ozone levels have been dropping since the 1970s because of destruction by halocarbons (primarily chlorofluorocarbons or CFCs). In the Antarctic "ozone hole" that appears each fall, ozone levels are 50–70% lower than they were in this region during the

1960s. (Ozone also exists in the lower atmosphere or *troposphere*, where it's a pollutant, a constituent of smog and a health hazard—and it causes warming there, too.)

Greenhouse gases are not the only thing that affects global temperatures. Another important player is *aerosols,* tiny particles created by both natural processes (dust storms and volcanic eruptions) and human activities, including burning fossil fuels and biomass (plants and trees). Most of these particles, especially those containing sulfur, are believed to cause cooling by scattering the sun's energy and also by altering the properties of clouds (which themselves have complex and poorly understood climatic effects, contributing to both warming and cooling). But some aerosols, such as dark soot particles, absorb sunlight and cause warming. To further complicate matters, some scientists now suggest sulfate particles may also cause some warming by reflecting infrared radiation back to the earth's surface.

Aerosols from volcanoes, the major natural source, can have an impressive climatic impact, reducing incoming solar radiation by as much as 5 to 10%. The eruption of Mount Pinatubo in the Philippines in 1991 blasted huge amounts of ash and gaseous sulfur dioxide high into the atmosphere, creating particles that circled the earth, scattered the sun's energy, and caused a two-year global cooling that temporarily interrupted what might otherwise have been an unbroken string of record-breaking hot years during the 1990s.

A key issue in climate science concerns the extent to which cooling caused by aerosols offsets warming caused by greenhouse gases. It's difficult to evaluate because, while greenhouse gases are fairly evenly mixed throughout the atmosphere, aerosols are more concentrated near their sources. In some places where aerosol levels are particularly high, they appear to have more than counteracted greenhouse gas warming on a local scale. On a global scale, scientists estimate very roughly that aerosol cooling may have offset or "masked" perhaps a third to half of the greenhouse warming that has occurred in the past century. This isn't particularly good news, however, since aerosols also create pollution, reduce air quality, and cause environmental damage and human health problems, notably respiratory ailments.

There's another reason why aerosol cooling is not an antidote to global warming: particles generally stay in the atmosphere for only a few days before being removed by precipitation (the exception being those created by violent volcanic explosions, which can sometimes remain aloft for several months to years). This quick removal explains why aerosols are not spread evenly worldwide, but are concentrated near their sources; they simply don't have time to travel far before they're washed out of the atmosphere.

This shorter atmospheric lifetime means that, even though we keep creating aerosols, they don't accumulate as they would if removal mechanisms didn't exist or worked more slowly. And if we reduce or stop aerosol emissions, their atmospheric concentrations drop rapidly in response. This is not the case with greenhouse gases; since their sinks operate more slowly, they accumulate for much longer periods. This has several implications for our efforts to combat global warming:

- Even if we stabilize greenhouse gas *emissions* (i.e., don't increase the amount we're putting into the atmosphere each year), atmospheric *concentrations* will continue rising because the gases are not being removed as rapidly as they're being added and thus will accumulate. To reduce *concentrations*, we must cut emissions considerably.

- Even if we greatly reduce or stop greenhouse gas emissions, atmospheric concentrations will not stabilize or drop for some time because it takes carbon sinks a long time to remove the gases that have accumulated. It could take a century or more for greenhouse gas concentrations to respond fully to cuts in emissions.

- Because of the time lag in the response of greenhouse gas concentrations to emission reductions, their warming effect is expected to overwhelm the cooling effect of aerosols, *even if aerosol emissions continue to increase.* If aerosol emissions are reduced, their masking of greenhouse warming will be weakened accordingly.

Climate Research: Searching for the Smoking Gun

Figuring out what's happening to the climate is among the most daunting scientific challenges humanity has faced. Although scientists have been studying the climate for centuries, it's only within the last two decades that major national and international research programs were started to study global climate change and its impacts. The first World Climate Conference was held in 1979 and led to the creation of the World Climate Program, sponsored by the International Council of Scientific Unions (ICSU), the World Meteorological Organization (WMO), and the Intergovernmental Oceanographic Commission (IOC) of UNESCO. These programs coordinate global research focusing on the atmosphere and oceans, the extent to which climate and climate variability can be predicted, and how human activities are affecting the climate.

Many countries also have national research programs. For example, the U.S. established its US$1.8 billion Global Change Program in 1990 with the goal of "observing, documenting, understanding, and predicting global change; assessing the consequences of these changes and the vulnerability of human and ecological systems to their potentially adverse impacts; and developing the tools and capabilities to conduct integrated assessments to synthesize and communicate this body of knowledge."

Canada has several research-oriented programs, many involving collaboration with university researchers. Environment Canada maintains a data-gathering network and archive with more than 150 years of climatic data. Another program provides researchers with remote-sensing data from satellites. Canada also has one of the most advanced computer models in the world, which is used for projecting future climate changes. The Climate Research Network (CRN) involves several groups in Canada, primarily in universities, that do research used to improve Canada's model. The Canadian Climate Program (CCP) coordinates research on climate change, impact studies, and data gathering and is also Canada's link to international research programs like the World Climate Program and the IPCC.

Under the umbrella of these programs, thousands of scientists around the world have been studying what happened to the earth's climate in the distant past, what's happening to it now, and what may happen to it in the future.

The past: Understanding past climate change is important for two reasons. First, it improves our understanding of how the climate system works, how it changes, and what causes it to change. Second, it provides a benchmark for comparison with the present and the future. After all, implicit in understanding change is knowing both before and after.

Since humans weren't around to monitor prehistoric climatic events, scientists examine what are described as the earth's "natural archives." These are vast "libraries" whose records of past climates are preserved in trapped air bubbles and microscopic fossils. For example, ice cores obtained by drilling deep into glaciers, some more than three kilometers long, contain trapped greenhouse gases, atmospheric dust and particles from volcanic explosions, and microorganisms that are indicators of climatic conditions in the atmosphere and oceans through the ages. They show a clear correspondence in the past between natural increases in atmospheric greenhouse gas concentrations and warming temperatures.

A similar kind of "proxy" data is found in the fossil remains of dust, pollen, and microscopic life forms in sediments collected from the bottoms of lakes and oceans. Since different species of tiny plants and animals responded in distinctive ways to local changes in temperature and precipitation, they tell a tale about climates past. Similarly, the skeletons of tiny sea creatures found in coral reefs provide an annual record of changes in ocean temperatures and local precipitation, while the spacing and chemical composition of tree rings provide a timeline of local changes in temperatures and rainfall. (Tree rings generally are spaced wider in warm years than they are in cool years.)

The work is painstaking, according to marine geologist and climatologist Thomas Crowley of Texas A & M University. The natural records are scattered across the globe and "can be recovered only piece by piece, and reassembled like a jigsaw puzzle, through the efforts of many people over many years."

Proxy data have enabled scientists to construct a relatively accurate climate history of the earth's cycles of heating and cooling. One intriguing aspect of this research, a field known as *paleoclimatology,* is what it tells us about how suddenly the earth's climate can change and how much of a nudge is required to precipitate rapid, dramatic climatic shifts. This helps us to understand the potential risks associated with the "nudging" we're doing today.

The present: While some scientists search for messages in the past, others deploy modern technology to monitor what's going on with the climate today. Modern data collection is relatively recent; temperature monitoring around the world began around 1860 and, since then, the network of land- and sea-surface monitoring sites has greatly expanded. By the 1940s, balloons were used to measure upper atmospheric temperatures and toward the end of the 1970s, satellites joined the network. In 1958 the first continuous and reliable measurements of atmospheric CO_2 concentrations started in Hawaii. Today there's a global network of instruments on land, at sea, in the air, and in space gathering data on temperatures, precipitation, wind speeds, greenhouse gas concentrations, and other important climatic variables.

Scientists are also studying changes in natural ecosystems that provide insights about the impact of climate change. Plant and animal species are moving and dying, glaciers are retreating, permafrost is melting, polar ice shelves are collapsing, insects, pests, and diseases are extending their ranges. These and other worrying changes, which are generally consistent with the environmental consequences expected from global warming, provide a growing body of indirect evidence about the potential risks of climate change.

Given the earth's size, and the fact that three-quarters of it is covered with water, comprehensive environmental and climatic data collection remains a daunting task, even with ships, aircraft, balloons, and satellites at our disposal. There are still large parts of the world, especially in the

Southern Hemisphere and over the oceans, where data are sparse. Sometimes, scientific debates arise over the accuracy of the data because of problems with sensors and instruments used to take measurements. However, as these technologies have been improved and more widely distributed, the information has become more comprehensive and reliable. Trying to take the earth's temperature doesn't come cheap, though, and environmental monitoring and climate research have been among the victims of government cutbacks in the 1990s. The IPCC notes that the decline in the data collection network threatens our ability to monitor the earth's changing temperature. Yet, we've never needed this information more; among other things, it's essential to validate the tools we use to project future climate change—computer models.

The future: Because climate models are the most powerful method we have for exploring the probabilities of future climate change—and because they've become the focus of intense political debate—it's important to understand how they work and what they can and cannot do. Climate models are complex computer programs that attempt to describe the relationships, interactions, and feedbacks among components of the climate system (atmosphere, oceans, ice and snow, land masses, etc.). They use thousands of mathematical equations to represent the physical laws of nature governing these interactions as well as data about these relationships obtained from observations in the real world. Models differ in complexity, some incorporating only limited components of the natural world or dealing with only one or two dimensions. The most sophisticated are three-dimensional general circulation models (GCMs), which run on supercomputers and are used to project future climate changes.

In essence, global models provide scientists with a "virtual" earth that can be manipulated at will. Once a model has captured the major components of the current climate system, it can be "run forward" to see how the climate evolves, both under natural conditions or if human-induced changes are added. Models are used for these experiments only when they're able to simulate with considerable accuracy the major features of today's climate; the more accurate this simulation, the greater trust that can be placed in the model's projections of future climatic changes.

Models allow scientists to ask "what if" questions, to simulate how temperatures, humidity, precipitation, wind speeds, and other elements of the climate system would respond to, for example, increases or decreases in atmospheric concentrations of greenhouse gases. If the models' equations could perfectly represent every interaction of the climate system down to the tiniest detail, they'd be very close to technological crystal balls. But that's not possible; scientists don't know enough about all those minute details to write such a complete and unerring set of equations and, in any event, the computing power needed to run such a model would be almost unimaginable. Still, the models have been greatly refined since the late 1960s, when they first began to produce fairly reliable climate projections. Many deficiencies have been identified and improved and different models now give consistent answers to the same questions, a measure of their increasing accuracy.

At present, models correctly depict many large-scale features of the current climate, such as seasonal temperature cycles, geographical variations, and temperature changes that occur vertically in the atmosphere. They can reproduce atmospheric fluctuations in the Pacific similar to those caused by El Niño, as well as some large-scale aspects of El Niño's global effects. They correctly replicated the global cooling caused by the particles ejected into the atmosphere by Mount Pinatubo. One study found that a phenomenon predicted by the models—the fact that plant growth would show a delayed response to warming and cooling episodes such as those caused by volcanic eruptions or El Niño—was confirmed by observations. The models also accurately represent some features of ancient climates deduced from proxy data. They have, for example, replicated the climatic changes resulting from past natural increases in greenhouse gas concentrations and aerosols.

However, there are still areas that need improvement. Models more accurately represent phenomena that extend over large areas—the entire globe or the hemispheres or continents—than they do smaller-scale events. Clouds are particularly troublesome because they're smaller than the smallest resolution of most models. And improvements are needed in describing the complex relationship between the atmosphere and the oceans. Because the oceans play a major climatic role, separate models of the ocean and the atmosphere have been combined to produce "coupled" GCMs to study their interactions.

It's particularly important for models to accurately reflect *feedbacks* that influence the extent and direction of climate change. A feedback is a response to a change in the climate that can either accelerate that change (a positive feedback) or slow it down (a negative feedback). Once a change is set in motion, a positive feedback tends to keep things moving in that direction while a negative feedback tends to move things in the opposite direction. A potentially serious positive feedback could occur if global warming causes melting of ice and snow in polar regions. Ice and snow reflect far more of the sun's energy than land or oceans do; in fact, regions of the Antarctic reflect about 90% of incoming solar energy, three times the global average. If melting exposes underlying land or ocean surfaces, this would cause increased absorption of solar energy at the surface, which in turn would increase warming that would further accelerate the melting of ice and snow. Most climate models indicate this positive feedback will occur with global warming, but its strength is uncertain. Another important feedback involves water vapor, the most powerful natural greenhouse gas. A warmer atmosphere holds more moisture and this would increase its tendency to trap heat from the earth's surface, thus accelerating further warming.

Despite recent improvements, climate models remain the target of relentless attacks from global warming skeptics. They argue that projections of future climate change can't be believed because the models are terribly flawed and thus don't accurately reflect what's happening in the real world. They note that when the models were applied retroactively to the climate of the past century ("backcasting"), the warming that occurred within the models was greater than what occurred in the real world, based on observational temperature data.

It turns out that cooling caused by aerosols and depletion of stratospheric ozone accounts for much of this discrepancy. Earlier climate models didn't account very well for aerosol cooling but when these effects were included, the models' ability to replicate past and current features of the climate greatly improved, showing results similar to the observed changes over the past century.

In assessing the accuracy of models in predicting future climate change, it should be remembered that flaws can come from the real world rather than the computer world. When models disagree with observational evidence, it's generally taken as evidence that something's wrong with the models but, given the limitations of data gathering in the real world, the discrepancies can sometimes result from inadequate data.

Models tell us what *could* happen under given conditions, but they provide no guarantees about what *will* happen. This doesn't mean the information is useless; models help us to assess the range of possible outcomes that flow from current and anticipated future behavior and they tell us what's likely to happen if we change that behavior. It's true that some data are uncertain—that's why it's wise to run a series of projections using high and low assumptions. But this is far better than not having even the slightest idea where we stand or what effect our actions—or inaction—are likely to have. This is precisely why we try, despite all the difficulties and uncertainties, to model the climate.

Today, model projections form the foundation of our awareness and concern about future climate change. But they're not crystal balls. They don't provide *proof* of what the future holds wrapped up in a comforting money-back guarantee. It's unreasonable to expect this; yet that's precisely what many people do seem to expect. It's what global warming skeptics have tried, with considerable success, to convince us we *should* expect. It's these unrealistic expectations that have catapulted climate modeling into the midst of a fierce political controversy that has diverted attention from the real issue raised by their projections: the urgent need to deal realistically with the *uncertainties* and *risks* associated with global warming, rather than wasting time waiting for scientists to meet an unachievable standard of certainty. (See Chapter Twelve.)

Evidence for a Warming World

By the end of the 1980s, climate scientists were generating enormous amounts of data from monitoring the real world and modeling it in computers. In 1988, the United Nations Environment Programme and the World Meteorological Organization established the Intergovernmental Panel on Climate Change (IPCC) with a mandate to assess the scientific information about climate change and its environmental and socioeconomic impacts and to formulate strategies to address what humans were doing to the climate. The objective was to offer relevant scientific information to policymakers around the world.

The IPCC has produced several major documents: its First Assessment Report in 1990, two intermediate reports in 1992 and 1994, and its Second Assessment Report in 1995. The first report played a critical role in negotiations leading to the creation of the 1992 Framework Convention on Climate Change. The second report, which summarized the current state of the science of climate change and its environmental and socioeconomic impacts, was the backdrop for negotiations through 1996 and 1997, culminating in the climate protocol signed in Kyoto in December 1997. A third assessment report is due in 2001.

The assessment reports, which synthesized peer-reviewed scientific studies from around the world, were put together by scientists and representatives from governmental and intergovernmental agencies and environmental organizations. (Peer review is a process by which papers submitted for publication in scientific journals are referred to other experts in the same field for comment, criticism, and recommendations for improvement. Even the most reputable scientists cannot publish papers in a refereed journal without undergoing this process.) An Environment Canada report described the IPCC reports as "the most authoritative statements on climate so far produced by the scientific community.

[They] were written and reviewed with the help of more than 2,000 scientific experts from more than 70 countries. There was a high degree of consensus among most scientists on the basic aspects of the climate change problem." As we'll see, none of this saved the IPCC from becoming the target of attacks on its scientific integrity and fodder in a highly charged international dogfight over climate change policy that escalated between the Rio Earth Summit in 1992 and the Kyoto conference in 1997.

The IPCC reports addressed several key questions: Is the earth getting warmer? If so, to what extent are human activities responsible? Will climate warming continue and how great will it be? What impact will future climate warming have on the weather, natural ecosystems, and human society?

We have better answers for some of these questions than for others. Some aspects of the climate system are well understood—for example, the fact that greenhouse gases trap infrared radiation from the earth's surface and thus heat the planet. Others are far less certain—for example, the role played by clouds and oceans. Equally complex, and even more uncertain, is the response of natural ecosystems and human societies to climate change. So, as we progress toward the ultimate question—what, if anything, should we *do* about climate change?—we must face the fact that, in the end, *we'll be forced to make decisions in the face of incomplete knowledge and substantial uncertainty about what the future holds.* This is precisely what makes climate change one of our more significant challenges, for if we knew better what was going to happen, our choices would at least be more obvious, if not necessarily easier.

But this dilemma is hardly unique to climate change. Being forced to make decisions in the face of uncertainty is pretty much standard operating procedure for most things in life. We gather the best evidence we can find, weigh the potential costs and benefits of different approaches, factor in how much risk we're willing to take, assess our options in light of our values and ethics, and then cross our fingers and get on with it.

Dealing with climate change should be no different. And the first step is examining the evidence.

Is the earth getting warmer? There's a growing and persuasive body of evidence that it is. The ice on Canadian lakes is breaking up about a week earlier than it did thirty years ago. Arctic permafrost is thawing in summer and polar sea ice is not as plentiful as it once was in winter. On mountaintops, glaciers are melting, receding, disappearing. Massive chunks of Antarctic ice shelves are breaking off. Sea level is rising. Over much of the Northern Hemisphere, spring is arriving earlier and the growing season has lengthened. Plants appear to be growing faster in spring and summer. In North America, Siberia, and Scandinavia, the tree line is pushing northward into the tundra. In the Alps, plants are climbing higher up the mountainsides. Canadian forests are experiencing record-setting losses to wildfires.

Along the Pacific coast of North America, species are on the move and whole ecosystems are being shaken up by the expansion of warm waters caused by the record-breaking El Niños of the 1980s and 1990s. Birds, fish, and marine mammals are faced with disruptions in the ocean food supply. Zooplankton, tiny sea creatures at the bottom of the food chain, have declined 80% off the California coast in the last 40 years.

Fish species are migrating. Warmer waters have encouraged the northward migration of mackerel, which prey on threatened cold-water Pacific salmon. Tropical fish are turning up far north of their usual haunts; in 1997, fishermen in Washington State were astounded to find themselves pulling in marlin, never before seen in their waters. The disappearance of fish and squid from warming waters off California has put sea lions and seals in dire straits; by late 1997, the sea mammals, with skin sagging from starving frames, swarmed onto California beaches and died in droves.

Tropical birds in Costa Rica are migrating to higher elevations, following the cloud cover driven up mountainsides by warmer temperatures below. Amphibians are spawning earlier with the early arrival of spring. In the Pacific and Indian oceans and in the Caribbean, coral reefs are being "bleached" to death by spiking ocean temperatures. In Antarctica, small shrimp-like creatures called krill have declined about 90% since 1980, victims of the loss of their food source, algae, because of melting sea ice. Some Antarctic penguin populations that feed on krill have been cut in half. Mosquitoes that carry dengue fever and yellow fever are now found at higher elevations than ever before in Central and South America. Outbreaks of pneumonic plague, Hanta virus disease, and cholera have accompanied unusually hot weather in Asia, India, and the U.S.

None of these events individually proves conclusively that global warming is happening. Some, notably those occurring in the Pacific Ocean off North and South America, are most likely related to El Niño—but, then, El Niño itself has recently become more intense and long-lived and some scientists suggest this could also be a symptom of global warming. (See Chapter Four.) However, taken together, this evidence can't easily be ignored; it defies logic to dismiss as pure coincidence so many events happening all over the world that are consistent with the projected consequences of global warming.

However, we need not rely only on the indirect testimony of krill and coral, seals and salmon, or mosquitoes and marlin. The global temperature record goes back to about 1860 and it clearly shows that the earth's surface temperature is rising. The last two decades of the 20th century have been the warmest on record, including 12 of the hottest years since record-keeping began—1981, 1983, 1987 through 1991, and 1994 through 1998, which became the 20th consecutive year in which the annual average temperature exceeded the "benchmark" against which comparisons are made (the average temperature between 1961 and 1990). The hiatus in 1992–93 was due largely to the cooling effects of the Pinatubo volcanic explosion in June 1991. Even with this cooling, however, the 1990s have been hotter than the 1980s.

The record set in 1997 got some help from El Niño, which warmed Pacific waters in some locations up to 6°C above normal. Otherwise, 1997 wouldn't have beaten 1995. When ocean and land temperature were considered separately, scientists found that the average ocean temperature in 1997 was the warmest on record, but the average land temperature was not, although it was within the top five. At a news conference, Tom Karl of the NCDC, which has the world's largest collection of weather and climate data, said 1997 would have made the top 10 warmest years even without El Niño's help.

In fact, 1997, along with 1995 and 1990, was warmer than any year since 1400, according to researchers at the University of Massachusetts, who reconstructed average annual temperatures going back 600 years using data from tree rings, coral, ice cores and historical records. They concluded that greenhouse gases were the dominant cause of warming in the 20th century.

The story in 1998 was even more dramatic. In October, the NCDC reported that *each* month of the year had set an all-time global surface temperature record—an unprecedented pattern that had persisted for 17 months. Until July, both land and ocean surface temperatures remained "far above both last year's record high levels and all other years." Remarkably, ocean temperatures during the first half of 1998 set records despite the waning of El Niño; they were actually warmer than at the same time in 1997 when El Niño was building rapidly. (By fall, however, they'd cooled below 1997's record levels, due to La Niña's influence.) Land temperatures from January through September were particularly hot, exceeding the long-term mean temperature between 1880 and 1997 by a record-breaking 3°C. "During the past few decades, global temperatures have persistently broken previous record highs every few years, but never to the extent observed in 1998," the NCDC noted.

The warming of the past two decades is part of a longer trend dating back to the latter part of the 19th century. Since then, the earth's mean surface temperature, based on both land and sea measurements, has climbed by about 0.3 to 0.6°C, with a best estimate of 0.5°C. More than half of this increase has occurred within the last three decades.

To most people, this may not seem like much; after all, we experience temperature changes far greater than this all the time—from summer to winter, from day to day, even from day to night. But while large temperature swings can occur locally over short periods of time, when you average temperatures, both hot and cold, from around the world over a year, you get much smaller numbers. This doesn't refute the fact that it takes a tremendous amount of energy to warm the average surface temperature of the *entire earth* by what seems to be a small amount. In fact, the warming of 0.5°C since the late 1800s is greater than any that has occurred in the past 600 years, according to the IPCC, which notes as a further point of comparison that it's unlikely that global mean temperatures have increased by 1°C or more during a century at any time within the past 10,000 years.

The observational record of warming during the 20th century is buttressed by proxy data, such as tree ring data from forests in North America, Europe, and Russia. A study of rings from 300- to 500-year-old trees in remote areas of Mongolia confirms that this previously undocumented region was unusually warm during the 20th century relative to past centuries. The 10 highest growth intervals in the rings occurred after 1920 and the largest 25-year growth period occurred between 1944 and 1968.

On a global scale, the earth's temperature did not climb upward steadily during the past century. The two sharpest and most sustained increases occurred between 1910 and 1940 and then again from the mid-1970s to the present. The period between 1940 and the mid-1970s was characterized by fluctuations—slight cooling in the 1940s, leveling off and then rising again in the 1950s, dropping in the 1960s, and leveling off again until the mid-1970s when the current, very sharp, climb began.

This mid-century "plateau" in the warming trend occurred when CO_2 concentrations were still climbing significantly, so the question arises whether its existence contradicts the global warming theory. Certainly, skeptics argue that it does. However, most climate scientists say this is not the case because the warming trend caused by the buildup of CO_2 from human activities was still in its early stages and had not yet had much effect. Kevin Trenberth of NCAR said it was only in the late 1970s that greenhouse warming started to emerge from the background of natural variability.

The earth's temperature tends to wobble up and down naturally on time scales of about a decade or so because of natural forces not only in the atmosphere but in the oceans, including events like El Niño and periodic temperature oscillations in the North Atlantic. Small changes in the sun's energy output may also have played a role in the temperature trends during the first half of this century. The warming caused by greenhouse gas emissions would be superimposed on these natural dips and swells, sometimes being reinforced by them and sometimes being counteracted. In the early stages of global warming, it would not be surprising that these natural wobbles might be large enough at times to mask the emerging greenhouse warming. It would be like the path of the stock market during this century—a roller coaster ride that trends generally uphill but with corrections and crashes along the way.

There's also aerosol cooling to consider. It may be no coincidence that the plateau occurred just after World War Two, according to Jerry Mahlman, director of NOAA's Geophysical Fluid Dynamics Laboratory at Princeton University. During the industrial expansion after the war, there was a sharp increase in the use of fossil fuels that emitted increasing amounts of both CO_2 and sulfur dioxide, the precursor of sulfate aerosals. The atmosphere reacted to each one in a markedly different way. The pulse of new aerosols was very large compared with the amount of aerosols already in the atmosphere; it may have doubled existing concentrations, Mahlman said. As a result, the atmosphere reacted immediately and strongly by cooling off.

On the other hand, the CO_2 entering the atmosphere was only a small fraction of the natural CO_2 already there. "CO_2 was being slowly added to a big bathtub; it was like filling it up a drip at a time," said Mahlman. It would take the atmosphere the best part of a century to respond significantly to this injection of CO_2. "With sulfates, the immediate effect is very sharp, so in the short term, cooling dominates." Mahlman's modeling studies indicate the post-war aerosol pulse is a likely explanation for the mid-century leveling-off in global warming.

Other scientists, such as University of Toronto geography professor Danny Harvey, have reached similar conclusions. His models show the best fit to observed temperatures of the 20th century when greenhouse gases, aerosols, and solar factors are all included. "The fact that the cooling occurred in no way undermines our view of the climate's sensitivity," he said.

Wallace Broecker, a noted paleoclimatologist at Columbia University, has said that while we can't eliminate the possibility that greenhouse warming has been held back by a natural cooling, neither natural nor human-induced cooling can continue to compensate for long. Natural cooling will give way to natural warming and since aerosols are short-lived compared with greenhouse gases, "the aerosol cooling will soon be eclipsed."

Some skeptics continue to question whether the global warming trend is real. In fact, they argue that the earth has actually cooled over the past two decades, based on two claims: first, that satellite data since 1979 show a slight cooling in the lower atmosphere (troposphere) and, second, that the surface temperature data is incomplete and also skewed by the fact that many measurements are taken in large cities that generate a lot of heat, thus creating a "heat island" effect. They contend these data cast doubt on the reliability of computer models used to predict future climate change.

The debate over the satellite data is highly technical, focusing on issues such as problems with sensors, instruments, and the satellites themselves, and the fact that satellite records are only two decades long—so short that they may have been biased by transient events like volcanic eruptions and El Niño.

The data that caused this brouhaha indicated that the atmosphere a few kilometers above the surface cooled about 0.03 to 0.05°C per decade since satellite monitoring began in 1979. Even though the surface warming trend is expected to lessen with altitude, these findings were inconsistent with climate model predictions and they became pivotal to the skeptics' claims that the models are so flawed that their projections cannot be trusted. On the day that NOAA scientists announced that 1997 had broken another global temperature record, the Science & Environmental Policy Project (SEPP), run by global warming skeptic Fred Singer, issued a statement headlined: "1997 Registers on the Cool Side, According to Satellite Global Temperature Data." It argued that satellite data, described as "the most reliable and only global temperature data," indicated that 1997 was "among the coolest years since satellite-based measurements began in 1979.... Satellite readings continued to show the slight downward trend seen over the past two decades, in contrast to ground-based data, which are strongly affected by the so-called 'urban heat-island' effect, and show a warming."

It now appears, however, that the satellite data were flawed. In mid-1998, Frank Wentz and Matthias Schabel of Remote Sensing Systems in California published a paper in *Nature* that showed that the observed tropospheric cooling resulted from a failure to correct for the decay in the satellites' orbits caused by atmospheric friction. They calculated the satellites dropped more than a kilometer per year, which affected the data gathered by the instruments; correcting for this slippage produced a tropospheric *warming* of about 0.07°C per decade, which is more consistent with the climate models.

Although there are still uncertainties associated with both the models and observational data, this finding, if it withstands further investigation, has been hailed as a killing blow to the skeptics' claim that the earth's atmosphere is cooling, not warming. *New Scientist* noted that "one of [their] last props...is crumbling." James Hansen, a leading proponent of the global warming theory, and his colleagues at the Goddard Institute for Space Studies, wrote in *Science* that both surface and troposopheric warming trends "are now sufficiently clear that the issue should no longer be whether global warming is occurring, but what is the rate."

However, the scientists who collected the satellite data do not accept this assessment. According to Roy Spencer, a climate scientist with NASA's Marshall Spaceflight Center, Wentz and Schabel provided the "first convincing evidence" that corrections were needed in the satellite temperature record but, after reanalyzing the data, Spencer and John Christy of the University of Alabama concluded that other adjustments partially offset the orbital decay effect and the net result is still a slight cooling of about 0.01°C per decade. Spencer noted that, given the uncertainties, this is too small to be called a trend—but he also points out that data from weather balloons show a stronger cooling. The upshot is that debate over temperature trends in the lower atmosphere will likely continue, as will the skeptics' claims that the earth is cooling, not warming.

In fact, many skeptics argue that the satellite data are more accurate than the surface temperature record that shows the earth is warming. They say surface temperature data have been skewed by the urban heat-island effect, citing studies showing that more of the surface warming has occurred at night than during the day. (Urban heating is stronger at night.) Recent research refutes this claim. While urban heating may affect local temperatures around big cities, much of the global temperature record consists of data taken in remote regions on land and from ships at sea. Moreover, when scientists examine the effect of *removing* temperature data collected from urban centers, they found this does not greatly affect the warming trend.

For example, an analysis by scientists from the U.S., Britain, Russia, Australia, and New Zealand found that urban heating has a "negligible" impact on observed global and hemispheric warming trends. The study, published in *Science*, was the most comprehensive analysis to date of global temperature data; it examined measurements taken between 1950 and 1993 from 5400 weather stations covering more than half the earth's land area, including data from locations in the Southern Hemisphere that had not been analyzed before. The researchers found evidence of global warming everywhere, even in undeveloped areas of the Southern Hemisphere.

The study confirmed that minimum (nighttime) temperatures are rising faster than maximum (daytime) temperatures—globally, about twice as fast. This means that the spread of temperatures over the 24-hour period—the daily or "diurnal" temperature range (DTR)—is decreasing. When the researchers did a separate analysis eliminating data from 1300 weather stations in urban areas (cities with 50,000 people or more), they found only "slight differences" from the one that used all 5400 stations. David Easterling of the NCDC, the lead author of the paper, expressed confidence that urban heating is not responsible for the warming trends over global or hemispheric scales. However, he said more research is needed to determine the causes of the day/night differences in global temperature increases.

The IPCC report concludes that, in general, urbanization "could have contributed only a small part...of the overall global warming." Moreover, it points out that indirect indicators independently support the surface temperature data; for example, measurements of shrinking glaciers, rising sea levels, a reduction in Northern Hemisphere snow cover, and temperature measurements taken from holes bored deep underground in remote locations are all consistent with the surface data.

Just as global warming varies by season and time of day, it also varies geographically. Some regions have warmed more than others. For example, eastern North America, parts of the eastern Mediterranean, and parts of Asia, particularly in China, have remained relatively cool, perhaps because of the local effects of aerosols. Land masses have warmed more than oceans—the north Atlantic has remained cool—and high latitudes have warmed more than equatorial regions. The greatest warming occurred over the continents at mid-latitudes (between 40° and 70°N) in winter and spring. Canada, for example, has warmed about 1°C in the past century, roughly twice the global average. Some northern regions have warmed even more; the average temperature in the Mackenzie Basin, an area of nearly two million square kilometers that includes parts of the Yukon and Northwest Territories and the provinces of British

Columbia, Alberta, and Saskatchewan, has increased about 1.5°C. The greater warming in the Arctic is consistent with model predictions that high latitudes will show signs of global warming earlier and more strongly than regions closer to the equator.

Scientific debates over the details of global warming will persist and scientists trying to make sense of the earth's climate system will continue to be kept on their toes by its incredible complexity. Nevertheless, most of the world's climate experts are convinced that when all the evidence is weighed—proxy data from past climates, temperature measurements taken from all over the world, and observations of the responses of natural ecosystems, plants, and animals—on balance, there's little room for doubt that the earth's surface temperature is warmer now than it was a century ago and that it's continuing to warm at a rapid rate.

four

GOING TO EXTREMES:
The Link Between Global Warming and Severe Weather

For millions of people, a Spanish phrase meaning "the boy child" has become the harbinger of disaster. El Niño, named after the Christ child because of its tendency to peak around Christmas, has been wreaking havoc with the world's weather, particularly in 1982–83 and in 1997–98.

El Niño is a huge swath of abnormally warm surface water that appears roughly every three to seven years in the eastern tropical Pacific Ocean off the coast of Peru. Water temperatures may rise 1 to 5°C above the average sea surface temperature of 28°C. (During the 1997–98 event, water temperatures in some places were up to 6°C above normal.) Its alter ego—known as La Niña ("the girl child")—occurs when the surface waters turn cold. There's a link between these changes in ocean temperature and changes in atmospheric pressure known as the Southern Oscillation—hence the acronym ENSO (El Niño/Southern Oscillation) by which scientists refer to the combined ocean/atmosphere phenomenon.

During an El Niño, the normal distribution of warm and cold water in the equatorial Pacific is reversed. Usually, winds blowing from east to west push warm water to the Asian side of the ocean where it sits in a huge pool around Australia and Indonesia. As the warm surface waters are pushed away from the eastern Pacific, cold water wells up from the depths off the coast of South America, bearing nutrients that sustain fisheries in the region. Every so often, the winds slacken and may even reverse direction and the warm water sloshes back across to the eastern Pacific. The upwelling of cold water is shut off and the South American fisheries take a direct hit. But we now know that the effects are more far-reaching; El Niño spreads heat to other parts of the world via ocean currents and the atmosphere. Next to the seasons, it is the strongest natural climate signal on earth.

The intense preoccupation with the 1997–98 El Niño makes it hard to believe that, until recently, there was relatively little scientific interest and almost no public interest in the phenomenon. There's evidence of El Niño in the paleoclimatological record going back at least 5000 years and in folk tales going back to the 16th century, but scientific record-keeping began around the 1880s. Even so, scientists didn't pay much attention to a phenomenon that was viewed as a regional event that primarily affected South American fisheries.

The first awareness of El Niño's worldwide reach came in the 1970s, when it caused drought and famine throughout the tropics. Then came the huge event of 1982–83, which caught scientists by surprise and was not even recognized until it was well underway. It caused an estimated 2000 deaths and US$13 billion in damages worldwide and led to the creation of the Tropical Oceans and Global Atmosphere Program (TOGA) in 1985. Buoys were deployed across the Pacific to measure ocean temperatures, which greatly improved the ability to predict El Niño. In fact, the 1997–98 event, which developed unusually early in the spring of 1997, was predicted about six months in advance, allowing people in vulnerable regions to plan for its destructive effects. These forecasts proved to be highly accurate.

During an El Niño, the warm surface waters heat and moisten the air, mostly through increased evaporation, and this affects atmospheric circulation, causing changes in wind, precipitation, and storm patterns throughout the tropics and even into the mid-latitudes. This sets the stage for extreme weather events. El Niño brings drier conditions and drought to Australia, Indonesia, Southeast Asia, and the Philippines, as well as to Hawaii, parts of Africa, Brazil, and Columbia. Droughts in India, usually related to the failure of monsoon rains, may become more intense during El Niño years. Meanwhile, the west coast of South America and the southern U.S. are drenched in rain and pummeled by storms. During the 1982–83 event, parts of Peru that normally get about 15 centimeters of rain a year got more than 300 centimeters. The situation was even worse in 1997–98, when normally dry regions of the country experienced the worst flooding in 50 years.

El Niño gives the Caribbean and the U.S. southeastern coast a break from hurricanes, but it increases tropical storms and typhoons in the Pacific. Evidence suggests that the number of intense winter storms in the Pacific also increases during El Niño years. There's also evidence that climatic conditions associated with El Niño and La Niña can affect the frequency and location of tornadoes in the eastern two-thirds of the U.S., increasing them in some regions and reducing them in others, according to Mark Bove of Florida State University. He noted that the El Niño–caused conditions that prevailed over Florida in the winter of 1998 "can lead to tornadic activity" and there were many more winter tornadoes there than normal.

El Niño can affect temperature and rainfall patterns and storm tracks as far as the northern U.S. and Canada. It strengthens and distorts the North American jet streams, currents of air high in the atmosphere that influence continental weather systems. An extreme distortion of the northern jet stream in December 1997 brought snow to the southern U.S. and Mexico while, in western Canada, people were golfing on normally snowbound golf courses and fighting unprecedented winter forest fires caused by unusually warm, dry weather.

El Niño may also affect storminess on the Great Lakes, according to a University of Michigan study, which found that strong storms and elevated waves in the Great Lakes followed strong El Niño events in the past. For example, in 1984 and 1985, following the 1982–83 El Niño, Great Lakes storms caused about US$130 million in damage.

Dalhousie University climatologist Owen Hertzman has described El Niño as the climatological equivalent of a motorcycle gang: When you see it coming, you know there's going to be trouble, you're just not sure what kind.

El Niño and Global Warming

The public and media attention devoted to the 1997–98 El Niño wasn't surprising; the photo ops provided by floods, mud slides, tornadoes, and ice storms are made for TV. What's more surprising was how little of the coverage explored the possible relationship between El Niño and global warming. In some respects, the conditions El Niño creates can be viewed as a dress rehearsal for global warming. "The lesson we learn from El Niño is that when you move towards a warmer, wetter world, you're going to see it through increased storminess," said James Baker, head of the U.S. National Oceanographic and Atmospheric Administration (NOAA).

The 1998 ice storm is one example. El Niño didn't cause the storm, but it helped to make it as bad as it was. Henry Hengeveld, Environment Canada's senior climate change adviser, said it took the confluence of four unusual events to make the storm so extreme: 1) a large mass of warm, wet air coming up from the Gulf of Mexico; 2) higher-than-normal temperatures for the time of year, hovering around the freezing point; 3) a high pressure region over Bermuda that deflected the warm moist air northward and kept it from moving over the Atlantic for nearly a week; 4) a shallow layer of cold air trapped in low-lying regions in the Ottawa and St Lawrence valleys. El Niño was the dominant influence on the first two factors, he said.

Environment Canada scientists emphasized that not every El Niño will create the conditions for such extreme ice storms. Bad ice storms occurred in Montreal in the past during a non–El Niño year. Nevertheless, icing conditions might become more frequent as the earth's atmosphere warms up, regardless of El Niño's influence. According to Kevin Trenberth of NCAR, if climate warming pushes the transition zone between snow and rain farther north to land that's very cold, one of the results is more freezing rain.

Questions about a link between El Niño and global warming began to surface in the mid-1990s with evidence that its behavior was changing. Since the mid-1970s, but particularly during the 1990s, El Niño has been bigger, hotter, and more persistent than usual. Of the 11 El Niños that occurred in the past five decades, eight have been in the past 20 years and within those 20 years, two (1982–83 and 1997–98) were dubbed the "El Niño of the century." Of the two, the 1997–98 event started earlier and lasted longer at a higher peak. By mid-1998, it appeared El Niño was making way for La Niña, but it didn't give up easily; in September, scientists reported that El Niño was "lingering" and apparently had weakened the developing La Niña. With the Pacific "running hot and cold," there was great uncertainty about weather forecasts for the winter of 1998–99.

From 1990 to 1995, the longest ENSO event on record occurred. There's usually a break of several years between El Niños, so this extended warm period, plus El Niño's increased frequency since 1976, "are unexpected given the previous record, with a probability of occurrence about once in 2000 years," according to a study by Kevin Trenberth and Timothy Hoar of NCAR. The researchers concluded it's highly unlikely these changes are caused by natural variability and said this raises the possibility that global warming may be at least partly responsible for ENSO's changing behavior.

Certainly, the changes in El Niño in the past quarter-century coincide with the rapid rise in global surface temperature attributed to global warming. So the question arises: Is global warming intensifying El Niño? Is this intensification an indirect mechanism by which climate change could influence the frequency and intensity of extreme weather events? Intuitively, it makes sense that global warming should affect a phenomenon whose defining feature is the warmth of surface waters in the ocean.

Scientific opinion is still mixed, however, as are the results of modeling studies. Although some indicate that El Niño could be intensified by global warming, different models project different outcomes. According to Trenberth, one projects that El Niño will get bigger and stronger with global warming, while another indicates it will becomes more frequent, and still another suggests that conditions will become more "El Niño–like" on average. Francis Zwiers of Environment Canada said some modeling studies suggest the possibility that rising sea surface temperatures caused by global warming could create a permanent state of El Niño. These studies all suggest El Niño will change, but it's not yet clear how.

On the other hand, there are computer studies that suggest global warming will not cause major changes in the great variability that El Niño already exhibits. "We have gone out to 1000 years with doubling and quadrupling of carbon dioxide and we find that the structure of the El Niños and their natural variability doesn't change noticeably," said Jerry Mahlman, a modeling expert at Princeton University. "You can get 40- or 50-year periods when they hardly happen at all." This also applies to the extreme weather events influenced by El Niño; they're highly variable because El Niño is variable and "we find that in a high CO_2 world, that's still true."

One scientist has suggested a mechanism by which global warming could intensify El Niño. In a scientific paper, De-Zheng Sun of NOAA's Climate Diagnostic Center argues that El Niño is driven by the temperature difference between the surface of the tropical Pacific Ocean, which is warmed by the sun's energy and by global warming, and the cold deep ocean, where temperatures change much more slowly. Using a computer model, Sun found that when the *difference* between the surface and deep-ocean temperatures reaches a certain point, it triggers the oscillatory (up and down) pattern characteristic of El Niño. Further increasing the surface temperature intensifies this effect, which suggests that increased global warming would result in increasingly severe El Niños. He concludes that enhanced global warming would strengthen El Niño using the same mechanisms that cause it to occur naturally.

However, Sun suggests that after a certain point, this intensification might also be indirectly reined in by global warming. It's expected that greater warming will occur in polar regions than at the equator; interactions between these warmer polar regions and the ocean may eventually cause temperatures to rise in the deep ocean at the equator more rapidly than temperatures are rising at the surface. This would reduce the difference between the two sets of temperatures and cause a weakening of El Niño. But this would probably take a few hundred years and, in the meantime, El Niño would intensify.

Even if global warming doesn't intensify El Niño, however, the two phenomena might work independently but in concert to create unprecedented weather extremes. "El Niños in a greenhouse-warmed earth could still have an amplified impact because of their signals being superposed on distinctly warmer tropics," says Mahlman.

According to Trenberth, the main way in which global warming and El Niño may reinforce each other is through their similar influence on the global water cycle. Global warming intensifies the hydrological cycle and thus affects rainfall, flooding, and droughts. This intensification is likely to exacerbate the floods and droughts caused by El Niño, even if global warming doesn't change El Niño itself. El Niño influences the distribution of floods and droughts and if, in addition, global warming alters the hydrological cycle, this is likely to "make the manifestation of these events worse in both cases," he said.

In short, even working separately, global warming and El Niño could deliver a devastating one-two punch. And if it turns out that global warming does intensify El Niño, the resulting weather extremes could become that much worse.

Global Warming and Extreme Weather

The intense focus on extreme weather caused by El Niño in 1997–98 overshadowed questions about the link between extreme weather and global warming. Until recently, media coverage of extreme events rarely even mentioned climate change. However, this link is a key factor in assessing the social and economic harm that global warming may cause, because the most serious climate-related problems we face will likely be caused by changes in climate variability and the incidence of extreme events. Extreme weather is how climate change will hit us where we live. These events, not gradual climate change, may determine whether humanity can adapt to climate change or will be exposed to "unacceptable economic and physical danger," said Henry Hengeveld.

Understanding the relationship between climate change and weather extremes is inherently difficult because the weather is naturally so variable and extremes are, by their very nature, rare events. Detecting trends requires comprehensive and accurate weather records going back a long time. There are relatively long records in some locations for some kinds of weather—for example, temperature records in England go back to the mid-1700s—but in many parts of the world, particularly the Southern Hemisphere and the oceans, records are inadequate or lacking. However, global weather monitoring indicates that some types of extremes are already increasing in many parts of the world. "We're starting to see weather conditions and climate conditions that are foreign to us," Hengeveld commented during a public forum on understanding weather disasters. "It suggests something is happening that's different than what we've been accustomed to in the last 50 years."

Projecting what will happen with extreme weather in the future is also difficult because many of the most damaging events (e.g., tornadoes and thunderstorms) are short-lived and small in scale, affecting only local or regional areas. Global-scale climate models do not resolve events this small and hence do not "see" them very clearly, making it difficult to project how they'll change as the climate warms. However, other methods, based on known physical principles of how weather systems work, can be used instead. For example, analyzing low pressure systems, which are associated with storms but are much larger, may help detect trends in severe storms. Understanding the role of temperature in forming thunderstorms and tornadoes can be used to infer how global warming might affect these storms.

Two other factors affect our ability to determine if weather extremes are actually increasing or not: improvements in monitoring and reporting and the effect of socioeconomic factors such as population growth, increased urban development and concentration, and the increasing economic value of property exposed to weather extremes.

Better reporting: As monitoring technologies improve and are deployed more widely, the amount of information available about weather extremes increases accordingly. But scientists are cautious about interpreting these data as evidence of real changes in the climate because the relative lack of records from earlier times makes comparisons difficult, particularly given the natural variability of the weather. It could be that what's changed is not so much the weather itself but the amount and quality of weather data. This is why paleoclimatology has been so valuable; by reconstructing climate changes going back millions of years, scientists are better able to calibrate more recent changes.

As the instrumental record continues to improve, the extent of the human influence on climate should become clearer over the next few decades. Many climate scientists believe the mid-1990s may prove to be the transition zone—the point where convincing evidence of humanity's impact on the global climate first began to emerge from the "noise" of natural climate variability.

The public, however, is more influenced by media coverage. There's no doubt that reporting of extreme weather events has increased, and it's possible that the public's perception that the weather is getting worse is based primarily on the sheer volume of in-your-face coverage. These images also have political consequences. In the U.S. since 1989, the percentage of denied requests for a presidential declaration of disaster has dropped from one-third to one-quarter of all such requests. Richard Sylves of the University of Delaware attributes this to the "CNN syndrome"—increased pressure on the president to act when a disaster attracts a lot of media attention.

However, saturation coverage can't wholly explain the perception of increasing weather disasters. The media are attracted to the new, the unprecedented, the spectacular; the mere fact that the weather is providing them with a steady diet of disasters deemed worthy of coverage says something about the escalating and unusual nature of these events. Weather coverage would not be sustained at this level if extreme events were not having more serious social and economic impacts.

Socioeconomic factors: There's no question the weather has taken an enormous and growing financial toll recently; the cost of property damage, insurance losses, and rescue and recovery operations has climbed steadily worldwide—much of it piled directly on top of the public debt. During the 1990s, most regions of the U.S. have at one time or another been declared disaster areas, which qualified them to receive funding from the Federal Emergency Management Agency (FEMA). According to FEMA, the 168 major disasters declared between 1988 and 1992 covered 43 states, plus D.C. and offshore territories. Between 1993 and 1997, the number rose to 219 major disasters covering 48 states, plus D.C. and offshore territories. This is a 30% increase in the number of declared disasters and a 12% increase in the number of states with declared disasters. All were weather-related except for two large earthquakes.

FEMA payments for disasters in 1988–92 were US$6.12 billion. During 1993–97, the payouts rose to US$14.44 billion—all for weather disasters except for about US$1 billion in earthquake relief. Between 1992 and 1997, FEMA's annual costs averaged US$1.88 billion (excluding earthquake costs). Only once before, in 1989, did their annual costs exceed US$1 billion.

According to the NCDC, from 1988 to 1997 there were 25 natural disasters that cost at least US$1 billion, 21 of which occurred from mid-1992 to mid-1997 and accounted for two-thirds of the total cost. Another study by NOAA found that insurance claims for damages from wind, hail, snow, and tornadoes in the U.S. increased 570% between 1986 and 1995. Hurricane Andrew alone caused about US$27 to $30 billion in damages, of which about US$15 billion were insured losses.

In Canada, the losses have been smaller in dollar amounts, but they have been rising rapidly, and by the late 1990s, Canada had joined the billion-dollar club for weather disasters. A study done by David Etkin of Environment Canada's Environmental Adaptation Research Group (EARG) and his colleagues found that nearly all of Canada's most costly natural disasters in the 20th century were weather-related, and more than three-quarters of the 23 events occurred in the 1980s and 1990s. The Saguenay flood in 1996 was the first billion-dollar weather event. Estimates of the total cost for the 1998 ice storm range from $2 to $3 billion.

Insured losses to natural disasters in Canada, mostly weather events, have risen sharply over the past two decades, according to insurance industry figures. From 1983 to 1989, there were nine catastrophic events that cost insurers $425 million; from 1990 to 1997, there were 41 events that cost $2.15 billion. According to the Insurance Bureau of Canada, the ice storm cost a record-breaking $1.44 billion in claims — but it also generated about $2.2 billion in new business and created 16,000 jobs.

Globally, economic losses to natural disasters have jumped more than fiftyfold in the past three decades. Weather-related disasters accounted for roughly 70% of those losses. According to Germany's Munich Reinsurance Company, annual losses remained relatively stable between 1960 and 1975, exceeding US$10 billion only twice in the early 1970s. Until 1988, the global insurance industry had never experienced a loss of more than US$1 billion in any single natural disaster; between 1988 and 1996, 15 such events occurred. Around the mid-1990s, the losses became frightening; Munich Re estimated total losses to major natural disasters at around US$110 billion in 1994, around US$180 billion in 1995,[*] and around US$60 billion in 1996. In 1997, while insured losses dropped considerably, they were still much higher than the average between 1970–1988, according to Swiss Reinsurance. The upward trend resumed in 1998; losses in the first half of the year exceeded the total for 1997.

It might seem logical to conclude that these statistics prove that extreme weather events are increasing in frequency and intensity, but

---

[*] This figure includes about US$125 billion for an earthquake in Japan.

other factors influence these figures. There's no doubt that increasing losses stem in part from increased exposure to extreme weather, as more people move to vulnerable areas such as storm-prone coastal regions and river flood plains. Roughly half the U.S. population already lives within 80 kilometers of either coast, and coastal migration is expected to grow as aging baby boomers retire. Increasing urban concentration is also a factor; since 1950, the portion of the world's population living in cities rose 15% and the number of cities with more than a million residents has quadrupled. Population growth, increased density of population, and the value of property exposed to risk are all factors that have increased losses to extreme events.

This is clearly shown in a study of U.S. losses to hurricanes done by Roger Pielke, Jr. of NCAR and Christopher Landsea of NOAA's Hurricane Research Division. When they eliminated differences due to inflation and changes in wealth and population over the years, they found that damages actually *decreased* during the 1970s and 1980s, compared with earlier decades. Damages increased during the first half of the 1990s, but were not unprecedentedly large. The scientists concluded that social factors such as population growth and increasing wealth, rather than climate change, were responsible for the increased losses. Far from making things worse, the climate actually gave the U.S. a break during the years that coastal development increased most rapidly. However, Pielke noted that the study shows only that past losses were not caused by an increase in the number or severity of hurricanes; it does not in any way refute suggestions that climate change could intensify storms in the future.

Because the public erroneously believes that Atlantic hurricane intensities have been increasing, people may be dangerously underestimating the potential for future damages, Pielke and Landsea said. Even if global warming doesn't enhance their frequency or intensity, a return to the levels of storm activity experienced earlier in the century could cause huge losses because of the greater exposure caused by coastal development and the higher value of vulnerable property. They warn it's only a matter of time before a storm costing US$50 billion occurs.

However, other researchers and many in the insurance industry aren't convinced that the huge increases in all weather-related losses in recent years can be written off solely to socioeconomic factors. Statistics on losses to natural disasters in the past three decades lend support to the idea that weather variability and extremes may be increasing. From the mid-1960s to 1990, global GNP increased about threefold and the world's population nearly doubled—far below the increases in the cost of weather-related disasters that occurred over the same period. The

Canadian Global Change Program (CGCP) reported that "populations in hazardous areas, especially coastal zones, have increased more rapidly than the general population but only in a few locations (e.g., Florida) has this increase been sufficient to explain much of the disaster loss."

An analysis of disaster losses compiled for a conference on natural disaster reduction found that between 1963 and 1992, the rates of major weather-related disasters not only grew faster than those of other natural disasters such as earthquakes, but also outstripped global economic and population growth. While the economic damage from earthquake disasters doubled or tripled over the period, drought losses increased five to seven times, flood losses increased eight to twelve times, and losses to tropical storms increased about fourfold. As for the number of people affected, floods and droughts topped the list, followed by tropical cyclones.

Is there more to this than just greater exposure of people and property to risk? James Bruce, a Canadian climate expert who was a senior official of the United Nations International Decade for Natural Disaster Reduction, concludes it's unlikely that differences in development patterns can account for the fact that climate-related disasters increased so much more than earthquake disasters. These trends suggest that climate-related hazards "may be increasing in frequency and severity," he said.

Many in the insurance industry agree. Since the mid-1990s, major international insurance organizations have argued that while socioeconomic factors are significant, they can't fully account for the rapidly escalating losses to weather extremes and that climate change must be at least partly to blame. A 1996 report by Munich Reinsurance said that in many places, "increasingly discernible" environmental and climatic changes "are leading to a greater probability of new extremes in terms of temperatures, amounts of precipitation, water levels, wind velocities, and other parameters.... This is why the Munich Re has long been pleading for measures to be taken with a view to curbing man-made changes in the environment." Several European insurance companies made this point at international conferences in 1995, 1996, and 1997 where negotiations on the treaty to limit greenhouse gases were being conducted.

North American insurers have been less politically active, but they too are concerned about the potential impact of climate change on their bottom line. In both Canada and the U.S., research institutes have been established to examine ways to reduce losses to natural disasters. Frank Nutter of the Reinsurance Association of America warned that if the insurance industry is unprepared for increasing weather extremes, there could be "major insolvencies." A 1997 study by Swiss Re found that most primary insurance companies in major world markets hadn't sufficiently insured themselves against catastrophic losses to natural disasters and "the equity base of many insurers could therefore come under considerable strain."

Paul Kovacs of the Insurance Bureau of Canada concurs that socio-economic factors alone can't account for the growing losses. "We've done our best to take those factors out. Maybe you can explain half of the increase that way. The amount of losses is growing faster than those factors.... It's getting very frightening." Within the insurance industry, it's largely accepted that as the world gets warmer, there will be more extreme events. "Our view is that there's no question it's warmer and there are more severe events, more people killed and property lost—and we're getting the bills."

Variability and Extremes: Climate at the Edges

Before examining scientific evidence that indicates extreme events are increasing, it's important to understand what's meant by climate variability and its relationship to averages and extremes.

Variability is a measure of departures from average conditions, a measure of the range between highs and lows. It tells us how far things are likely to go in either direction. *Extremes* are events at the far ends of the expected range of climatic variability or that exceed that range altogether. Scientists typically define extremes as events in the top or bottom 10%, 5%, or 1% of the range. For example, a storm that dumps more rain in 24 hours than 99% of other rainstorms would be defined as an extreme event.

Extremes, *by definition*, have a low probability of happening under current climatic conditions. At issue is whether global warming will change those probabilities by causing existing weather extremes to occur more often or by creating events so extreme they've never happened before. Such record-breaking events often involve the coincidence of two or more unusual or rare weather conditions.

Looking at variability and extremes, rather than just at averages, provides a more realistic picture of the day-to-day impact of climate. Two locations could have exactly the same average annual temperature, yet experience very different weather. Say that in the first case, day-to-day temperatures generally deviate only a few degrees to either side of the annual average, providing a relatively uniform (i.e., low variability) climate with few nasty surprises. In the second case, however, the temperature swings are much larger, resulting in more very hot or very cold days. This climate is more variable than the first and, clearly, far more difficult and expensive to cope with. Yet both have the same average annual temperature.

These examples show that while variability and extremes are related, they are not the same thing. This is important when considering the potential impact of global warming. For example, *it is possible for extremes to increase even if variability does not,* solely because of a shift in the average temperature. Small changes in the average can cause large changes in the frequency of extremes. If *both* the average and variability increase, the extremes would be that much worse.

The chart below illustrates this point in a simplified way. It shows what would happen to extremes if global warming causes an increase in the mean temperature with *no* increase in variability (Case 2) compared with what would happen to extremes if there was an increase in the mean temperature *plus* an increase in variability (Case 3).

	low extreme	mean	high extreme
Case 1: Original climate	10———25——— 40		
Case 2: Increased mean temperature, no increase in variability	15———30———45		
Case 3: Increased mean temperature *plus* increased variability	5—————————30—————55		
Case 4: Increased variability; no increase in mean temperature	5————— 25 —————45		

We see in Case 2 that if the mean temperature increases without an increase in variability (the *range* of the extremes), there is an increase in extremes at the upper end but not the lower extreme. In 1995, England experienced a high average summer temperature that, under current climatic conditions, would be reached or exceeded only once every 75 years. If the earth's average temperature warms as much as projected, the probability of this hot summer event occurring would increase twenty-fivefold and the return period would drop from 75 years to three years.

If variability also increases, as in Case 3, the result is unprecedented extremes at both ends—a double whammy. The IPCC report notes that small changes in either the mean or in variability can cause large changes in the frequency of extreme events; however, changes in variability have a greater effect on the incidence of extremes than just a change in the mean.

Unfortunately, the fuss over whether the *average* global temperature is rising has obscured the public's understanding of its real significance, which is its impact on *the frequency and severity of weather extremes*. As geography professor Barry Smit of the University of Guelph put it, "the average is probably irrelevant, it's not important by itself." On either side of the average are all the weather conditions we actually experience and to which our activities are adjusted. It's when an event exceeds our "coping range" that we have problems, Smit said. "We haven't experienced it, we don't expect it, and we're not structured to accommodate it."

Climate change can create conditions beyond our coping range even without a change in variability, he said. "Just assuming a change in the average condition and no change in variability, you can still get a change in the frequency and magnitude of extremes." Any change in variability on top of this would just make matters worse.

The flip side of this coin is also bad news. An increase in variability would cause more extremes even if there is *no* increase in the average temperature (Case 4). If the world became 5°C hotter in summer and 5°C cooler in winter, its average global temperature wouldn't change, but, on a local level, these temperature changes would still have an enormous impact.

These examples illustrate why we shouldn't be overly concerned about what's happening with the earth's *average* temperature, except for what it tells us about how global warming may affect weather variability and extremes.

Linking Global Warming to Climate Variability and Weather Extremes

What *is* the likelihood that global warming will increase either the variability of the climate or the frequency and severity of extreme weather events, or both? Has it already done so? Can we link the weather extremes of the 1980s and 1990s with the concurrent rise in greenhouse gas concentrations and global temperatures during the same decades? Does the future hold increasingly wild weather if global warming continues as the computer models project?

Answering these questions is hampered by the weather's natural variability and the lack of comprehensive historical global weather data. Climatologist D. E. Parker of the British Meteorological Centre has noted that "long records are essential if the rarity of recent events is to be properly assessed." Changes in instruments and measuring techniques also make it difficult to compare data taken at different times to calculate long-term trends. Even when trends can be detected, they often vary considerably by region and season, which makes it difficult to determine global trends. Global figures average regional trends that go in both directions and, as we've seen, this can create a misleading impression that nothing much is going on.

It's not surprising, therefore, that the IPCC concluded there was insufficient evidence to determine whether consistent changes in climate variability or extremes have occurred in this century *on a global scale*. Some skeptics have claimed this demonstrates the global warming theory is flawed but this is not the case; the fact that no globally averaged trend is yet apparent does *not* imply that extremes have not or will not increase in frequency. Some scientists have suggested it may not even be possible to determine whether the frequency of extreme weather events is increasing on a global scale but even if this proves true, it doesn't mean that different regions won't experience increasingly difficult weather.

Indeed, the IPCC states there's "clear evidence" of *regional* changes in extremes, with some areas experiencing increases and others experiencing decreases in variability and extremes. "Potentially serious changes have been identified, including an increase in some regions in the incidence of extreme high-temperature events, floods and droughts, with resultant consequences for fires, pest outbreaks, and ecosystem composition, structure and functioning."

El Niño's sometimes aggressive intervention in the global weather system is a confounding factor here. As we've seen, while El Niño has had a growing influence on extreme weather events in recent years, it's still uncertain how much this is related to global warming.

Since it's not possible to attribute any single weather event directly to global warming, scientists look for long-term trends that show departures from average or "normal" conditions. The job of climatologists, according to Tom Karl of the NCDC, is to "separate any meaningful signals from ever-present noise and to discern, if possible, whether there is indeed at work the sometimes slow and subtle hand of significant change."

To that end, Karl and his colleagues have developed a Climate Extremes Index (combining five temperature and precipitation indicators), which shows that the U.S. climate has become more extreme. Between the mid-1970s and mid-1990s, the index averaged 1.5% higher than the average from 1910 to 1976. The researchers cannot say whether this means the U.S. has permanently moved into a new state of extreme climate or whether the fluctuations are part of natural variability; however, the current state has persisted for about two decades, roughly twice as long as two similar decade-long episodes that occurred in the 1930s and the 1950s.

A second index indicates that the trends in extremes during this century are "suggestive of a climate driven by greenhouse warming," according to the NCDC researchers. The data indicate there's only a 5% chance the changes could be due to natural variations. They say the case for linking the observed changes in extremes to greenhouse warming is strong enough for a case in civil court (requiring a preponderance of evidence) but not enough for a criminal conviction (requiring proof beyond a reasonable doubt). Some indicators used in the two indexes increased abruptly during the 1970s and "have more or less remained at these levels....Real changes in climate remain the most likely explanation for the most conspicuous changes."

Canadian scientists are currently developing similar indexes with special emphasis on indicators relevant to northern climates (e.g., snowfall). Global warming is expected to be more pronounced at high latitudes than in tropical regions, including greater warming and larger increases in precipitation.

Circumstantial evidence about weather extremes is accumulating in other parts of the world. Although information is still sparse and many uncertainties remain, there are disquieting observations about changes in temperature extremes, precipitation, floods, droughts, storms, and sea level.

Temperature Extremes

The earth's average surface temperature has increased by about 0.5°C in the past century. The IPCC report says there's no evidence that temperature variability has also changed on a global scale, though there are signs of increased regional temperature variability. Computer models indicate there will be an increase in global mean temperature but *not* an increase in the range of temperatures around the mean, that is, in variability. (This corresponds to Case 2 in the variability diagram on page 64.)

The implication is that a warmer world will experience an increase in the frequency of extremes at the upper end of the temperature scale (hot events), but a decrease in those at the lower end (cold events). The IPCC report concluded that warming will "lead to an increase in the occurrence of extremely hot days and a decrease in the occurrence of extremely cold days." For example, modeling studies indicate that daily maximum temperatures over much of North America—especially in southern regions—

will cross the 33°C threshold more often and stay above it for longer periods. In Toronto, the number of days above 30°C each summer would be expected to increase from 10 to 53 days if atmospheric CO_2 levels double. Other Canadian cities could expect three- to sevenfold increases in the number of days above 30°C.

These data indicate that heat waves will likely be more common in summer. The decrease in cold extremes would mean fewer frost days in spring and fall. Reductions in the length of the frost season have already been reported in several places; in the northeastern U.S., the frost-free season now starts 11 days earlier than it did three decades ago.

This doesn't mean that record-breaking cold days are a thing of the past. Despite the rise in global temperature during the 1980s and 1990s, some bad cold snaps occurred. One in December 1989 broke 250 minimum temperature records across the U.S., and cold spells in January 1994 and February 1996 were among the ten worst such events in the U.S. Midwest in this century. Ironically, El Niño can cause unusually cold conditions at times, as it did in December 1997, when a shift in the jet stream delivered cold and snow to the southern U.S. and Mexico. Generally, however, exceptionally cold days will likely become less frequent. While most people living in cold climates welcome this, it does have a potentially vicious downside, as the 1998 ice storm demonstrated.

One of the major reasons for the freezing rain was that temperatures were hovering around the zero-degree mark, warmer than usual for the time of year. Henry Hengeveld says that as winters get warmer, the number of times the temperature crosses the zero mark during winter will probably rise. This, combined with a predicted increase in the intensity of precipitation, could increase the probability of severe mid-winter freezing rain storms. If spring and fall get warmer too, they'd likely experience less freezing rain, but this may not compensate for the difficulties associated with mid-winter ice storms; recovery efforts after the 1998 event were hampered by the return of low temperatures, high winds, and heavy snow.

Scientific evidence shows that the frequency of high temperature extremes has been increasing while low temperature extremes have been decreasing, just as expected in a warming climate. Writing in *Science*, an international group of scientists analyzed temperature data from more than 4000 stations worldwide and found that in most places, the maximum daily temperature had increased between 1950 and 1993. Temperatures also increased in the Northern and Southern Hemisphere individually and in all seasons except autumn in the Northern Hemisphere.

A study by the British Meteorological Office shows that, around the late 1970s, more regions of the earth began experiencing the top 10% of temperature extremes, while fewer regions experienced the bottom 10% of temperature extremes. The divergence between these trends, which greatly increased after 1985, demonstrates that hot extremes are increasing and cold extremes are decreasing, consistent with computer projections of global warming.

Studies in England, Australia, and the U.S. have all found increases in warm days and decreases in cold days. Modeling studies by Australian scientists project that by 2030, Australia and New Zealand could experience as much as a doubling in the number of extremely hot days (over 35°C) in summer and equally large decreases in the number of cold days (below 0°C) in winter. They also projected increases in heat spells and decreases in cold spells lasting five days or more.

U.S. researchers have found decreases in the number of days below freezing and a faster retreat of snow cover in the spring. According to David Easterling of NCDC, the Northern Plains and the Great Lakes region tend to have the greatest decrease in days below freezing "and that could extend into Canada as well." Spring and summer snow cover in the Northern Hemisphere has dropped by about 10% since the late 1980s.

Environment Canada has just started to analyze extreme temperature trends in Canada. According to William Hogg of Environment Canada's Climate Research Branch, two kinds of temperature extremes are most significant in northern countries: extreme warmth in winter and spring that melts snow and ice, and extreme cold in fall, winter, and spring that damages crops and endangers life. While extreme warmth in summer can cause problems in southern regions, "it's not a nation-wide problem." In Canada, so far there have been only minimal changes in summer temperature extremes.

A significant temperature trend detected in recent years is a greater increase in daily minimum (nighttime) temperatures than in daily maximum (daytime) temperatures in many places worldwide—a phenomenon predicted by computer simulations to occur with global warming. Although both are rising, nighttime temperatures are warming faster. This has caused a decrease in the average daily range of temperature or daily temperature variability—a shift indicative of a more tropical climate.

Precipitation Extremes

Some of the most compelling data linking global climate change to weather extremes relates to precipitation, particularly heavy rainfall and snowfall. Climate models project an increase not only in total precipitation, but also in the intensity of precipitation, creating a greater likelihood of more frequent extreme rainfall events. Increases in precipitation are expected to be most pronounced in mid- to high latitudes during the cold season.

A warmer world is expected to experience increased evaporation of surface water and an increase in precipitation on a *global* scale to balance it (although local and regional precipitation patterns will likely change in different ways and some places may experience droughts even if global precipitation rises). It's been estimated that global precipitation will rise on average about 1.5 to 2.5% for every degree of warming. A warmer atmosphere could also hold more moisture, so its water vapor content would rise; concentrations in the lower atmosphere are projected to increase at a rate of about 6% for every 1°C of warming. (Since water vapor is a powerful natural greenhouse gas, this would further accelerate warming.)

Kevin Trenberth emphasizes that most of the increased heat from global warming goes into evaporating moisture. Because of this, he says, places that normally get little or no rain likely will experience more intense droughts that start earlier and last longer, while in areas that normally get rain, increased moisture in the atmosphere likely will cause intense rainfall, snowfall, and flooding. Weather systems that create thunderstorms and snowstorms will feed on the extra moisture in the atmosphere, so these storms likely will be more intense.

Worldwide observations reveal precipitation trends consistent with model projections of a warmer climate. It's estimated that global precipitation over land has increased about 1% during the 20th century. There's also strong evidence that moisture in the atmosphere is rising; from the mid-1970s to the early 1990s, it climbed by about 5% per decade over the U.S. and the Caribbean.

Precipitation in high northern latitudes is expected to increase substantially as warm air carrying increasing amounts of moisture moves north and cools, releasing its water content. There's evidence that since the 1970s, precipitation has increased in the mid- to high latitudes of the Northern Hemisphere and decreased in the tropics, subtropics, and mid-latitudes of the Southern Hemisphere. In 1997, high latitudes were wetter than normal and tropical regions were drier than normal—a pattern that's continued for most of this century and is consistent with computer projections of global warming.

In Canada north of 55°N, annual snowfall has increased by about 20% since the middle of the century, while Alaska has seen an increase of about 11%. On the whole, however, snow cover over most of the

Northern Hemisphere has declined about 10% in the past two decades because of higher temperatures. In southern Canada and the northern U.S., this has resulted in a higher ratio of rain to snow.

In the U.S., precipitation has increased roughly 10% since 1910, with the largest increases occurring in spring and fall, according to Tom Karl and his colleagues at the NCDC. (The first three months of 1998 were the wettest on record.)

Increases in both the number and the intensity of precipitation events contributed to an increase in total precipitation. Not only did the annual number of rain days rise, but there was also a large jump in the proportion of total precipitation that fell in the heaviest rainfalls. The increase in heavy rainfall came at the expense of more moderate precipitation events. The data also show about a 20% increase since 1910 in the total area of the U.S. that receives much higher than normal amounts of total annual precipitation during extreme one-day events (i.e., days with more than 50 millimeters of rain). Karl calculates there's less than a one in a thousand chance this is caused by natural processes and says it's a sign of a changing climate: "Not only is precipitation increasing, but the proportion of precipitation derived from extreme and heavy events is increasing."

In 1998, the NCDC reported that during the 1990s, 4% more of the U.S. territory experienced severe or extreme wetness compared with any other decade this century.

It's more difficult to sort out what's been happening to precipitation in Canada as a whole because there's almost no data for the northern part of the country before the 1950s, while the record in southern Canada goes back to 1900. However, from mid-century to the present, there's been an increase in the proportion of total precipitation falling in extreme events.[*] This is almost entirely confined to the North, according to William Hogg. In southern Canada since 1900, the fraction of precipitation falling in heavy events has declined.

Total annual precipitation has increased about 10 to 15% throughout the country, except for the prairies, and "that's because of more rain on rainy days rather than more days with rain," said Hogg. The available data therefore indicate that both southern and northern Canada have experienced increases in the average intensity of rainfall. However, Hogg said the increase in total precipitation appears to have resulted from increases in the number and intensity of moderate rather than extreme

---

[*] Extreme rainfall events are defined as above 25 millimeters per day because Canada almost never gets the 50-millimeters-per-day events used in the U.S. index.

events, which is "not quite the way we expect the climate to change in association with increased greenhouse gases, but is consistent with observations to date in Canada and elsewhere."

Evidence of increases in the frequency of extreme rainfall events during this century has also been found in Japan and parts of northern Europe, South Africa, Australia, and India, as well as around the periphery of the north Atlantic Ocean. In Australia, studies have shown that most of the continent experienced increases in the intensity of precipitation in summer and fall, while other seasons show a decrease. Overall, heavy rainfalls within the top 10% in intensity increased by about 20%. Combined with a decrease in dry days, this resulted in an increase in total precipitation, though some regions experienced decreases in both rain intensity and total rain. Australian researchers says it's too soon to tell if the increasing rainfall intensities are due to natural or human-enhanced warming. El Niño also exerts a strong influence in this region. "Continued monitoring of extreme rainfall trends may provide an early signature of climate change," they conclude.

Large areas of the tropics and subtropics from northern Africa to Southeast Asia and Indonesia have experienced decreases in precipitation. The IPCC notes these are areas where droughts are typically caused by El Niño, so the observed decreases probably reflect the influence of the more frequent and intense El Niños since the late 1970s. Precipitation in South and Central America, which has increased or decreased in different areas, is also strongly affected by El Niño.

As for the future, computer studies indicate that human-induced climate change—specifically a doubling of CO_2 levels—will increase rain intensities almost everywhere, especially at the upper end of the extremes. A study by the British Meteorological Office indicates that the heaviest rainfall events would more than double. Similarly, an Australian computer study found that doubling atmospheric CO_2 concentrations might not greatly change the total amount of rainfall there, but the number of days with heavy rainfall would increase markedly, while the days with light rain would decrease.

Computer models predict similar increases in intense rainfall and decreases in light rainfall in other regions, particularly in mid-latitudes. A study using the Canadian climate model indicates that heavy rains that would, at present, be expected to occur only once every 20 years would start occurring once every 10 years. U.S. scientists calculated that increases in average precipitation cause disproportionate increases in the frequency of extremes. For example, they found that in southern Canada, a 10% increase in average summer precipitation would be manifested as a 25 to 45% increase in the number of days with extreme precipitation. For Mexico, the frequency of extreme rainfall days would rise by 20 to 40%. In the U.S. and Russia, a 5% increase in mean summer precipitation yields roughly a 20% increase in the probability of days with extreme precipitation.

The fundamental message in these studies is consistent: global warming is likely to give new meaning—and an ominous spin—to the old cliché that "it never rains but it pours."

Droughts and Floods

Changes in temperature and precipitation patterns, especially extremes, are likely to change the incidence and severity of floods and droughts, but the effects will vary widely. Some regions will experience more severe floods and droughts, while others will experience less severe ones, but it's not possible to predict how any given region will be affected.

The IPCC notes that relatively small changes in temperature and precipitation can cause relatively large changes in runoff, especially in dry regions; therefore, places that already have inadequate water supplies will be particularly vulnerable. In many regions, the tendency of rain to fall in heavier downpours means more water will be lost as increased runoff rather than sinking into the soil.

Because global warming is expected to increase evaporation as well as precipitation, in some places it will exacerbate naturally occurring droughts. Climate models project that, if CO_2 levels double, the interiors of continents will experience increased soil dryness and more droughts because of increased evaporation. This could also cause levels in the Great Lakes to drop by as much as one-half to one meter and reduce water flow from the St. Lawrence River by as much as 20%, despite increased precipitation. This would put considerable strain on water supplies in this highly populated area.

In many countries, droughts are closely related to El Niño and their incidence has increased with its growing intensity and frequency. Some climate models suggest that global warming could worsen El Niño–induced droughts.

Shifting precipitation patterns under global warming conditions are expected to influence the frequency and intensity of dry spells as well. Several studies indicate that in some areas, such as central North America, the shift to more intense rainfall events could be accompanied by an increase in the number of dry days in between. This could lead to an increase in the length of dry spells (consecutive days without precipitation) and an increase in both droughts and flash flooding. The NCDC found that during the 1990s, the area of the U.S. experiencing both severe wetness and severe drought was greater than during any previous decade this century. In 1998, four southern states experienced their driest April through June since record-keeping began more than 100 years ago.

In areas where average precipitation drops, dry spells could increase even more dramatically. One study found that with a doubling of CO_2 levels, precipitation over southern Europe would drop about 22% while the probability of dry spells lasting at least a month would soar by 200 to 500%.

Modeling studies have also shown that greenhouse warming would lead to a substantial loss of soil moisture in summer. U.S. scientists who examined what a quadrupling* of CO_2 levels would do to soil moisture found the greatest increases in summer dryness would occur between about 40 and 50°N latitude, suggesting the possibility of more severe droughts in regions of North America and Asia where most cereal crops are grown.

Flooding patterns are also likely to change with global warming. Current precipitation trends (e.g., more rain in heavy downpours) will likely increase flooding in some regions. Kevin Trenberth suggests that the trend toward more intense precipitation in the U.S. helps to explain the huge amounts of rain and snow that fell in the winter of 1996–97, causing massive spring flooding throughout the West and Midwest. The Red River flood in 1997 occurred mainly because North Dakota and Manitoba received two to three times the normal amount of snow during winter and then were hit with a massive spring storm. Trenberth said that increased moisture in the atmosphere probably enhanced the severity of these events by as much as 10% over what would have occurred in the previous two decades.

According to a Canadian Global Change Program report, in regions where melting snow causes flooding, there's evidence that the early arrival of spring due to climate warming has caused more rapid melting of the winter snowpack and higher peak flows of water. This is likely to contribute to increasing flood disaster losses unless countermeasures are adopted. According to one study, global warming could increase spring runoff in Quebec by roughly 2 to 13%, which could increase the risk of spring flooding. On the other hand, regions where snow cover is reduced by global warming could experience water shortages; Environment Canada notes, for example, that water supplies in the Canadian interior

---

* Quadrupling of CO_2 is not expected to occur before the end of the next century, but a tripling is possible.

could be strained and "more frequent and severe droughts would be likely."

According to Roger Street, director of Environment Canada's Environmental Adaptation Research Group, extremes related to precipitation are the major concern for Canada. "The big issue, the big vulnerability, is water—too much or too little, at the wrong time in the wrong place."

Storms

Because storms are among the most destructive types of weather, there's considerable interest in whether they'll become more frequent or severe as the climate warms. There are different kinds of storms, ranging from small-scale but highly destructive events like tornadoes and thunderstorms to large cyclonic systems covering a thousand kilometers or more in which winds move in a circular path around a central low pressure core. Tropical storms, hurricanes, and typhoons are examples of cyclones that originate in tropical and subtropical regions though they can move into middle latitudes. In mid-latitudes, winter storms ("extra-tropical" cyclones) are large-scale systems that drop rain, freezing rain, or snow over wide areas. For all their destructive potential, however, many of these weather systems are too small for climate models to "see" very well, so there are many uncertainties associated with computer projections of future changes in storminess caused by global warming.

David Phillips, Environment Canada's senior climatologist, said storms caused an estimated 88% of worldwide natural disasters from the mid-1980s to the mid-1990s. During that period, 10 of Canada's most costly natural disasters were storm-related. Yet only a minority of storms actually cause economic damage; according to geography professor David Chagnon of Northern Illinois University, damage results from only about 5 to 10% of thunderstorms, 10 to 15% of freezing rain storms, 10 to 25% of hail storms, and 30 to 40% of tornadoes. All hurricanes cause damage, but only a few produce major damage. Chagnon estimated that thunderstorms account for about 40% of all storm-related property damage in the U.S., while hurricanes account for about half.

Scientists are trying to measure whether storminess has already increased and whether it's likely to increase more as the earth warms. Since heat rising from the surface into the atmosphere is a major factor driving the formation, development, and movement of storms, there are sound theoretical reasons to suspect that warming may increase the frequency and severity of some storms. The IPCC reports says there's no clear evidence of a uniform change in storminess worldwide, but there's evidence of increased storminess in some places and decreases or no changes in others.

Studies done since the IPCC report was released provide growing evidence that global warming may increase the potential for more frequent and/or intense storms. Since there are significant differences between the three major types of storms, they should be considered separately.

Summer convective storms: These include thunderstorms as well as related hailstorms and tornadoes. These storms are too small to be resolved in global climate models, so it's hard for computer simulations to predict what effect global warming will have on them. However, there are other grounds for arguing that the number and intensity of these storms will increase in a warmer climate. Hengeveld notes that warmer surface temperatures will cause air to become more buoyant, rising more rapidly and setting up the conditions that create such storms. A warmer atmosphere will also hold more moisture, so the rain intensities of storms are expected to increase.

A study of "thunder days" in the U.S. found increasing numbers from 1901 to 1940, then a decreasing trend until around 1970, after which the incidence rose again through 1980—a pattern that parallels rising global surface temperature during this century. Another study found a strong correlation between convective storms and the U.S. annual temperature. More storms occurred after 1970, mainly in southern and eastern regions. The researchers concluded that the eastern two-thirds of the U.S. could experience more frequent intense storms as the climate warms.

There's also evidence in both Canada and France that increasing summer temperatures, especially at night, "have been related to more frequent severe thunderstorms, hail and tornado events," according to the CGCP report. "Warm summers will be more frequent in a greenhouse enhanced world."

Tornadoes may also become more common, according to David Etkin of EARG, who found that tornado frequency in the Canadian prairies increased during warmer springs and summers. Since the mid-1980s, there's also been a dramatic increase in reported tornadoes in the U.S., but researchers caution this may be due to better observations.

Extra-tropical storms: Extra-tropical (mid-latitude) storms, also known as winter storms, are large cyclonic systems that can develop over land or water. Scientists are still debating whether these storms will become more intense or frequent in a warming climate. The IPCC report said evidence that such changes have already occurred was inconclusive; while some regions had experienced changes in these storms, others had not. It also said there was little agreement among computer models about possible future changes in storm frequency or intensity. There is, however, evidence of increased storminess in some parts of the world, according to several studies, including some done since the IPCC report was released.

For example, several studies indicate that severe winter storms over the eastern Atlantic and western Europe abruptly increased in the late 1980s. British researchers found an increase in the intensity of severe gales hitting the U.K.; some reached unprecedented intensities in the late 1980s and early 1990s, exceeding anything in records going back to the 1880s. However, the total number of gales did not break records.

There's recent evidence of an increase over the past century in the intensity of cyclonic storms over the Great Lakes during November and December. November is typically the most active month for storms; the lake waters are warmest then—warmer than the air—and the transfer of heat and moisture to the atmosphere fuels passing storms. James Angel of the Illinois State Water Survey and Scott Isard of the University of Illinois found that the number of strong storms per year has more than doubled since 1900. "There was a decrease in the total number of cyclones, but the ones we're getting are stronger," said Angel.

The researchers say this is the first study to document an increase in strong cyclones over the Great Lakes and it contradicts assumptions in previous climate change studies that cyclones in the region have not or will not change over time. Angel said they didn't examine the impact of global warming in detail, but he speculated that increasing temperatures and further warming of lake waters could lead to more intense storms, particularly if it delays the formation of surface ice. The lakes' ability to intensify storms "goes away once the ice cover is there. If the ice cover is delayed, there's more opportunity for this kind of situation to set up."

Several studies of large-scale cyclonic storms in the U.S. have shown an increase in frequency in the first half of the century, then a decrease from the mid-50s to the mid-1980s, followed by an increase through the mid-1990s. The IPCC notes that the prevalence of the most destructive storms over the eastern part of North America, though erratic, has increased overall; seven of the eight most intense storms of the past half-century are clustered in the last 25 years. The incidence of severe winter storms over the Northern Hemisphere has increased as much as three times in the past century.

There's also evidence of increasing storminess over the North Atlantic in the past three decades. A study by Munich Reinsurance found that cyclone activity in 1995 was more than twice as high as the average over the previous 45 years. Observations show that North Atlantic wave heights have increased, an indicator of stronger winds and more stormy conditions.

Modeling studies by the British Meteorological Office suggest that, as the climate warms, Northern Hemisphere storm tracks will move northward and increase in intensity, especially in the eastern Atlantic and western Europe. A Canadian modeling study agreed with the projected increase in storm intensity but did not agree that the storm track would shift northward, according to Steven Lambert of Environment Canada's Centre for Climate Modelling and Analysis. Lambert's research indicates that, if atmospheric CO_2 levels double, there may be a decrease in the total number of winter storms over the Northern Hemisphere, but the number of very intense storms may rise; storms that are among the top 5% in intensity could increase by 17%.

After making this projection using a computer model, Lambert examined observational data to see if there's already been an increase in the number of intense storms in this century. Weather records going back to 1900 indicated the annual number of intense storms remained relatively constant until about 1970, then rose sharply during a period when global temperatures were rising. "In the 1990s, they were 60% higher than in the 1960s," Lambert said. "As the surface temperature goes up, evaporation increases, the moisture in the atmosphere increases and this plentiful supply of water vapour encourages storms to intensify."

However, Jerry Mahlman urges caution in attributing regional changes in storminess to global warming caused by greenhouse gases. Identifying trends in a phenomenon with such large natural variability is difficult. "You may be finding things, but what's causing it is the question. It's not that such signals do not exist—signals of global warming could exist—but they can easily be overwhelmed by the noise of natural variability."

Tropical storms: Tropical cyclones—otherwise known as hurricanes and typhoons—are the deadliest and most costly storms. Featuring high winds moving in a circular pattern around a low pressure core, they develop over warm water, deriving their energy from evaporation and heat rising from the ocean surface. Their intensities are categorized according to wind speeds; in the North Atlantic, hurricanes are considered intense if they belong to Categories 3, 4, and 5 with wind speeds of at least 180 km/h.

There's much debate about what effect global warming is likely to have on tropical cyclones. Some studies predict increases in frequency and intensities but others do not. The IPCC report concluded there's insufficient knowledge to say whether the occurrence or distribution of severe storms will change.

Because tropical storms are too small to be resolved by computer models, predictions of their behavior have been based on other factors—for example, the fact that global warming is likely to increase sea surface temperatures (SSTs), a critical factor in the formation of tropical storms. Meteorologist Kerry Emanuel of the Massachusetts Institute of Technology has argued that a doubling of atmospheric CO_2 will warm SSTs enough to increase the intensity of tropical cyclones. (He suggests that even higher SSTs could trigger incredibly powerful hurricanes known as "hypercanes.")

A study by Thomas Knutson and colleagues at NOAA's Geophysical Fluid Dynamics Laboratory at Princeton University found that greenhouse warming would create more intense typhoons in the northwestern part of the Pacific basin. This region, near the Philippines, is where the strongest typhoons currently occur. The study involved a special high-resolution weather model used for predicting hurricanes that can resolve smaller-scale weather phenomena than global climate models. When the researchers examined the effects of increasing atmospheric CO_2 levels, they found that maximum wind speeds would increase and the low pressure core of the typhoons would get lower (a sign of increased intensity). Writing in *Science*, they concluded that the maximum intensity that typhoons could possibly attain would increase in a warmer climate.

Most typhoons or hurricanes don't reach their maximum possible intensity; they weaken or die as they move over land or cooler waters. What the Knutson study indicates is that the storms that survive are likely to reach higher maximum intensities in a warmer climate. Once a cyclone forms, the maximum potential intensity it can reach depends on how warm the sea surface temperature is, according to Jerry Mahlman. Since sea surface temperatures will be higher in a greenhouse-warmed world, the maximum potential intensity is greater. However, while the latest study indicates hurricane *intensities* might increase with global warming, Mahlman said there's no evidence that the annual *number* of tropical storms, hurricanes, or typhoons will increase in a warmer climate.

Christopher Landsea of NOAA's Hurricane Research Division analyzed global data on cyclone and hurricane trends. Unfortunately, for most regions, measurements go back only 25 to 30 years; the longest records are for the Atlantic basin, where measurements began in the mid-1940s. The evidence suggests that some regions have experienced increases in the number of cyclones, while others have experienced decreases and still others show no significant trend. In the oceans around Australia, the number of moderate cyclones has decreased substantially since 1970, likely because of El Niño's influence, but the number of intense tropical cyclones increased slightly. These findings imply that while El Niño reduces the frequency of cyclones in the region, it doesn't affect their intensity once they form.

As for Atlantic hurricanes, we've seen that the growing economic losses in recent years is primarily due to socioeconomic factors, not an increased frequency or intensity of the storms themselves. In fact, records going back to the mid-1940s show that while there's been no major change in the total annual number of Atlantic hurricanes, there's actually been a decrease in the number of *intense* hurricanes.[*] During the five decades studied, there was no change in the highest intensity achieved by the strongest hurricane in each year. In fact, the period between 1991 and 1994 was quieter than any other time in the five decades studied, partly because strong and more frequent El Niños suppressed Atlantic hurricane activity. However, this was overshadowed in the public mind by the record US$27 to 30 billion in damages caused by Hurricane Andrew in 1992.

In 1995 and 1996, however, hurricane activity picked up dramatically. In fact, 1995 was the most active year since 1933; eleven of its nineteen storms became hurricanes, five were classified as intense (Category 3 or higher) and three of those made landfall. Nineteen ninety-six was also active; nine of the thirteen storms were hurricanes; six were classified as intense and two of those made landfall. This increase in Atlantic activity has been attributed to the fact that El Niño gave way to La Niña in 1995 and 1996, which were the most active back-to-back Atlantic hurricane years on record. Although 1997 was not an above-average year, as had been predicted, the period from 1995 to 1997 was the most active three-year stretch on record.

According to a 1997 paper by hurricane expert William Gray of the University of Colorado and his colleagues, "a new era of major hurricane activity appears to have begun with the unusually active 1995 and 1996 seasons." Although they don't blame global warming, suggesting these changes may be related to natural climate fluctuations, they nevertheless warn that there could be a significant increase in major landfalling Atlantic hurricanes over the next couple of decades, which could cause unprecedented damages in the U.S.

In 1998, with El Niño's influence waning, the Atlantic hurricane season again intensified. There were 14 tropical storms, nine of which reached hurricane strength. On September 25, four hurricanes were in progress at the same time, "the first time such an event occurred in this century," according to the U.S. National Hurricane Center. In late October, Mitch, a rare Category 5 hurricane, rampaged through Central America, killing at least 10,000 people, with many more missing. This

---

[*] Intense hurricanes are of particular interest because, while they make up only about 21% of hurricanes, they cause about 83% of hurricane damage.

was the deadliest Atlantic hurricane season in 200 years. Forecasters predict 1999 will be just as active. The United Nations said development had been set back by decades.

Sea Level Rise

Global sea level has risen about 10 to 25 centimeters in the past century and the average rate of increase has been significantly higher than at any time in the last several thousand years, according to the IPCC report, which attributed the rise largely to the concurrent increase in global temperature.

There are two mechanisms by which this could happen. First, heated water expands, so warmer temperatures increase the ocean's volume. Second, warming accelerates melting of land-based ice (i.e., glaciers and ice sheets) and the water flowing into the oceans causes them to rise. (However, ice already floating on water does not cause a sea level rise when it melts, just as ice cubes melting in a glass of water don't cause the level in the glass to rise.)

The IPCC estimates that roughly half or less of the observed rise in sea level can be attributed to thermal expansion of the ocean and melting of glaciers and ice caps. Most of the rest likely comes from melting of the huge polar ice sheets that cover Greenland and Antarctica. However, there are large uncertainties associated with the ice sheets because global warming may have contradictory effects. In a warmer climate, more snow would fall on interior regions of the ice sheets, causing them to grow; at the same time, warmer temperatures would cause iceberg calving and melting at the edges. The IPCC report says melting is expected to dominate over ice growth in the Greenland ice sheet (causing a rise in sea level) but the opposite would happen in Antarctica (causing a drop in sea level). It suggested that observational data were not good enough to tell for certain whether these ice sheets have grown, shrunk, or remained balanced over the past century.

However, more recent research indicates the Greenland ice sheet has been shrinking by about 2.5 centimeters per year. In the Antarctic some of the floating ice shelves at the edges of the ice sheet have also been shrinking and a large part of one collapsed into the ocean in 1995. According to Henry Hengeveld, however, the interior Antarctic ice cap itself "may be slowly building due to a moister climate."

There's little doubt sea level has risen over the past century, but scientists are debating about how much of it can be attributed to global warming. John Shaw of the Geological Survey of Canada, who studied the impact of rising sea levels in Atlantic Canada, doubts we've seen the human signal yet. Atlantic Canada experienced a large and rapid sea level rise of about 30 to 40 centimeters over the past century, but that's largely due to sinking of the earth's crust in the region. "About 10 to 15 centimeters is unexplained," he said. Though some take this as a sign that sea level is rising because of global warming "no one has concretely linked the two." Natural fluctuations are a more likely explanation, he argued. "Global sea level goes up and down every few centuries; the rise we're seeing now has been going on for 300 to 400 years." But this doesn't belie the possibility that global warming could boost natural sea level rise in the future and if that happens, "we'll be in a fairly bad position in Atlantic Canada."

Canadian studies indicate that the Atlantic region could face increased flood risk and coastal erosion from sea level rise. Environment Canada estimates that storm surges and flooding that now hit parts of the Maritimes every 100 years might occur every 20 years. Agricultural lands in diked regions of Nova Scotia's Annapolis Valley could be threatened. These projections indicate British Columbia and the Yukon also face rising waters, perhaps as much as 30 to 50 centimeters in northern regions by 2050.

Sea level along much of the U.S. coast has been rising at about 2.5 to 3 millimeters per year (or 2.5 to 3 centimeters per decade), and this has already had an impact on coastal communities. According to James Titus of the EPA, beaches are eroding, floods are coming further inland and water tables are rising. By some estimates, beaches along the U.S. Atlantic coast have receded about 60 to 90 meters in the past century.

Projections of a substantial sea level rise in the future are "very probable" (a 9 out of 10 chance), according to Jerry Mahlman. Using a variety of scenarios with different assumptions about greenhouse gas emissions, the IPCC's "best estimate" indicates sea level would rise about 50 centimeters by 2100. However, because of the uncertainties involved in these scenarios, the projections range from as low as 13 centimeters to as high as 110 centimeters. The IPCC says sea level is expected to continue rising "over many centuries even after concentrations of greenhouse gases are stabilized." Estimates in the second IPCC report are lower than those in 1990, largely because the projected rise in global temperature is smaller. "If global warming were to occur more rapidly than expected, the rate of sea level rise would consequently be higher," the report notes.

An EPA study projected that global warming is most likely to increase global sea levels by 15 centimeters by 2050 and 34 centimeters by 2100. Because other factors such as land subsidence are also causing rising seas, the total will actually be higher. EPA has estimated there's a fifty-fifty chance that by 2100 sea levels will rise between 48 and 63 centimeters on the U.S. east coast, between 33 and 48 centimeters on the west coast, and as much as 140 centimeters on the Gulf coast. EPA calculations also indicate that by 2100, climate change could greatly increase how *fast* sea level is rising. It said stabilizing greenhouse gas emissions by 2025 could cut the rate of sea level rise in half.

Perhaps the most significant aspect of the projected rise in sea level is that it will be driven, especially during the first half of the 21st century, by the temperature increase caused by greenhouse gases that are *already* in the atmosphere. This is because the oceans and ice sheets take longer than the atmosphere to respond to rising temperatures; during the early 21st century, they'll still be adjusting to temperature increases that have already occurred. Greenhouse gases emitted during the 21st century will therefore primarily affect sea level rise beyond the year 2100. Model simulations indicate sea levels will continue rising for many centuries, even after greenhouse gas concentrations stabilize and, in fact, even after global *temperature* stabilizes. This is significant when assessing the consequences of delaying action to deal with global warming.

The uncertainties about whether the Greenland and Antarctic ice sheets are growing or shrinking affect estimates of their future impact on sea levels. There's also concern about the possible collapse of the huge West Antarctic Ice Sheet, which by some estimates could increase sea level a disastrous six meters. The IPCC concluded this is unlikely to happen in the 21st century and that the contribution of ice sheets to sea level changes before 2100 will probably be "relatively minor."

As for extreme events, rising sea levels present two threats. The first is increased flooding or inundation of low-lying coastal areas, including some major cities. Deltas of many major rivers could become unlivable, including the Mississippi in the U.S., the Yellow and Yangtze in China, the Nile in Egypt, the Ganges in Bangladesh, the Rhone in France, and the Po in Italy. Even regions protected by dikes and sea walls could be threatened. Incursions of sea water could also contaminate coastal freshwater supplies and threaten drainage and sewage systems.

The second major threat from higher sea level is an increase in the size of storm surges, elevated water levels caused by winds and pressure changes during storms. A huge wall of water, typically several meters higher than normal, builds up ahead of the storm and when it crashes onto land it often causes more death and destruction than the storm itself. The IPCC estimates that a half-meter sea level rise would more than double the world's population at risk from storm surges, from about 46 million to about 92 million. A one-meter rise would increase that to 118 million.

New Orleans, already facing problems because of land subsidence, would likely be the most threatened urban area in the U.S. During storms, water levels already rise to within a meter of the top of levees that protect the city from storm surges. Computer projections indicate that with a one-meter sea level rise, a severe hurricane could put most of the city under water.

Climate Surprises

There are two others ways in which climate change may be considered "extreme." The first is how *fast* climate change is happening and the second concerns the potential for sudden surprises—the possibility that, at some unpredictable point, global warming will pass a threshold that suddenly throws the earth into a new climatic state.

The rate of change: We've seen that the earth's climate has been relatively stable for the past 10,000 years. Most of human civilization has developed during that time—agriculture, urban infrastructure, technology. In the last thousand years, when these trappings of civilization advanced at a pace unprecedented in human history, the average temperature of the earth has neither increased nor decreased by more than 0.5°C in any one century.

It appears we're about to change that. According to model projections, we'll usher in the new millennium by warming its first century at least 1°C and quite possibly twice to three times as much. This rate of warming would be unprecedented in the past 10,000 years. By way of comparison, this temperature change is equivalent to about half or more of the change (in the opposite direction) that led to the last ice age, but it will occur over just one century, whereas ice ages typically take many centuries or millennia to develop.

Humans, plants, animals, and ecosystems can adapt to a changing climate, but they need time. According to Gordon McBean of Environment Canada, "while most species can migrate in response to slow climate change, paleo studies suggest that rates of change in excess of 0.1°C/decade (i.e., 1°C/century) are almost certainly too rapid to avoid major disruption." This is the minimum increase expected to occur over the next century due to global warming. In 1997, leading U.S. ecologists issued a statement saying that "rapid climate change is more dangerous to plant and animal communities than gradual climate change even if the total amount of change that eventually occurs is exactly the same....The rate of projected change is enough to threaten seriously the survival of many species."

Human societies will also be vulnerable. Adaptive measures require time to implement, particularly those that involve transforming infrastructure, energy use, and lifestyles. Perhaps the greatest danger of global warming is that some of its most serious consequences will come upon us before we know it—and long before we're ready for them. Our continuing reluctance to take precautions only increases our vulnerability.

Climate thresholds: There are those who argue we should wait for proof of global warming before taking costly steps to avoid it. A common counterargument is that the time lag associated with climate change means the problem will continue to get worse long after we have such proof. In short, climate change is like a ship that can't be quickly turned around. But perhaps a more important question is whether it can be turned around at all. Scientists are raising concerns that we may at some point unknowingly cross a threshold, a trip wire, that will suddenly throw the climate into a state that will be irreversible on human time scales.

This could happen because of positive feedbacks, mechanisms that speed up changes in the climate system. A positive feedback is like an accelerant in a fire. One key example involves water vapor, the most important natural greenhouse gas. A warmer atmosphere can hold more water vapor, which will accelerate further warming, leading to even higher water vapor content, and so on. Other potential feedbacks are more ambiguous—for example, changes in cloud cover could cause both positive and negative results.

One of the more ominous phrases used by climate scientists is the term "nonlinear." A nonlinear response is one that's disproportionate to the force that caused it. It is, in essence, an overreaction. As Henry Hengeveld put it, "a nonlinear response happens when a response is not smooth. For example, a door that's stuck will not open when pushed with moderate force but then bursts open when the push is strong enough to unstick the door." Perhaps the most dramatic example of this was the sudden and unpredicted appearance of the ozone hole in the 1980s.

According to the IPCC, the climate system's "nonlinear nature" makes it subject to unexpected behavior. As the climate changes beyond the boundaries of what we know from the past, it's likely we'll be confronted with "surprises and unanticipated rapid changes."

Paleoclimatic data indicate that rapid and dramatic shifts happened in the past. Ice cores from Greenland show that the climate throughout much of the North Atlantic region experienced both rapid cooling and rapid warming that occurred over just a few decades or even, in some places, a few years. A study of ocean sediments by paleoclimatologists at Columbia University indicates that Africa regularly experienced abrupt and dramatic shifts between cold/wet and warm/dry conditions, often within the span of a human lifetime.

These findings have astonished many scientists who previously believed it took millennia for such large shifts to occur. The evidence has persuaded paleoclimatologist Wallace Broecker of Columbia University's Lamont-Doherty Earth Observatory to reject arguments that natural mechanisms will kick in to counteract global warming. "I find absolutely no support for the self-regulation concept....The climate system has the bad habit of undergoing large and abrupt jumps from one mode of operation to another." In the journal *Science*, he wrote: "The paleoclimate record shouts out to us that, far from being self-stabilizing, the earth's climate system is an ornery beast which overreacts to even small nudges."

Scientists are currently studying whether global warming could trigger one of these abrupt changes—one that could plunge most of Europe into a deep freeze and might even set off a little ice age if it were extreme enough. The system in question is an ocean current known as the "thermohaline circulation" (THC) or, more popularly, the "conveyor belt" that redistributes heat between the equator and the Arctic. This conveyor belt is powered by the sinking of dense, cold, salty water in the North Atlantic, where it plunges to the depths of the ocean and streams back toward the equator. This plunging action draws warm surface waters northward from the equator along the path of the Gulf Stream; winds blowing across this current pick up heat and give Europe a mild climate compared with places at the same latitude on the other side of the Atlantic. As the current moves north, it cools and becomes increasingly salty because of evaporation; by the time it gets to Greenland, it's denser than the surrounding waters and starts to sink, keeping the conveyor going.

According to Broecker, who first proposed this concept with George Denton of the University of Maine, if the conveyor belt shuts down, temperatures in the north Atlantic and surrounding land areas would drop by about 3°C within a decade. In fact, some scientists have said a complete shutdown of the THC could trigger an ice age. Studies of ocean sediments done by researchers at the Massachusetts Institute of Technology indicate that a rapid reduction in the THC may have ended the last "interglacial" (a ten thousand-year warm period similar to the one we're in now) and brought on the earth's last ice age.

Ironically, one of the things that could slow or shut off the conveyor belt is global warming. If, as projected, global warming increases precipitation in northern latitudes and melts Arctic ice, this would inject large amounts of fresh water into the Arctic that would dilute the salinity of the incoming conveyor current. If the North Atlantic region became too warm and if fresh water overwhelmed the incoming salty water, the current might not cool and sink; this would effectively slow the conveyor belt or possibly even stop it. This is exactly what some scientists think happened about 12,000 years ago, when the earth did an abrupt about-face while coming out of the last ice age and plunged back into cold conditions.

This could happen again. According to the IPCC, most computer simulations show global warming could reduce the strength of the THC. Several modeling studies have produced disquieting results. A German study found that it would not take large changes in the amount of fresh water entering the North Atlantic to alter the THC and cause major climatic changes. Another study done by Swiss researchers indicated that the conveyor belt could shut down permanently if atmospheric concentrations of greenhouse gases double by the end of the 21st century, as most scientists expect. If concentrations grow more slowly than expected, the conveyor belt might slow down, but it would not entirely stop, they concluded.

Mahlman has estimated there's about a two in three chance that the THC will be suppressed because of increased precipitation in higher latitudes caused by global warming. However, he doubts this would occur suddenly or that it would have worldwide consequences. The sudden shutdowns in the past occurred during cold periods, and climate models suggest that during warm periods, "we see the thermohaline circulation very slowly shut down. If that were to happen, it would have a major effect on Europe, but it would not likely have a profound effect globally."

How long the conveyor shutdown would last is a matter of debate. Some scientists suggest the cooling that would result from the loss of the conveyor belt would set up conditions that would eventually kick-start it again, though this might take a thousand years or so. In the interim, life could become decidedly unpleasant in a large area surrounding the North Atlantic Ocean.

The lesson to be learned from abrupt climate changes in the past is that they *can* happen. Most people think of climate change as a gradual process, not something that we can experience in a human lifetime. But both historical evidence and computer modeling suggests that rapid climatic about-faces are not impossible. More to the point, there's little we can do about them once they begin.

There is no simple answer to the question of whether extreme weather is increasing all over the world. At present, we can't say there's a single global trend, and it may be some time before a clear signal emerges from the noisiness of day-to-day weather—if it ever does. Some scientists have suggested it may not; because of the natural variability of weather, a decrease in extreme events in some regions may, in a statistical sense, "cancel out" an increase in events in other regions and thus create the impression that little is happening on a global scale.

Even if this were to prove true, it doesn't eliminate the potential for increased danger from extreme weather events as the world warms. It's the local and regional weather events, not global averages, that have a direct impact on our lives.

Sherwood Rowland, a Nobel prize–winning University of California scientist who was one of first to identify that CFCs destroy ozone, has suggested that the weather disasters caused by the 1997–98 El Niño might galvanize public concern about global warming much as the discovery of the Antarctic ozone hole did for ozone depletion. The ozone hole is widely regarded as the "smoking gun" that convinced everyone that human activities were destroying the ozone layer. David Runnalls of the International Institute for Sustainable Development says the history of environmental action indicates that "there has to be a smoking gun." For climate change, "extreme weather could be it."

five

FINGERPRINTS IN THE GREENHOUSE:
Detecting the Human Influence

For people who make it their life's work to explore the unknown, scientists are surprisingly conservative when it comes to reporting back to the rest of us what they've found at the frontier. They live in a world of hypothesis, probability, and uncertainty. They pounce on each other as a matter of course, prodding for the slightest weakness in theory, observation, or logic. They're uncomfortable with speculation but have a deep-seated aversion to the word "proof," resisting demands for definitive statements about the way things are. This is especially true of climate scientists, whose lot in life is wrestling with what is undeniably one of the Big Questions facing humanity: Are we warming the climate in a way that is dangerous to life on earth?

It would be nice if we could put them on the witness stand and bark, as lawyers often do, "Just answer the question—yes or no?" Actually, that is, in effect, what we often expect of scientists.

The IPCC's now-infamous statement that "the balance of evidence ...suggests a discernible human influence on global climate" comes as close to a definitive answer as we're likely to get for now. Cautious as it is, it's still remarkable, coming from an international organization involving more than 2000 scientists worldwide. The statement reflects a growing—though not universal—consensus among climate experts that they've detected suspicious human fingerprints in the climate record of the past century, if not—yet—a "smoking gun."

Predictably, the IPCC statement precipitated an intense political and scientific firestorm because, if we accept that human activities *are* warming the climate, we must consider what to do about it, even if only to make a conscious decision to do nothing. There are many divergent and often incompatible views about what we should or shouldn't do. The costs and benefits of different options, as best we can determine them, fall on different people in different places at different times, including future generations who have no say in the present debate. In the short term, there may be "winners" and "losers" and we can't rule out the possibility that, in the end, everyone will lose. It has all the makings of a messy, fractious, intractable battle—and that's exactly what it's become.

Previous chapters examined whether the world has already warmed. More significant, however, is whether this warming is simply part of natural climate variability or the result of human intervention. Few questions have generated more scientific effort, or more intense controversy. Indeed, the mandate of the IPCC is to assess the "relative magnitude of human and natural factors ...[and] whether a human influence on present-day climate can be detected." It notes the difficulty of doing this: "In observed data, any 'signal' of human effects on climate must be distinguished from the background 'noise' of climate fluctuations that are entirely natural in origin." At present, the impact of human activities is still small compared with the great natural variability of the climate. As the report notes, "our ability to quantify the human influence on global climate is currently limited because the expected signal is still emerging from the noise of natural variability and because there are uncertainties in key factors."

The obsession with whether we've already seen the human signature is puzzling in some ways. After all, there is *no question* that human activities emit greenhouse gases into the atmosphere, or that atmospheric concentrations of greenhouse gases are increasing as a result, or that greenhouse gases cause warming. Therefore, logically, there is *no question* that human activities contribute to warming the climate. And since there is also *no question* that greenhouse gas emissions from human activities are currently increasing every year, we know that their climatic influence is bound to grow.

In 1995, a forum on climate change convened by the U.S. Global Change Research Program concluded that it's known "with great certainty how greenhouse gases affect the energy balance of the earth, and with similar confidence that concentrations of these gases are now increasing due to human activities, and that these increases should result in global warming."

So where's the debate?

"At issue is not whether the Earth will warm, but by how much, where, when and with what consequences for society and ecosystems," the USGCRP forum elaborated. The unanswered questions focus on the magnitude and rate of change: *How much* and *how fast* have human activities already changed the climate? Can we actually detect this change against the backdrop of natural climate variability? *How much* and *how fast* will human activities change the climate in the future? What *impact* will this have on the natural environment and human society in different places at different times? And, most importantly, will these impacts be large enough to matter?

The ultimate concern, as it's framed in the Convention on Climate Change, is preventing "*dangerous anthropogenic interference*" with the climate system. The Convention does not, however, define what's meant by "dangerous," because this involves political, social, and economic factors as well as scientific ones. As for scientists, what preoccupies them is the question of "interference"—detecting it, measuring it, assessing its present and future magnitude.

As we've seen, there's strong evidence that the earth has warmed over the past century. There's also strong evidence that atmospheric concentrations of greenhouse gases are rising as well. Ice core studies show that throughout the earth's history, warm periods are well correlated with high CO_2 levels and cool periods with low CO_2 levels.

However, large changes in greenhouse gas concentrations in the past were caused by natural processes, so the question is whether the recent temperature rise is attributable to human activities. Here again, there's strong evidence that it is. From ice cores and atmospheric measurements, we know there've been sharp increases in important greenhouse gases—CO_2, methane, and nitrous oxide—since the Industrial Revolution began in the late 1700s. Carbon dioxide, for example, has increased almost 30%, from 280 ppmv (parts per million by volume) to more than 360 ppmv today—higher than any level found in the ice core record for the past 220,000 years.

Carbon dioxide concentrations are now growing at a rate of about 0.4% a year or 4% per decade. This takes on added significance given findings from ice cores that during the current millennium prior to industrialization, CO_2 concentrations remained much more stable over long periods, varying by only about 3.5% around the 280 ppmv mark. During that time the climate was also quite stable.

Methane, though present in smaller quantities than CO_2, has grown even more since pre-industrial times, rising from about 700 ppbv (parts per billion by volume) to 1720 ppbv, an increase of about 146%. It's increasing at a rate of about 0.6% a year or 6% per decade. Nitrous oxide has increased by more than 13%, from 275 to 312 ppbv, and is rising at a rate of 0.25% a year or 2.5% per decade.

The IPCC report says there's no doubt the increases in atmospheric concentrations of CO_2, methane, and nitrous oxide during the past century are largely due to human activities, primarily fossil fuel burning, agriculture, and land-use changes such as deforestation. It estimates these gases are responsible for about 80% of the greenhouse warming caused by human activities since pre-industrial times, with CO_2 alone accounting for about 60%. More recently, the World Watch Institute estimated the combined contribution of the three gases at 90%, with CO_2 accounting for 65%.

Tropospheric ozone and halocarbons (which include chlorofluorocarbons or CFCs) account for most of the rest of the warming. In the lower atmosphere, ozone, a constituent of smog, is created by reactions between the sun's energy and chemicals such as nitrogen oxides, hydrocarbons, and carbon monoxide, produced by fossil fuel combustion as well as natural sources. It's estimated that tropospheric ozone has perhaps doubled in the Northern Hemisphere, but the relative contribution of human sources is unknown. There's not enough data to determine if concentrations have increased in the Southern Hemisphere.

The climatic impact of stratospheric ozone, which is formed by natural processes, is more complex. Levels have dropped by roughly 4 to 5% per decade since the late 1970s, largely because of destruction by halocarbons. Since ozone both absorbs the sun's energy and acts as a greenhouse gas, the stratospheric losses of ozone contribute to cooling in the upper atmosphere and to a lesser extent at the surface.

The increase in halocarbons is, of course, entirely caused by human activities, since these chemicals didn't exist before industrialization. Halocarbon concentrations are in the parts per trillion, much lower than those of other greenhouse gases, but some are far more powerful heat-trappers and will remain in the atmosphere for decades or centuries—even millennia in some cases. Some halocarbons have been banned to protect the ozone layer, so their growth rates have started to slow down or reverse. Others, however, continue to grow; chemicals used as substitutes for those banned have increased by annual rates ranging from 5 to 100% in recent years.

The climatic impact of halocarbons is complex. They're greenhouse gases, so they cause warming, but they also destroy stratospheric ozone, so they indirectly contribute to cooling. Scientists have estimated cooling caused by stratospheric ozone depletion may have counteracted as much as 30% of the global warming that should have occurred since 1979. However, it appears so far that the direct warming caused by halocarbons has exceeded their indirect cooling effect.

Ozone "holes" further complicate this picture. These recurrent holes have appeared over Antarctica and sometimes over the Arctic; the most severe have occurred in the 1990s, with depletions of up to 99% in some parts of the stratosphere. The hole that developed over Antarctica in the fall of 1998 was the largest ever, covering an area the size of Canada and the U.S. combined.

A surprising number of people believe ozone holes are what's causing global warming, but in fact, the opposite may be closer to the truth. While greenhouse gases cause warming in the lower atmosphere, they cause cooling in the stratosphere and this appears to accelerate the destruction of ozone by halocarbons. Scientists at Columbia University suggest that greenhouse gases may have been partly responsible for ozone holes in the Arctic in recent years, including record-breaking losses of ozone in 1997. The implication is that increasing greenhouse gas emissions will contribute to continuing destruction of ozone *even after CFC levels in the stratosphere drop.* The scientists calculated that by 2020, stratospheric ozone loss will be roughly twice what it would have been without the increases in atmospheric greenhouse gases expected to occur over that period.

Greenhouse Gas Emissions from Human Activities

It's sometimes claimed by skeptics of the global warming theory that human emissions of greenhouse gases cannot be that significant because they're much smaller in volume than natural sources of these gases. For example, R. J. Eaton, chairman of Chrysler Corporation, argued in the *Washington Post* that eliminating all cars and trucks worldwide would reduce CO_2 emission by less than half of one percent. He noted that nearly 97% of CO_2 comes from natural sources and that cars contribute "only one-eighth of that small remaining fraction of CO_2 attributable to man."

This argument is flawed, since it's the "small remaining fraction attributable to man" that's precisely the problem. This concept was more easily grasped by John Browne, CEO of British Petroleum of America, a company that broke ranks with the fossil fuel industry in accepting the need to act against global warming. "Human activity accounts for a small part of the total volume of emissions of carbon—but it is that part which could cause disequilibrium," Browne said in a 1997 speech.

Even though it's true that natural emissions of greenhouse gases are currently much greater than human emissions, this is irrelevant to the problem of global warming. The reason is simple: natural emissions and natural sinks are nearly equally balanced and thus produce little net change in atmospheric concentrations of these gases. In fact, during this millennium prior to industrialization, CO_2 concentrations were very stable, varying only a few percentage points around the 280 ppmv mark. This small variation indicates that nature had achieved a remarkably good equilibrium between the forces that release CO_2 and those that remove it from the atmosphere. Even though human emissions are small in absolute terms compared with natural forces, they have still been large enough to throw this delicate natural balance out of whack. The 30% rise in atmospheric CO_2 levels since pre-industrial times is clear evidence of this.

There's another important fact to remember: greenhouse gases are very powerful at trapping heat. They keep the planet 33°C warmer than it would otherwise be, yet they comprise less than 1% of the volume of the atmosphere. Small changes in their concentrations are not trivial.

Emissions of greenhouse gases continue to rise. About half of all CO_2 added to the atmosphere throughout human history has been emitted in the last three decades. The major human sources of CO_2 are fossil fuel burning, deforestation, and changing land use. The IPCC estimated that during the 1980s, fossil fuel burning released an average of 5 to 6 billion tonnes of carbon into the atmosphere in the form of CO_2 each year. At the 1992 Earth Summit, developed countries made voluntary commitments to reduce their emissions to 1990 levels by 2000, but virtually all failed to achieve this goal and emissions continued to climb throughout the 1990s. In 1996, according to the World Watch Institute, carbon emissions from fossil fuels jumped to an all-time high of 6.35 billion tonnes, up from 6.2 billion tonnes in 1995, which also set a record. In 1997, CO_2 emissions from fossil fuels rose to 6.4 billion tonnes. When emissions from deforestation and other land use changes are added, the global total comes to more than 7 billion tonnes a year.

Not all the carbon emitted by human activities remains in the atmosphere; it's estimated about half is fairly rapidly removed by sinks, primarily absorption in the oceans and increased uptake by plants. The remainder is removed more slowly and thus accumulates year after year. This is why it takes a long time for CO_2 levels in the atmosphere to respond to reductions in emissions.

Human sources of methane include rice cultivation, fossil fuel production, transportation, biomass burning, and landfill sites. The IPCC report estimates that human activities are responsible for an estimated 60 to 80% of total annual methane emissions, which averaged about 535 million tonnes a year in the 1980s; the production and burning of fossil fuels accounted for about a fifth of the human emissions. Compared with CO_2, methane is removed more rapidly and completely by sinks, primarily chemical reactions in the atmosphere; it's estimated that less than 10% of annual emissions remains in the atmosphere.

Emissions of methane and CO_2 from natural processes could also be altered by global warming caused by human activities. For example, wetlands, which account for an estimated 20% of annual methane emissions, may increase their output if microbial activity is boosted by climate warming. On the other hand, if conditions become drier and water levels drop, this would reduce the amount of methane released from wetlands but it would increase peat decay and thus increase CO_2 emissions. Warming may also release methane now frozen in Arctic tundra or locked up in ocean sediments, further enhancing methane's contribution to the warming trend.

Nitrous oxide is produced by industrial processes, agriculture, biomass burning, and automobile catalytic converters. An estimated 3 to 8 million tonnes is emitted each year, roughly half of the amount released by natural sources in oceans and forests. The major removal process (being broken down by the sun's energy in the stratosphere) is slow, so most of the nitrous oxide emissions added each year remain in the atmosphere for more than a century.

The link between rising greenhouse gas concentrations and rising global temperatures is well established and accepted by most climate scientists. The idea that CO_2 emissions from burning fossil fuels could enhance greenhouse warming was first theorized in 1896 by Swedish chemist Svante Arrhenius, who estimated that a doubling of CO_2 would raise the global temperature by 6°C. In 1924, a U.S. physicist speculated that industrial activity would double atmospheric CO_2 concentrations from pre-industrial levels within 500 years (clearly underestimating the explosive growth in energy use that would advance the date of projected doubling by roughly three centuries.) In 1949, just as post-war expansion was taking off in industrialized countries, a British scientist, G. S. Callendar, argued there was a link between a 10% jump in atmospheric CO_2 between 1850 and 1940 and warming in North America and northern Europe that began in the 1880s.

More recently, ice core studies have shown a consistent relationship between temperature changes in the earth's past and natural variations in greenhouse gas concentrations. "The good correlation between atmospheric CO_2 and climate at multi-million-year time scales and longer is evidence that the climate is sensitive to variations in CO_2 concentrations," writes University of Toronto geography professor Danny Harvey in the journal *Climatic Change*.

It follows that since human activities emit greenhouse gases, they contribute to global warming.

The Problem of Natural Climate Variability

While most climate scientists don't dispute that increasing greenhouse gas concentrations cause warming, there are questions about the extent to which natural forces (either warming or cooling) have influenced the observed warming trend since pre-industrial times. There are several possibilities: natural variability could have been the dominant factor, it could have been equal to human-induced warming, or it could have been only a minor contributor. Or it might even have worked *against* the warming caused by human activities, resulting in a temperature increase that's less than it might otherwise have been.

In its 1990 report, the IPCC concluded that the evidence was not good enough to determine the likeliest scenario. While the observed warming was "broadly consistent" with model predictions of the effect of human activities, it was also about the same magnitude expected from natural climate variability and could, therefore, have been largely natural in origin. But, the report added, it was also possible that natural variability plus other human factors that cause cooling (e.g., aerosols) "could have offset a still larger human-induced greenhouse warming."

One contentious debate over natural causes of the warming since pre-industrial times concerns the role of variations in the sun's energy output. Direct measurements of the sun's radiation by satellites has been possible only since 1979, but indirect evidence derived from observing sunspot activity has been available since the 1600s. The satellite data indicate that, at present, the sun's energy currently varies by only a small amount (one-tenth of 1%) over the 11-year sunspot cycle.

Most climate scientists don't believe this is large enough to account for all the warming since pre-industrial times, especially the rapid warming since the late 1970s, although it may have caused some of it. Estimates range from as low as 15 to 20% of the temperature increase since the late 1800s to as high as 71 to 94%, though most scientists reject the higher figures. Analysis by Judith Lean, a solar physicist with the U.S. Naval Research Laboratory, and David Rind, a physicist at the Goddard Institute for Space Studies, indicates that solar changes might account for about half the warming from the mid-1800s to the mid-1900s but less than a third of the sharp warming between the 1970s and the 1990s. "The rapid warming since 1970 is several times larger than that expected from any known or suspected effects of the sun, and may already indicate the growing influence of atmospheric greenhouse gases on the earth's climate." They said that for "any significant fraction" of the warming since 1850 to be attributed to the sun, it would have to be more than twice as variable as indicated by recent satellite measurements.

Another study attributed much of the 1.5°C warming in the Arctic from the mid-1800s to the mid-1900s to solar variability, but said the additional warming since then is too large to attribute to natural causes and must be largely due to greenhouse gases. Raymond Bradley of the University of Massachusetts, a participant in the international study, said Arctic warming to date, unprecedented as it is, is small compared to what will happen if greenhouse gas emissions continue at their current pace.

According to Jerry Mahlman, the hypothesis that solar variability is primarily responsible for the warming since pre-industrial times would require a "double miracle"—a claim that the earth's climate is extraordinarily sensitive to small changes in solar output but extraordinarily *insensitive* to greenhouse heating, which is very much larger in magnitude.

Unless the sun's energy varies by considerably larger amounts than indicated by recent satellite measurements, its role in global warming is likely to be minor compared with that of greenhouse gases.

Detection of the Human Signal

Scientists are still trying to sort out the human signal in the natural noise of climate variability, but many feel the signal has been growing stronger since the late 1970s and that the 1990s provides strong evidence that it's emerging from the noise.

In 1990, the IPCC hazarded a guess that detection of the human fingerprint in global warming would take at least another decade or more. However, just five years later, in its second assessment report, it was willing to go farther, saying that the warming since pre-industrial times "is unlikely to be entirely natural in origin" and that "the balance of evidence ...suggests a discernible human influence on global climate."

One reason the IPCC was willing to make a firmer statement in 1995 was because of findings from so-called "pattern-based" studies that attempt to establish a cause-and-effect relationship between human activities and observed climate changes. The premise is that changes in the earth's temperature occur in different ways over time and over three-dimensional space (latitude, longitude, and vertical height through the atmosphere) depending on what's causing the change. In other words, different *causes* of climate change, whether natural or human, produce distinct *patterns* of temperature change that are different enough from each other to act as a kind of fingerprint.

The patterns of temperature change observed in the real world were compared with patterns generated in computer simulations that included changes in greenhouse gas concentrations and aerosols. These studies showed significant correspondences between the patterns the computer models said should occur if human activities were causing the changes and patterns that were actually observed in the real world. The match grows stronger over time, consistent with the fact that the human climate impact gains strength as greenhouse gas emissions increase.

The IPCC report says it's highly improbable that the match between real-world and computer-generated patterns could occur by chance due to natural climate variability. It notes that the patterns that occur vertically through the atmosphere are inconsistent with what would be expected if the changes were caused by natural factors like volcanic eruptions or changes in the sun's energy. It concludes that "increasing confidence in the emerging identification of a human-induced effect on climate comes primarily from such pattern-based work."

As Henry Hengeveld of Environment Canada put it, "the pattern of change in space and time is increasingly inconsistent with that expected of natural causes of change but broadly consistent with that projected by climate models simulating the 'human experiment'. The evidence clearly suggests that recent climate behavior is increasingly unusual relative to past climates and that human interference with the climate system is the most plausible explanation."

The IPCC statement was reinforced by a 1996 "state of the climate" paper by Robert Quayle and Thomas Karl of the NCDC, which summarized scientific evidence linking human activities with global warming. The fact that many different and independent forms of data show signs of greenhouse warming "is convincing evidence to most scientists that these signals are messengers of non-random variations in climate," they said. These data include sea and air temperature measurements from ships, balloons, satellites, and land stations, as well as glacial retreat, reduced snow cover, rising sea levels, decreasing daily temperature range, and changes in precipitation patterns. None of these sets of data is definitive and all have problems with uncertainties, but the scientists say it's important to focus on the overall message contained in all these kinds of data taken together. "Only by looking at the aggregated field of evidence can some level of confidence be placed in the results....But the *prepon-*

derance of evidence supports the notion of a real man-made climate variation in the record." (Emphasis added.)

The willingness of scientists to attribute global warming to human interference increased substantially after data showed that 1997 would set yet another global temperature record. NOAA officials explicitly attributed the warming at least partly to human activities: "For the first time, I feel confident in saying there's a human component," said Elbert Friday, Jr., NOAA's associate administrator for oceanic and atmospheric research. Karl added that scientific studies indicate there's only a 5 to 10% chance that human activities are not involved in global warming.

Similar sentiments were expressed by British scientists when they released their 1997 data. "We are beginning to see the fingerprint of man's impact on the climate," said Catherine Senior of England's Hadley Centre for Climate Prediction and Research.

Mahlman, who regards the IPCC's "balance of evidence" statement as "wimpy," takes a stronger stand. He says there's a nine in ten probability that the 0.5°C warming already observed was caused by human activities. While the temperature increase isn't a "definitive smoking gun" that proves the human influence, "we can't think of anything else that makes much sense." Arguments about natural variability or changes in the sun's output "fail various elementary credibility tests. I'll admit there's a one out of ten possibility of something weird out there, but it's not a something weird we can point to."

What's most ironic about the fuss over whether we've *already* detected a human influence in climate change is that the answer is not really important. Given what's known about the effect of greenhouse gases on warming and about the increasing amounts of greenhouse gases humans are releasing into the atmosphere, there can be little doubt that the human fingerprint will emerge at some point. "We understand the climate system well enough to realize that this must be having some effect even if we haven't detected it by now," said Gordon McBean of Environment Canada. "It's unbelievable to expect that we could do what we're doing over the next one hundred years and not expect some change."

Eric Barron of the Earth System Science Center at Pennsylvania State University and a senior official of the U.S. climate research program, has written that debates about the uncertainties of climate models and why the human signal hasn't yet been clearly seen have confused people and "clouded the clearer picture that increases in CO_2 will increase the global mean temperature." It has caused the whole global warming issue to be branded as "controversial" and created an erroneous impression that "the general concept, and not just the details, is in serious doubt."

The fact is that further increases in greenhouse gas concentrations are inevitable; a doubling of CO_2 by the end of the next century appears unavoidable and a tripling is within the realm of possibility.

How Much Will Human Activities Warm the Climate in the Future?

Predicting future climate warming depends critically on assumptions about the volumes of greenhouse gases that will be emitted. There are large uncertainties about this—not least of which concern whether humanity will take serious steps to cut back on burning fossil fuels and other activities that emit greenhouse gases. Despite past promises by developed countries to do something, emissions are rising, not falling.

Projections of future global warming are also affected by other uncertainties associated with estimates of the rate of population and economic growth and how the mix of energy sources will change. The extent to which there will be a shift from fossil fuels to other sources such as renewable and nuclear energy is still speculative.

To account for these possibilities, the IPCC report incorporates a range of assumptions about demographic, socioeconomic, and energy-use trends in six scenarios used to project growth in CO_2 emissions to the year 2100. Only two scenarios, which assume controls on population growth and a substantial switch to renewable energy, result in a leveling off or a drop in emissions by 2100. The other four, which include scenarios that assume current trends will continue or accelerate, all show large increases in emissions, ranging from roughly two and a half to five times the current level of about 7 billion tonnes annually.

The IPCC then projected what impact these different emission levels would have on atmospheric concentrations of CO_2. The results show that CO_2 concentrations *will continue to grow throughout the 21st century in all scenarios, even the ones that assume flat or decreasing levels of emissions after 2025*. This growth in concentrations reflects the effect of the slow removal of CO_2 from the atmosphere.

Of the six scenarios, only the one that involves a reduction in emissions after 2025 avoids a doubling of atmospheric concentrations over pre-industrial times (280 ppmv) by 2100. The CO_2 levels in this scenario would still be roughly 30% higher than 1990s levels. The highest emissions scenarios produce concentrations as high as three and a half times pre-industrial levels, or 170% of 1990s levels. In all scenarios except the lowest one, concentrations would continue rising after 2100, in some cases very sharply. For some scenarios, the concentrations reached would be unprecedented in the earth's history.

In considering what's likely to happen in the future, it's important to remember the distinction between *emissions* and *concentrations* and the implications for global warming:

- *Because of the long atmospheric lifetimes of most greenhouse gases, concentrations will continue to rise for a long time even after emissions stabilize or drop.*
- *If emissions continue increasing at the current pace, concentrations will soar for perhaps centuries to come.* The IPCC report notes that "if global CO_2 emissions were maintained at near current [1994] levels, they would lead to a nearly constant rate of increase in atmospheric concentrations for at least two centuries."
- *Large decreases in emissions are required to get atmospheric concentrations to stabilize or drop.* The IPCC examined several scenarios that could lead to stabilization and concluded that this would require emissions eventually to drop "well below current levels." In fact, immediate stabilization of CO_2 at 1990s levels could only be achieved by immediately reducing CO_2 emissions by 50 to 70%, with further reductions after that. The scenarios show quite clearly (and not surprisingly) that the longer emission reductions are put off, the more severe will be the cuts needed in later years to stabilize CO_2 concentrations.

Calculating the future course of global warming requires consideration of factors other than greenhouse gas emissions, and these unfortunately bring along their own uncertainties. Two factors are particularly important—aerosols and sinks. The IPCC report said the contribution of aerosols is "probably the most uncertain part" of predicting future warming. Since aerosols aren't mixed evenly throughout the atmosphere, it's difficult to compare their effects with those of greenhouse gases that are mixed evenly. The fact that aerosols are removed from the atmosphere more quickly than greenhouse gases also complicates the calculations.

Sinks that remove greenhouse gases also must be included in model simulations of future atmospheric concentrations and their effect on climatic warming. We saw earlier that various processes remove greenhouse gases, thereby influencing their atmospheric lifetimes. For CO_2 the two most important sinks are the oceans and plant life, and both are subjects of intense research at present.

The IPCC scenarios used to project changes in CO_2 emissions and concentrations assumed there would be little change in these sinks over time. However, more recent research by scientists at Princeton University suggests that global warming actually may *reduce* the ability of the oceans to absorb CO_2. In the journal *Nature*, they reported on a modeling study indicating that increased rainfall in a large portion of the southern ocean would freshen the surface waters, which would act as a barrier to downward movement of CO_2 and reduce the loss of heat from the ocean to the atmosphere. Both effects would decrease the ocean's uptake of atmospheric CO_2. There might be changes in the plant and animal life in the oceans that could counteract this effect, but not enough is known about their response. The researchers conclude these changes in the ocean may be occurring already, and "they could substantially affect the ocean carbon sink over the next few decades."

The effect of terrestrial plants on future CO_2 concentrations is also being studied and debated. An argument frequently made by skeptics is that if atmospheric CO_2 levels rise, this will stimulate more abundant plant growth. These plants will, in turn, sequester more CO_2, thus acting as a brake on global warming. Fred Palmer of the Western Fuels Association, an organization that supplies coal for electric generation, has argued that CO_2 is not a pollutant but a nutrient. "It's required for life on Earth. It's good for agriculture, forests, and plant life."

An article published on the Internet by the Oregon Institute of Science and Medicine, a small private research organization that supports the skeptical view of global warming, says that while it's reasonable to believe that humans are responsible for much of the increased CO_2 levels since pre-industrial times, the environmental impact of this is "likely to be benign. Greenhouse gases cause plant life, and the animal life that depends upon it, to thrive....Mankind is moving the carbon in coal, oil, and natural gas from below ground to the atmosphere and surface, where it is available for conversion into living things. We are living in an increasingly lush environment of plants and animals as a result of the CO_2 increase. Our children will enjoy an Earth with far more plant and animal life as that with which we now are blessed. This is a wonderful and unexpected gift from the Industrial Revolution."

There are problems with this argument. It's true that increasing CO_2 levels can increase plant growth, a process known as *carbon fertilization*. However, several studies indicate this doesn't necessarily offer us a bailout on global warming, because the phenomenon is more complex than it seems.

A study by researchers at the University of New Hampshire found that plants have a delayed reaction to atmospheric temperature increases. The scientist examined changes in atmospheric CO_2 levels and plant production after four warm spells that occurred between 1980 and 1991, including the strong 1982–83 El Niño. They found that roughly one to three years after these events, the increase in atmospheric CO_2 levels slowed down and plant growth sped up. The implication is that plants were soaking up extra CO_2.

The patterns of plant growth varied in different regions; plants in polar and temperate regions tended to grow faster at the peak of the warm spell, while plants in tropical rain forests and dry savannas declined. This suggests that the warm temperatures were putting additional stress on regions that were already warm or dry. However, one to three years after the warm peaks, the situation reversed—plant growth in northern latitudes declined while plants in tropical latitudes started to grow faster. This demonstrates that the response of plants to climate change is rather complex.

One question this study didn't answer was how long the increased plant growth would last. Research by Mingkui Cao of the University of Virginia and Ian Woodward of the University of Sheffield suggests there are limits to the ability of plants to take up excess CO_2. When they modeled the impact of climate change on ecosystems to the year 2070, they found that plant production would indeed increase significantly but "this response will decline as the CO_2 fertilization effect becomes saturated and is diminished by changes in climatic factors."

In fact, some studies suggest that global warming could, under some circumstances, convert plants from net *absorbers* to net *emitters* of CO_2. Writing in *Climatic Change,* Danny Harvey points out that when the climate changes, plants that are not adapted to the warmer temperatures die out before other, better-adapted plants take their place. The time lag could reduce the ability of ecosystems to store carbon for as long as decades to centuries. According to some calculations, forest "dieback" could release anywhere from four to six billion tonnes of carbon a year—an amount that rivals the annual emissions currently attributable to burning fossil fuels. Unrestrained global warming presents "a growing risk that what is probably a net carbon sink at present could significantly weaken and possibly become a net carbon source," Harvey said. "This may have already happened in some Arctic ecosystems."

Equally disturbing is a study that shows that plant growth enhanced by the increasing amounts of nitrogen we're putting into the environment actually may reduce the ability of ecosystems to store carbon. Nitrogen compounds are released by burning fossil fuels and by commercial fertilizers used in agriculture. Scientists have speculated that increased plant production resulting from this nitrogen fertilization would help take up additional amounts of atmospheric CO_2. However, research by ecologists David Wedin of the University of Nebraska and David Tilman of the University of Minnesota demonstrated this is unlikely to happen.

The researchers, who examined the impact of different levels of nitrogen fertilization on plots of grassland species over a 12-year period, found that increased nitrogen loading dramatically changed the composition of the species and reduced the diversity of plants by killing off some species. The grasses that contributed most to sequestering carbon were most vulnerable to being destroyed, and the weedier species that thrived under high nitrogen loading were not good alternatives. The reason for this has to do with a feature of carbon sequestration that's not widely understood outside scientific circles: in temperate climates, most of the carbon is sequestered not in the above-ground portion of living plants but in organic matter in the soil. This is particularly true for grasslands. Therefore, enhanced above-ground plant productivity is not the best indicator of how much carbon is being sequestered. "Most of the carbon is in the soil," said Wedin. "If you don't study that, you're really going to miss the point."

The amount of carbon stored by plants doesn't depend just on the amount of above-ground growth, but on how long the plant tissues live and how fast they decompose after they die. The faster the decomposition, the faster the sequestered carbon is released back into the atmosphere. In their experiments, Wedin and Tilman found that increased nitrogen loading favored faster-growing, weedy plants with tissues that decompose quickly once the plant dies. These plants grew more—in other words, plant productivity increased—but the carbon storage in the soil did not increase at all.

The bottom line is that nitrogen pollution from human activities actually reduces the ability of some natural ecosystems to sequester carbon. "The nitrogen-caused shifts in species composition limit the ability of temperate grasslands to serve as significant long-term carbon stores," the researchers wrote in *Science*.

Converting natural prairie grasses to agriculture significantly increases plant productivity but greatly reduces carbon storage in the soil, Wedin said. When they die, agricultural plants put less organic matter into the soil and repeated plowing accelerates decomposition. "A fertilized cornfield has at least four times the productivity of a native prairie, but less than half the soil organic matter. The productivity is higher, but the conversion of the prairie to agriculture has meant a huge contribution of carbon to the atmosphere. Once you break up that prairie soil, you lose carbon a lot of faster than you build it. In half a century, the carbon in the soil drops between 25 and 50%. It would take 500 years to rebuild the carbon that you can lose in 50 years."

Global warming itself may reduce the ability of soils to store carbon.

Wedin said that studies of forests ranging from northern regions to the tropics indicate that "with cooler temperatures, you get a much higher proportion of carbon stored in the soil." Temperature increases accelerate decomposition "so with global warming you'll get an increase in plant growth, but you'll decrease carbon in the soil much more."

He added that changes in the species composition of Canada's forests as a result of logging and increased nitrogen loading could also reduce the carbon storage capability of these forests. For example, black spruce, which is cut for pulp and paper, is not growing back very well, he said. "Part of the reason is that a lot of these black spruce stands are 100 years old. It was a cooler period when they got established; now it's warmer and there's more nitrogen floating around the atmosphere and it's hard to get the black spruce to come back." What often grows in their place are fast-growing deciduous trees like aspen or birch, which have a different effect on carbon storage in the soil. "Replacing the slow-growing conifer forests with fast-growing deciduous tree plantations may increase above-ground production, but it will also dramatically decrease soil organic matter pools," Wedin said. "If we change species composition because of a changing climate and nitrogen deposition, there might be considerable losses of carbon in the soil." He added that forest fires do not have as much of an impact as cutting trees because fires don't burn all the wood. "Some of the carbon in the soil and in the wood survives. It's completely different from logging."

Wedin said that "the most pressing issue is what's going to happen to the mass of soil carbon in the boreal forest. That's Canada's biggest carbon pool." He ranked the threats to this carbon pool in the following order: climate warming (especially when accompanied by a loss of soil moisture), more fires, increased logging, and higher nitrogen levels in the atmosphere.

More bad news comes from a study done by a group of U.S. scientists who studied the carbon balance in a 120-year-old black spruce stand in Manitoba. They found that when the normally frozen soil thawed, the decomposition of organic matter in the soil increased tenfold. They concluded that climatic changes that promote thawing of the soil are likely to cause a net release of CO_2 into the atmosphere.

The import of these studies is that global warming is likely to do more harm than good as far as carbon sequestration in living things is concerned. While the added CO_2 in the atmosphere will cause plants to grow more rapidly at first, it appears this effect saturates eventually. In addition, the warming caused by CO_2 promotes faster decomposition of organic matter in the soil and thus faster release of sequestered carbon to the atmosphere. Given these findings, it would be extremely unwise to count on carbon fertilization (or nitrogen fertilization, for that matter) to bail us out of our global warming problems.

Nevertheless, carbon sinks continue to cause controversy. In October 1998, scientists from Columbia and Princeton universities and NOAA published a paper in *Science* suggesting that North America harbors a large sink that's taking up most of the carbon estimated to be absorbed by land masses around the world. They speculate that forest regrowth on abandoned farmland and previously logged forests—perhaps aided by carbon and nitrogen fertilization and longer growing seasons in a warmer climate—is responsible for this CO_2 absorption. An accompanying commentary in *Science* notes that the suggested size of the North American sink is "straining belief among other scientists." An Australian study indicates the North American sink is only one-third as large as claimed in the new study, and British researchers found that tropical South America accounts for perhaps 40% of terrestrial carbon uptake, which is hard to reconcile with such a large North American sink. Moreover, the latest findings suggest that forests in Europe and Asia are absorbing only one-fifth as much as North American forests, which one ecologist described as "incomprehensible." Finally, critics note that the study covered a period (1988 to 1992) when global temperatures were cooled by the Pinatubo explosion, which would contribute to increased carbon uptake. A major question, therefore, is how long North American ecosystems would continue absorbing CO_2 at this rate even if these findings are true.

While the researchers who did the study stand behind their analysis, they acknowledge there are many remaining uncertainties. Carbon sinks are known to vary year by year and more research is needed to confirm the size and location of the North American sink. They also acknowledge that these mechanisms are temporary; at some point, such a sink would undoubtedly stop absorbing CO_2 at such a clip.

There's concern among scientists that these results will be used by opponents of the Kyoto treaty to cut greenhouse gas emissions. The U.S. wants to apply CO_2 absorption by sinks against its targets for emissions reductions and opponents of the treaty have already suggested that the North American sink is so large that the U.S. may not be producing any net emissions at all. Environmental groups reject this, saying the sink won't last forever. Even the researchers who did the study argue that while this research may be important for global management of CO_2 emissions, the findings do not justify arguments that sinks in a particular region should be used to offset that region's CO_2 emissions. Carbon dioxide from all sources is mixed globally long before it encounters a sink, so all emissions contribute to the global problem, and all sinks reduce global CO_2 concentrations.

It's important to emphasize that these findings, even if true, do not imply that atmospheric CO_2 levels are not rising. What's at issue here is the relative strength of different terrestrial sinks, which, along with the oceans, are estimated to remove perhaps 50 to 60% of the total annual emissions from human activities. The remaining amounts continue to build up in the atmosphere year after year.

Temperature Projections

The ultimate question is just how much warming will occur. Based on different scenarios for greenhouse gas emissions, and taking into account the effects of aerosols, sinks, and other factors that affect atmospheric greenhouse gas concentrations, the models cited in the IPCC report project a global warming between about 1 and 3.5°C between 1990 and 2100, with a mid-range "best estimate" of 2°C.[*]

The report gives a range of projected temperature increase because of the uncertainties associated with various factors in the calculations. On a global scale, the range between 1 and 3.5°C is actually fairly large, so it embraces quite a lot of uncertainty. At the same time, it puts some boundaries on the likely extent of the problem. This projected rate of warming will likely exceed anything that has occurred in the past 10,000 years.

These projections are somewhat lower than those in the 1990 IPCC report; the best estimate of 2°C is about two-thirds of the 1990 projection, primarily because of the addition of aerosol cooling effects to the models. There were also improvements in the way the models treat sinks and other atmospheric processes, as well as revisions in assumptions about greenhouse gas emissions. The revision of the temperature projections is completely consistent with the fact that such projections are critically dependent on the initial assumptions and the atmospheric and biological processes included in the model equations. As scientific knowledge of these processes improves, it will be used to further refine the models.

Predictably, however, the effort to improve the accuracy of the models elicited only contempt from the skeptics, who charged that the new

---

[*] The IPCC calls these temperature increases *projections* rather than *predictions* because they're "what if" scenarios based on assumptions about emissions and other important socioeconomic factors that could change in the future. The scenarios project how the climate will respond *if* forced by a given range of factors. If those factors change (for example, if substantial cuts in greenhouse gas emissions are implemented), the response of the climate will also change.

projections demonstrated not only that the models were flawed but also that scientists had previously been guilty of alarmism designed to generate increased research funding. A newsletter with the misleadingly impartial-sounding name *World Climate Report*[*] expressed little surprise that scientists were trying to "explain away the lack of warming with some competing compound. After all, no one wants to publish a letter that says, 'Dear World Leaders, we are sorry but we goofed and really didn't understand how the greenhouse effect would change the climate after all. Anyway, it looks like no big deal (or at best a moderate one), so ta ta, Yours Truly, the Consensus of Scientists. P.S. Thanks for the billions. Hope you get your carbon tax.'"

Sarcasm aside, the lower projections do not suggest there'll be a "lack of warming" and they're nothing to cheer about, since they still represent a rate of warming unprecedented in the last 10,000 years. Moreover, warming would continue beyond 2100; because the oceans take longer to respond than the atmosphere, only about 50 to 90% of the total eventual warming would occur by 2100 and temperatures would continue to rise afterwards, *even if atmospheric concentrations of greenhouse gases were stabilized by then.*

Modeling studies have continued since the 1995 report and at least one has indicated recently that the upper limit on the projected warming may in fact be higher than that cited by the IPCC—around 4.5°C. These results were obtained using the Canadian climate model, among the most advanced in the world.

The warming is not expected to occur uniformly but will vary geographically and seasonally. A 1996 IPCC statement said that virtually all modeling studies project "greater surface warming over land versus oceans in winter, a maximum surface warming in high northern latitudes in winter [and] little surface warming over the Arctic in summer." In his analysis in *Science,* Jerry Mahlman says it's "very probable" (i.e., a 90% probability) that by the middle of 21st century, "the higher latitudes of the Northern Hemisphere will experience temperature increases well in excess of the global average increase and substantial reductions in sea ice."

---

* At one time, WCR's web site said the newsletter was funded by the Western Fuels Association and associated energy companies. Subsequently, however, WCR's funding was attributed to the Greening Earth Society, an organization created to "spread the good news about the beneficial impact of carbon dioxide on earth's biosphere." Fred Palmer, head of Western Fuels, is chairman of the Greening Earth Society and the two organizations share the same address.

There's debate about the impact of natural variability on the future course of global warming, just as there was about its impact on global warming to date. The question of variations in the sun's energy arises again. Most researchers who have looked at this question say that, even if solar variability affected the warming observed so far, it's unlikely to dominate future warming. Judith Lean and David Rind calculated that if, over the next century, the sun varies by no more than the 0.1% observed by satellites in the past two decades, this will have "a negligible effect" on global temperatures relative to the warming projected to occur from greenhouse gases.

In any event, increased solar warming, no matter how "natural" it may be, would not be particularly helpful if it's added to warming caused by greenhouse gases. On the other hand, it's possible the sun's output could *decrease*, causing cooling that might slow down global warming. Scientists who've analyzed this question have concluded it would not be sufficient to counteract global warming from greenhouse gases. The message, in short, is that we shouldn't count on the sun to bail us out, any more than we should count on growing plants.

Global concentrations of greenhouse gases will assuredly increase for most of the next century, no matter what we do, and they will assuredly cause warming. Danny Harvey says that if "business-as-usual" levels of emissions continue, their influence will grow relative to that of natural causes and "will dominate natural sources of climatic variability during the coming century."

In *Science*, Jerry Mahlman commented that the knowledge scientists have accumulated about how the climate works means that greenhouse warming cannot "rationally be dismissed or ignored....None of [the] recognized uncertainties can make the problem go away."

A November 1997 *Washington Post* editorial commented that "overlying the uncertainty [about global warming] is a broad scientific consensus on the fundamentals of the warming forecast." It quoted James Baker, head of NOAA, saying: "There's a better scientific consensus on this than on any issue I know—except maybe Newton's second law of dynamics."

six

DANGEROUS INTERFERENCE:
Environmental and Socioeconomic Impacts of Climate Change

The Convention on Climate Change calls for humanity to stabilize atmospheric greenhouse gases at a level that would prevent "dangerous" interference with the climate. But it doesn't define "dangerous" beyond saying that we should avoid changes that threaten natural ecosystems, food production, or sustainable economic development. We need to be more specific than this before we can fully assess the threats we face.

We do know some things about the dangers associated with climate change. They will vary from place to place and some populations—particularly those in developing countries—will be more vulnerable than others. The potential for danger also changes with time. Though global warming has been underway for more than a century, its socioeconomic and environmental effects are still emerging from the noise of natural variability and we've barely begun to detect and document them. As greenhouse gas concentrations grow, so too will human interference with the climate, increasing the probability of adverse impacts that could be deemed dangerous. This means that the dangers confronting future generations will almost certainly be greater than those we face today.

What we don't know yet is exactly what these dangers are, how bad they will be, when and where they will occur, and who will be most affected. Answering these questions is a tall order and when you add the complexities of natural ecosystems and human societies to those of the climate system, the uncertainties multiply. Much remains to be learned about how plants, animals, and humans will respond and adapt to a changing climate, particularly at the regional and local levels where the effects will be felt most strongly.

To answer these questions, scientists around the world are now doing "impacts" studies and several national and international programs have been established. The 800-page second volume of the IPCC report synthesized thousands of these studies to arrive at a series of "plausible impacts" of climate change on natural ecosystems and human societies. This was intended to help policymakers evaluate whether these "plausible impacts" constitute a dangerous interference with the climate but, in the end, "interpreting what is 'dangerous' involves political judgement," the report emphasizes.

According to Ian Burton, former head of Environment Canada's Environmental Adaptation Research Group (EARG), impacts studies are needed because while the public and policymakers want to know how serious climate change is and what actions should be taken to protect against it, the unknowns make them hesitant to act. "They are frequently warned that the threat is real and potentially very serious, but the warnings are clouded in uncertainty, and policymakers are reluctant to take major decisions without a clearer sense of the reality and potential severity of the risks." Moreover, powerful groups such as the fossil fuel industry and the business community argue that climate change is less of a threat than the measures governments may take to avoid it, and their opposition will be all the stronger in the absence of evidence of adverse impacts. "The cost of curbing greenhouse gas emissions is potentially very high, and can only be justified if impacts are similarly high," Burton noted. "In order to justify such costs, and the policies that would make them mandatory, it is essential to know what the costs of inaction would be."

Impacts research is also essential in developing strategies for adapting to climate change. "It is important to know where the greatest impacts can be expected and where adaptation measures can do the most good," Burton said.

Adaptation (which is examined further in Chapter Eleven) involves measures that reduce the vulnerability of natural and human systems to climate change. It's one of three key elements considered in the IPCC's impacts studies, along with sensitivity and vulnerability to climate change. The *sensitivity* of a system is a measure of how strongly it would respond to changes in the climate. Its *adaptability* refers to the adjustments that can be made to cope with climate change. *Vulnerability*—a measure of how a system may be damaged or harmed by climate change—depends on both its sensitivity and its adaptability. All three factors are affected by both the *magnitude* and the *rate* of climate change.

We saw in Chapter Five that the rate of change is particularly important, though it's often overshadowed by a preoccupation with how large the changes will be. Both the rate of warming over the next century and the exposure of human populations and property to its effects are expected to be unprecedented, suggesting that the impacts associated with climate change in the next 100 years could be unique in human history.

The more quickly the climate changes, the harder it is for systems to adapt and the higher the probability that damage and disaster will occur. According to Henry Hengeveld of Environment Canada, "historical and paleoclimatological studies show that it's unlikely that economic losses and threats to natural ecosystems will be significantly greater than those caused by natural variability if global warming does not exceed 0.1°C/decade, if the absolute amount of warming remains below 1°C, and if sea level rise is less than 20 centimeters. Above these thresholds, ecosystems are likely to become stressed, regional food production is more likely to fail and island and coastal regions will be threatened. Hence these values could be considered as a cautious threshold for danger." If these values are doubled—that is, a 0.2°C/decade warming with an absolute increase of 2°C and a sea level rise of nearly half a meter— "the likelihood of large scale ecosystem dieback, major regional food shortages and coastal flooding...is quite high, thus perhaps providing an upper criteria for danger thresholds." These "upper criteria" for danger are about equal to the IPCC's "best estimate" of what's likely to happen over the next century if greenhouse gas concentrations continue to rise at current rates.

Impact Research

The impacts studies done so far have examined the effects of climate change on subsets of the natural environment and socioeconomic activity—agriculture, forests, fisheries, energy production, water supplies, etc., which will be discussed later in this chapter. Chapters Seven and Eight examine the impact of climate change on human health and security, which, surprisingly, has not been studied much until recently.

It's important to emphasize that impacts studies are *not* detailed predictions of what will happen in the future. As with climate change itself, there are no crystal balls. Scientists are concerned that the public, the media, and policymakers tend to interpret impacts studies as firm predictions, but the results should be viewed more as "what-if" scenarios. Whether these scenarios come true depends on the social and political decisions we make over the next few decades.

These studies are really a first cut at identifying areas where scientists can most productively concentrate further research effort. Henry Hengeveld emphasized that they're "a *qualitative* indication of the direction that climate change will take us....They identify where the highest risks or problems will be and where we're likely to have fewer problems, but they do not tell us in detail what the magnitude of the problems will be." Because it takes several years to do an impacts study, it's difficult for them to use the most current climate models, which are constantly evolving at the same time. Consequently, "we need to be cautious about hard numbers," Hengeveld said.

Still, knowing the general direction of future change is useful. For example, it's helpful for engineers who design dams to know whether river flows are likely to increase or decrease in a warmer climate. Reforestation programs could also benefit, Hengeveld said. "We're planting trees today that are going to be around for three or four decades at least, so we should be planting species that do well today and in a drier, warmer climate. There are many long-term policy decisions we're making today which we'll have to live with for decades to come. Many of those are made on the assumption that climate is a constant. That's no longer the case. So it's important that we have an idea of what direction it's going in. Impacts studies can provide that."

Speaking at a climate conference, Joel Scheraga, director of the climate and policy assessment division of the U.S. Environmental Protection Agency, said: "We have very imperfect information and a very complex planet. Does that mean we can't say anything? No—we can draw insights; we can identify the vulnerable systems. We don't know exactly what's going to happen in 20 years, but all we need is a probability that there will be adverse impacts to start worrying about those impacts."

Accentuate the Negative

The emphasis on "adverse impacts" is often criticized by those who believe the dangers of global warming are overblown. They argue that climate change will produce many benefits that are largely ignored by "Chicken Little" researchers and environmentalists.

It's true that global warming may produce some benefits. Cold countries like Canada may enjoy an extension of the growing season or a reduction in winter heating bills. Climate change also may alter the environmental range of plants and animals, perhaps creating new agricultural opportunities. In some regions, forestry may benefit from an expected northward expansion of the tree line and shifting ranges of boreal and temperate forests. Shipping may be extended as warmer weather reduces the ice cover on lakes, rivers, and the ocean. There may be fewer cold-weather illnesses and injuries; in Canada at present, these health problems outstrip those associated with warm weather.

It's certainly wise to be alert to the opportunities that climate change may send our way, but this doesn't mean that placing greater emphasis on the negative impacts is inappropriate, for two reasons. First, it's the negative impacts that will give us the most trouble and will be the most costly to cope with, especially if we're unprepared for them. But preparing can be costly too, if we decide to make precautionary investments to protect ourselves from their worst effects. So it makes sense to direct our efforts and our dollars to the aspects of climate change that are likely to have the most serious costs and consequences.

The second reason for focusing on negative impacts is that no region is likely to experience *only* beneficial consequences of global warming. Climate change is not a simple binary situation—either you win or you lose. While some regions may be overall "winners" and others may be overall "losers," every region will experience a complex and shifting mixture of both good and bad impacts. A longer growing season may be accompanied by severe drought or excessive rain or the invasion of new insects and pests. A reduction in cold-weather health problems may be offset by an increase in illness and death caused by hot weather or by the migration of insects or disease-causing organisms. Reduced heating costs may be offset by increased air conditioning costs. A reduction in ice cover on lakes or rivers may be accompanied by an increase in shoreline damages caused by storms.

How these good/bad equations work out will vary from place to place. In Canada, the benefits of reduced winter heating costs may more than offset increases in summer air conditioning costs but, as you move south, this cost/benefit balance changes. At some point, reduced winter heating costs will roughly equal increased summer cooling costs, and as you move further south, increasing temperatures will greatly increase the cost of cooling without any countervailing benefit associated with lower winter heating costs. And since many poor tropical countries can't afford air conditioning, the increased temperatures in these regions could have deadly consequences.

It's worth remembering also that countries will not be affected only by the direct impacts of climate change that occurs within their own borders. In an era of economic globalization, even regions that experience benefits from global warming will find themselves dealing with negative impacts occurring elsewhere. Everything from export markets and food prices to currency values and immigration patterns could be affected.

Finally, in assessing the potential benefits and costs of climate change, it's important to factor in the role of weather extremes. Events that, in moderate doses, might provide benefits can turn into disasters if they're too extreme. Most impacts studies to date have focused on changes in average conditions, but it's the extreme events that push even the strongest socioeconomic system to its limits. How close to those limits will climate change take us? Impacts research may give us some answers.

Socioeconomic and Environmental Impacts of Climate Change

There isn't much about modern life that climate and weather do not affect. Agriculture, energy production and consumption, transportation, recreation, industry, and the entire infrastructure of society are all vulnerable to variations in climatic conditions and to weather extremes. So are the natural ecosystems—water resources, forests, fisheries—on which human economic activity depends. While it's true that technology has buffered us from some of the worst the weather has wrought, it's a dangerous conceit to assume that technology has, or ever can, make us immune. Projections of future climate change force us to examine whether our technology will continue to provide the same level of protection in the future. We're tinkering with an intricate, complex climate system that will undoubtedly react in ways that are hard to predict. The fact that we're not certain exactly how it will react, far from being a cause for complacency, should make us take notice.

"We are performing a global experiment on our natural ecosystems for which we have little information to guide us," a group of leading U.S. ecologists commented in an open letter to President Clinton. "While plant and animal communities may be able to eventually adapt to a stable climate system that is warmer than the existing one, many species may not be able to survive a rapid transition to that new climate....We believe that this situation constitutes a dangerous anthropogenic interference with the climate system."

We've seen that it's very difficult for scientists to forecast how the climate will change at the regional or local level because computer models aren't sophisticated enough to provide this kind of detail. This, in turn, makes it difficult to predict the impact of climate change on natural resources and economic activity in different regions. Social factors, such as agricultural practices, urban growth and development, and energy use, are also highly variable. Therefore, the only things that can be said with certainty are that the socioeconomic and environmental impacts will vary and that each region will likely experience both good and bad effects.

Regional Impact Studies

Many countries have established research programs to study the regional impacts of climate change. Canada, a country so large that it embraces seven distinct "ecoclimates," has been particularly active in this area. During the 1990s, several regional studies were done, focusing on the Mackenzie Basin, the Great Lakes–St. Lawrence Basin, and the Niagara–Toronto region. In 1996, Environment Canada started the Canada Country Study (CCS), which synthesized existing research on the impacts of climate change on different regions and economic sectors of Canada. Several regional reports were released in 1997 and 1998.

The CCS was the first national review of existing research on the impact of climate change on Canada's socioeconomic and biological/ environmental systems. As a northern country, Canada is expected to experience warming several times the global average—perhaps as much as 5 to 10°C during the 21st century. The first phase of the study, which focused on assessing the current state of research, produced six reports on different regions of Canada, as well as a series of studies on the potential impact of climate change on social and economic sectors (e.g., agriculture, energy, fisheries, forestry, human health, insurance, transportation, etc.). Future studies will add to this knowledge base.

In 1997, the U.S. Global Change Research Program began a similar national assessment program to determine the consequences of climate change for the U.S. and to examine adaptive mechanisms. It focused on two time frames—the next three decades and the next 100 years. A series of workshops were held around the country in 1997 and 1998 and a synthesis report, along with regional and sectoral reports, will be submitted to Congress in 2000.

Both countries collaborated on the Great Lakes–St. Lawrence River Basin Study (GLSLBS), which started in 1992. It examined the impact of climate change on a region that contains one-fifth of the world's supply of fresh water and serves a population of more than 42.5 million people in two Canadian provinces and nine U.S. states. Managing and protecting the Great Lakes Basin is extremely complex because two countries, several provinces and states, and many municipalities all have a stake in the resources this area offers. The study focused on four main topics: water use and management, land use and management, ecosystem health, and human health.

The Mackenzie Basin Impact Study (MBIS) was especially significant because it was one of the first to examine regional impacts of climate change in northern latitudes, which are expected to experience more warming than southern regions. The Mackenzie Basin has warmed about 1.5°C in the past century, about three times the global average, and is projected to warm 4 to 5°C between the "baseline" period of 1951–1980 and the middle of the 21st century. The six-year study, which concluded in 1997, explored the potential effects of climate change on the environment and the people in a region of 1.8 million square kilometers that includes parts of the Yukon, the Northwest Territories, British Columbia, Alberta, and Saskatchewan. Research indicates that warming has already caused thawing of the permafrost, increases in landslides, a lowering of water levels in lakes, more forest fires, a longer growing season, and other ecosystem changes. "The region may well be going through the beginning of significant changes to its climate as a result of global warming," the report notes.

The Toronto–Niagara Study (TNS), started in 1997, is looking at the effects of climate change on the Toronto–Niagara region, a highly urbanized area containing the largest concentration of population and economic/industrial activity in Canada. The study examines the sensitivity of the region's diverse natural ecosystems, social systems, and economic activities to expected changes in the climate with the aim of suggesting ways in which these impacts can be managed.

A summary of some of the major changes expected in several vital socioeconomic sectors follows. This is a representative sample only, since there are far too many studies to cover in detail. Most impacts studies examine the effects of a doubling of atmospheric CO_2 levels over the next century. These findings should be taken as *qualitative* estimates of the direction of change rather than precise predictions.

These studies strongly suggest that the *rate* of climate change may be more important than the magnitude of climate change, at least over the next century. A too-rapid acceleration in warming could stress many of these systems beyond their ability to adapt, particularly when other stressors, such as population growth and increasing urbanization, are added to the picture.

Water Resources

Fresh water is one of our most precious and limited resources; more than 97% of the water on earth is salt water and most of the rest is locked up as ice, so only a small fraction is available for nearly 6 billion people to use for domestic and industrial activities and food production. We depend on water not only as a necessity of life, but as a critical component of our agricultural, power generation, transportation, and recreational activities. Water also sustains natural ecosystems important to human survival, such as forests and fisheries.

The IPCC comments that "the illusion of abundance of water on the earth has clouded the reality that in many countries, renewable freshwater is an increasingly scarce commodity." Water systems around the world are under increasing stress from population growth, urbanization, and other human activities such as building dams and diverting rivers. Since the mid-1800s, a fivefold increase in population has decreased the water available annually from 43,000 to less than 9000 cubic meters per capita. The Worldwatch Institute estimates that by 2025, 40% of the world's population, some 3 billion people, will be living in places experiencing water stress or chronic water scarcity. "Although fresh water is renewable, it is also finite," notes a WWI report.

Global warming will likely exacerbate water availability problems already being experienced around the world. It can affect water levels and flows in lakes, rivers, and streams, as well as groundwater. Changes in precipitation are expected to make some areas drier and others wetter. An increase in intense precipitation events, with longer periods of little or no precipitation between them, could intensify both flooding and drought. Changes in the mix of rain and snow could alter the timing and volume of spring runoff, which could also cause increased flooding and severe water shortages at critical times.

Rising temperatures are also expected to change the rate of evaporation—something that's likely to cause a drop in water levels in lakes and rivers even where precipitation increases. Reduced water levels and flows would affect domestic and industrial water, shipping and navigation, power generation, recreation, wildlife habitat, and coastal structures. Lower water levels could also increase the concentration of pollutants and reduce the effectiveness of urban drainage and sewage systems. Disputes over the removal of water from lakes and rivers for human consumption and competing industrial, agricultural, and recreational activities could be further exacerbated if water volume drops substantially.

Several impacts studies that examined the effects of climate change on Canadian water resources found that precipitation is expected to increase in northern regions, but the interior of the country will likely experience increased soil dryness and drought in summer. As the climate warms, Canada is also expected to experience an increased percentage of precipitation as rain rather than snow. In some places, increased precipitation is likely to be offset by increased evaporation, resulting in a reduction in water supplies.

Several U.S. and Canadian studies indicate that the Great Lakes–St. Lawrence Basin (GLSLB) will experience reduced water levels and flows, mostly because of increased evaporation in a warmer climate. Modeling research indicates that the basin's average temperature could rise by about 4.5°C by 2055, with decreased precipitation in the southern regions and increased precipitation in northern regions (mostly in spring). It's estimated that water levels could drop by one half to one meter and the outflow of the St. Lawrence River could be reduced about 20%.

Lower water levels in the basin will have a profound effect on the 42 million people who live nearby. More than two-thirds depend on the system for drinking water and there are competing requirements for shipping, power generation, sewage, and recreation. Changes in the St. Lawrence would have a profound impact in Quebec; an estimated 97% of the province's population lives within its watershed and nearly 70% live within 10 kilometers of the river.

Climate change is a "wild card" that will only aggravate existing problems with water withdrawals from the Great Lakes, according to Sarah Miller of the Canadian Environmental Law Association (CELA). At a 1997 climate conference in Toronto, she noted that only about 1% of the water in the basin is renewable (via rain and runoff), while 99% was deposited by melting glaciers some 20,000 years ago. There are increasing demands to withdraw water from the lakes, and users are not being charged enough because of a widely held belief in surrounding regions that "we can always turn to the Great Lakes," Miller said. People in this region "continue to be the largest per capita wasters of water in the world. We pay less for water than almost anyone in the world." She said that if the combination of water withdrawals and climate change results in consumption in excess of the renewable portion of the water, "we are no longer using a resource sustainably. Most of us do not recognize that we are irretrievably eroding our precious water supplies."

On the prairies, drought is a major problem that may be worsened by climate change. During the 1988 drought, increased withdrawal of water from prairie basins caused widespread deterioration in water quality. According to the Canada Country Study (CCS), "evaporation rates from reservoirs and lakes have been abnormally high in recent decades." Future climate scenarios indicate that global warming will bring increased temperatures and reduced soil moisture, but estimates of precipitation and runoff are uncertain; some models show increased precipitation, greater frequency of extreme storms, increased evaporation, decreased snowpack, and an earlier melting of the snowpack. Most climate scenarios indicate that semi-arid regions of the prairies will experience an increase in the frequency of droughts, except possibly for southern Alberta. Warmer weather will likely increase competing demands for high quality drinking water for livestock and humans and "more conflict will occur among users over the limited resource," the study notes.

Projections for British Columbia and the Yukon suggest an increase in the frequency and severity of spring flooding of rivers and streams and increased drought in the summer and fall because of reduced stream flow in southern regions. Many rivers and streams in southern B.C. and the Rocky Mountains get their water from glacier melting in the late summer and fall; many of these glaciers are expected to shrink substantially or disappear altogether in a warmer climate. Unless reservoir capacity in these southern regions increases, "water supply will be reduced in the dry summer season when irrigation and domestic water use is greatest," the CCS said. Landslides and "debris torrents" are expected to increase in mountainous areas that become unstable because of increased precipitation, melting permafrost, and retreating glaciers. This will increase the risk to water quality, fish and wildlife habitat, roads, and other manufactured structures.

Researchers are projecting similar effects in the U.S., particularly in the West where water scarcity is already a significant problem in many areas. The western part of the U.S. receives only about one-fifth of the national average freshwater runoff and about three-quarters of the water used goes to agriculture. A study by Resources for the Future notes that in the western U.S., water flow in streams and rivers comes largely from melting snow in spring and summer. If global warming alters the rain/snow mix and the timing of snowmelt, the resulting changes to runoff patterns could increase the likelihood of flooding and reduce the amount of water available during spring and summer when it's most needed for irrigation. The author of the report, Kenneth Frederick, notes that there is already concern about water supplies because "demands are outpacing supplies, water costs are rising sharply, and current uses are depleting or contaminating some valued resources." He says that, on balance, the impact of climate change on these resources is "likely to be adverse because the existing water infrastructure and uses are based on an area's past climate and hydrology."

A 1993 study by the Environmental Protection Agency found that the Colorado River is "extremely sensitive to climatic changes that could occur over the next several decades." One of the most significant rivers in the U.S., it provides the main source of water in a basin covering nearly a quarter of a million square miles over seven states. The EPA study said that a 2°C temperature increase, without added precipitation, would reduce the annual runoff 4 to 12%, while a 4°C increase would cause a drop of 10 to 20%. "Increases in temperatures decrease winter snowfall and snowpack, increase winter rainfall, and accelerate spring snowmelt. These changes would increase winter runoff and decrease spring runoff, creating a greater potential for spring flooding in some regions, reductions in water storage and deliveries and increases in average annual salinity." The report notes that runoff is even more sensitive to changes in precipitation. Depending on whether precipitation increases or decreases, it could either reduce or increase the effects of higher temperatures on runoff.

Because climate change is amplified at higher latitudes, the North is expected to experience more dramatic changes than southern regions. Precipitation increases of up to 25% are expected, and an increasing fraction of it will likely fall as rain rather than snow in fall and spring. An earlier and longer snowmelt period and an earlier peak to the spring thaw are also expected. It's anticipated that warmer temperatures and the increased thaw period will increase evaporation enough to largely offset the increased precipitation, so northward-flowing Arctic rivers may experience decreased levels and flows. The Mackenzie Basin study projects a slight drop of about 7% overall in water runoff from rain or melted snow, although some regions within the basin may experience increased runoff because of melting permafrost. Evaporation is likely to decrease water levels and flows in Great Slave and Great Bear lakes.

It's not only the quantity of water in lakes and rivers that will be affected by climate change, but the quality as well. Lower water levels and flows would result in higher concentrations of pollutants, toxic chemicals, and sewage. The reduction in water runoff and discharge from rivers and streams would reduce the flushing of bays and the dilution of organic waste and chemicals. Dredging done to offset the lower water levels could resuspend pollutants that had settled into sediments. Higher water temperatures could decrease the amount of oxygen dissolved in the water and encourage added growth of bacteria and algae.

In coastal regions, freshwater systems and groundwater could become contaminated with salt water from rising sea levels. "A rise in the water table associated with higher sea levels may increase the salination of groundwater over a long period, with consequent effects on agriculture and water supply," according to an impacts analysis by Britain's Climate

Change Impacts Review Group, a panel of academic, government, and industrial experts. Groundwater provides almost a third of Britain's water supply. The report noted that even if precipitation increased in winter, higher summer temperatures and increased evaporation could decrease the water-holding capacity of soils, increase the likelihood of droughts, and put added pressure on water supplies, forcing greater investment in the water resources infrastructure.

According to the IPCC report, similar threats to freshwater systems would occur all over the world, particularly in arid and semi-arid regions of developing countries, where water systems tend be simple and based on a single, isolated source (e.g., a single river or reservoir), which increases their vulnerability. Studies indicate that these water systems are extremely sensitive to climate change and less able to adapt than the larger, sophisticated, and more managed systems found in developed countries. Water systems in developed countries are probably resilient enough to cope with the effects of climate change, the IPCC report concludes, but adds that little is known about what this adaptation would cost.

The World Health Organization says there are large cities around the world that already have inadequate water supplies and cooling systems. "Under heat wave conditions, the availability of water will become extremely critical for human survival."

One of the most potentially volatile side effects of these changes in water systems would be to increase existing conflicts over water resources among countries and user populations. There's a considerable disparity in the allocation of fresh water around the world; for example, Canada (as well as Iceland, Norway, and New Zealand) gets more than 100,000 cubic meters of water per person per year. The average in the U.S. is 14,000 cubic meters, while Africa gets about half as much. Billions are spent each year moving and storing water.

Tensions over river water are already common in places—disputes have occurred over the Ganges, the Jordan, the Nile, and the Tigris-Euphrates. "Worldwide, at least 214 rivers flow through two or more countries, but no enforceable law governs the allocation and use of international waters," according to a Worldwatch Institute report. The report's author, Sandra Postel, director of the Global Water Policy Project in Amherst, Massachusetts, said political leaders greatly underestimate the influence of water scarcity on food production, natural systems, and regional peace and stability. "Some unpleasant surprises may lie in store," she warned. According to Tad Homer-Dixon, director of the Peace and Conflict Studies Program at the University of Toronto, outright wars over water resources are unlikely because many countries are too poor to fight, but the social stresses associated with water scarcity are likely to get worse, leading to chronic unrest and violence. These problems are likely to occur regardless of global warming, but a changing climate may make matters worse in many places.

Water conflicts have also occurred in North America and are likely to grow. There have been disputes over water diversions from the Great Lakes—for example, during the 1988 drought, when water levels were too low for barge traffic on the Mississippi. According to Sarah Miller of CELA, water is increasingly being viewed as a commodity, and there are nearly three dozen companies in North America trying to tap into what they think will become a multibillion-dollar water market. Jack Lindsey, head of Sun Belt Water Inc. of California described water exporting as an "emerging industry." His company has been trying to export water from British Columbia, a move the province has resisted with a moratorium. In 1998, there was an uproar when the Ontario government issued a permit to a private company to withdraw water from Lake Superior for export to Asia. Concerns were raised that this would set a precedent that would make water resources a tradable commodity and force Canada to allow water exports in the future under international trade agreements. Miller, who said water will be to the next century what oil was to this one, warned of the political and environmental consequences if there's an "open season" on the Great Lakes when water shortages start to occur in the U.S.

Permafrost and Ice

Two of the more significant impacts that climate change will have on northern regions is thawing permafrost and altering the timing and extent of ice cover on lakes, rivers, and coastal waters.

Permafrost refers to regions where earth materials (e.g., soil, peat, clay, sand) remain permanently below 0°C. The northern two-thirds of the Mackenzie Basin sits on a foundation of permafrost. There are zones of continuous permafrost where all the ground is frozen and, in more southerly regions, discontinuous zones where patches of unfrozen ground are interspersed.

A report by Environment Canada's Environmental Adaptation Research Group says that tens of thousands of square kilometers of permafrost in the Arctic are "within one or two degrees Celsius of the melting point. Therefore, much of the permafrost environment would be profoundly affected by the transition to a warmer climate." It notes that permafrost is inherently unstable and warming could cause considerable destabilization of the ground throughout the North, affecting the safety of structures such as roads, bridges, dikes, and pit mines. There's evidence that warming has already increased the frequency of landslides and it's anticipated that a "poleward retreat of permafrost" could lead to even more landslides in the future, according to the Mackenzie Basin study.

Global warming is also expected to reduce ice cover on lakes, rivers and coastal waters. The CCS projected that the ice season for Arctic rivers could be reduced by up to a month by 2050; for lakes, the reduction would be up to two weeks. Ice on the Peace River is expected to form later in the fall and break up sooner in the spring, and there could be more than a 200-kilometer reduction in the distance it advances upstream. Residents in the Northwest Territories report that fall freeze-up already occurs later and spring breakup starts earlier than previously.

While a reduction in ice cover will have some benefits—for example, it should extend the shipping season—there's a downside. Ice keeps surface waters from being whipped up by winds and helps limit the damage caused by storm surges and pounding waves along coasts. Steve Solomon of the Geological Survey of Canada has found that in the past 50 years, coastal areas of the Beaufort Sea have receded by up to 100 meters, and this could accelerate if reduced sea ice causes increases in storm surges.

Solomon explains that storms in this region can create waves only if there's open water or fetch. "If you have an extensive fetch of 300 to 400 kilometers, quite large waves can form." The storm season in the Beaufort Sea starts in late August, but freeze-up usually occurs in October, so it's only open to severe storms for a month or so. If the water stays open longer, "the probability increases of a severe storm hitting the coast." One estimate suggests the Beaufort Sea's open water season could as much as double from around 90 to 150 days, considerably increasing the potential for coastal erosion and damage.

Forests

It's expected that global warming will shift the range of forests toward higher altitudes and latitudes and will alter the composition of tree species within forests. Many tree species cannot migrate fast enough to accommodate the habitat changes that may occur with warming, and slow-growing species may be replaced with faster-growing, more adaptable species. Overall, about a third of the world's forests may experience major changes in vegetation types; in some regions (e.g., northern boreal forests), as much as two-thirds of the forest area could be affected. Carbon fertilization is expected to increase tree growth, but these gains may be offset by an increased risk of forest fires in a hotter, drier climate and a higher risk of pest and disease infestations.

Climate change is likely to have different effects on forests in different climatic regions. The IPCC report says that tropical forests are likely to be more susceptible to changes in soil moisture than to temperature increases per se; in areas where water is already marginal, reduced soil moisture could cause the loss of forests. Tropical forests are likely to benefit from carbon fertilization and may be able to store more carbon as long as they're not being cut down. The IPCC notes that the high rate of deforestation is currently more of a threat to tropical forests than climate change. Tropical forests in South America, Africa, and Asia are burning at an incredible rate, both from natural fires caused by warmer, drier weather and from fires deliberately set to clear land for farming. In 1997, the El Niño–induced drought in Indonesia caused fires to burn out of control, creating a haze that caused health problems throughout the region for months. The World Wildlife Fund for Nature estimates that 88% of Asia's forests have disappeared.

Forests in temperate zones are also disappearing; for example, the WWF estimates that Europe has lost about 62%. In a warmer climate, it's anticipated that the mix of tree species in temperate forests is likely to change significantly. And while carbon fertilization will likely increase tree growth, these forests are unlikely to store more carbon overall because warmer temperatures will accelerate the decomposition of organic matter in the soil, releasing sequestered carbon. The IPCC notes that while temperate forests are currently a CO_2 sink, they could become a *source* of CO_2 due to losses caused by climate change or air pollution. An Australian study estimated that a doubling of atmospheric CO_2 would increase forest productivity 27% at first, but these gains would drop to 8% within less than a decade.

Canada's northern boreal forests will likely be most affected by climate change, both because they're more sensitive to climatic changes and because changes in the North are expected to be larger. It's anticipated the forests will expand northward, but this may be curtailed by poor soils and thawing permafrost. Some forests may benefit from carbon fertilization but net biomass and carbon storage may still drop because of increased drought, fires, pest infestations, and accelerated decomposition of soil organic matter. Studies of the boreal forest carbon budget that have examined the risk of increased losses to forest fires project than Canadian forests will be a net *source* of atmospheric CO_2 for the next few decades.

Forests are extremely important to Canada; it contains about 10% of the world's forested area, covering more than 400 million hectares or nearly half the total landmass of the country. Forestry is one of the mainstays of the Canadian economy, contributing $19 billion to the GDP in 1994 and providing more than 300,000 jobs. In 1993, forest exports added more than $22 billion to Canada's net trade balance, exceeding the contributions of energy, mining, and agriculture combined. Forests also play a key role in Canada's multibillion-dollar tourism and recreation industries.

Several Canadian studies have examined the potential impact of climate change on the country's forests. In general, rising temperatures, increased drought, reduced soil moisture, and increases in forest fires and pest infestations are expected to produce major changes. One study found that a doubling of atmospheric CO_2 would cause a 40 to 50% increase in Canadian forest fires.

In the prairies, gains expected from the northern expansion of forests are likely to be more than offset by losses caused by reduced soil moisture further south. According to an Environment Canada report, "low soil moisture stresses trees, making them more susceptible to pests, disease, and fire. It has been suggested that 170 million hectares of forest could be lost in the south and only 70 million hectares gained in the north, as they would be limited by poor soil or rock." The CCS notes that droughts in the 1980s caused reduced wood volume and increased insect outbreaks and fire damages. Areas affected in the 1970s and 1980s were greatly increased compared with the previous 50 years.

The Mackenzie Basin study found that commercial timber regions in northern B.C. and Alberta and southeastern Yukon could experience large changes in the Fire Weather Index (a measure of weather and forest conditions conducive to fire). Without countermeasures, the average number of hectares annually burned could double and yields from commercial stands could be cut in half. The study also found that insects common to southern Canada would likely migrate both northward and upward. Overall, forest losses caused by climate change in the Mackenzie Basin "will probably offset any potential benefits from a longer growing season," the study concluded. "Some of these changes have been observed during the recent 35-year warming trend."

In the U.S., commercial softwood forests in the southeast and south central states are expected to be hardest hit by climate change, according to a study done by the Environmental Protection Agency. Productivity in hardwood forests is projected to increase in northern regions and decrease in southern regions. The study projected a 20 to 45% drop in U.S. timber harvests, a 15 to 50% reduction in standing inventories, and a doubling or tripling of timber prices. The annual net loss was estimated at between US$3 and $12 billion.

Fisheries

Global warming influences many factors that affect the complex and diverse ecosystems that sustain fisheries, including water temperature and salinity, precipitation, winds, ice cover, and storms. However, the impacts will vary from place to place. "Global warming will likely cause collapses of some fisheries and expansions of others," according to the IPCC report. "It may be one of the most important factors affecting fisheries now and in the next few hundred years." While global fishery production might remain about the same or even increase with climate warming, there could be significant shifts at regional and local levels that could have serious economic repercussions.

Global warming is expected to alter fish habitats and shift the range of both ocean and freshwater species. Rising water temperatures will be a major mechanism causing these dislocations. Fish cannot regulate their internal temperature so they seek out waters that provide the optimum temperatures they need to thrive. Different species have different optimum ranges; sockeye salmon and brook trout, for example, grow best between about 10 and 15°C, while bass thrive in the mid-20s to the mid-30s. Changing their behavior is the only way most fish species can adapt to changing water temperatures, so they will shift their location in response to global warming.

The CCS notes that fish species in inland lakes and streams are expected to shift northward by about 150 kilometers for every 1°C rise in temperature. Cold-water species such as brook trout could be replaced by warm-water species migrating from the south. Similarly, northern species like Arctic char would find themselves competing with northward-shifting cold-water species like brook trout. A U.S. study found that freshwater fish with a preference for cold or cool water are vulnerable to climate change. A doubling of atmospheric CO_2 levels would cause losses of up to 50% of brown trout in more than half a dozen states and losses of 50 to 100% in another dozen states. Such changes are estimated to cost recreational fishing in the U.S. tens of millions of dollars annually.

On the Pacific coast, according to an Environment Canada report, "warmer river temperatures could cause severe prespawning mortality in some fish that go up rivers to spawn, such as the Pacific salmon. However, warmer ocean temperatures could create favourable conditions for species such as tuna, hake, and squid to migrate from the south." Similar shifts in the distribution and migratory routes of fish species are expected to occur along the Atlantic coast. The CCS projected that Atlantic fisheries are expected to decrease with climate warming.

The strong El Niños in 1982–83 and 1997–98 provided sneak previews of this effect. In the early 1980s, displacement of salmon migratory routes and habitat changes along the Pacific coast of North America led to increases in the price of salmon. In warm years, especially El Niño years, sockeye salmon take a more northerly route through cooler water to return to the Fraser River, causing a shift in the catch available to Canadian and U.S. fishers. Meanwhile, El Niño provided fishers off the coast of Peru and Chile with larger catches of scallops, tuna, and billfishes. During the 1997–98 El Niño, fishers in southern California hauled in one of the largest catches of squid they'd ever had, while fishers along the northwestern U.S. coast landed marlin, a tropical species never before caught in their waters. Warmer Pacific waters also caused a northward shift in the range of mackerel, which prey on salmon and became yet another stressor affecting the threatened Pacific salmon fishery.

Pacific salmon could be in for even more trouble, according to David Welch, head of the salmon research program for Canada's Department of Fisheries and Oceans. When he combined findings from research cruises during the 1990s and historical data going back 40 years, he found that salmon are stopped in their tracks by ocean temperatures higher than about 7°C in winter and 12 to 13°C in summer. Using climate models, Welch projected that by the middle of the 21st century, the north Pacific would no longer contain any regions cool enough to suit salmon; instead, they would likely be driven in summer to the upper half of the Bering Sea. By 2100, he suggests they may be pushed as far as the Arctic Circle.

What may be even more significant is that salmon would hit a high-temperature barrier much sooner during their southward trek to rivers along the Pacific coast. This could prevent many from making it back to spawn and thus could seriously affect reproduction rates. If this happens, the west coast salmon fishery could face extinction from climatic factors alone. To gather more data, Welch plans to implant salmon with computerized sensors that will record water temperatures and allow him to track where the fish travel.

Freshwater fisheries, especially at high latitudes, should benefit from climate warming, according to the IPCC report. Warming will extend the growing season for some species. Since most warming is expected to occur during the fall and winter and at night, it should improve the survival of fish that are not very cold-tolerant without damaging fish that are near their upper temperature limits. Many species will grow and mature faster in warmer waters, but this will also increase their need for food.

The CCS projected that dropping water levels and flows in southern lakes in the Canadian Shield and along the Pacific coast will reduce freshwater fisheries there, especially cold-water species like trout, char, and whitefish. In the Great Lakes, yields of whitefish, northern pike, and walleye are expected to drop. Species that are already living at their temperature limits are the most vulnerable. However, more northerly freshwater populations will likely expand because of longer growing seasons and smaller drops in water levels. The diversity of fish species may increase with the migration of species from the south, but in some cases resident species may be displaced by the newcomers.

Climate warming is expected to shift and expand the range of toxic organisms that produce "red tides," as well as microbes that cause cholera. Fish and shellfish (e.g., mussels, clams, and oysters) consume these toxins and species at the top of the food chain can contain concentrations lethal to humans. Consuming raw shellfish is a potential health risk for diarrheal and paralytic diseases and, in some cases, even cooking fish does not destroy the toxins.

Understanding the impact of climate change on fisheries is complicated by the fact that their fate also depends on socioeconomic factors and political choices. The IPCC report notes that most of the world's fisheries are already fully or overexploited and both the quality and quantity of fish have declined. The collapse of Canada's Atlantic cod fishery was influenced by environmental factors "but the additional burden of heavy fishing probably caused the stock collapse," the report says. Canada's salmon fisheries are also in serious trouble; in May 1998, the federal government banned fishing for endangered coho salmon in British Columbia, a move with major repercussions for the nearly $1 billion sport fishing industry as well as the commercial fisheries, which bring in about $400 to $800 million in a good year. A month later, the commercial fishery for Atlantic salmon was also shut down.

Declining fish stocks are already a source of growing conflict. In some cases there have been armed confrontations between fishing nations. The Pacific salmon crisis has escalated tensions between Canada and the U.S., while fishery closings on both coasts of Canada have created conflict between the provinces and the federal government. The situation is becoming increasingly volatile as coastal communities are devastated by the loss of their livelihood. In 1997, fishery officers wore bulletproof vests while confronting fishers determined to move into closed areas. There have been some examples of cooperation, however. In 1998, Greenland agreed to shut down its commercial salmon fishery to allow the fish to return to Canadian rivers to spawn.

Overfishing has left most fisheries around the world with little room to withstand added stress caused by climate change, the IPCC notes. "Climate change may magnify the effects of overfishing at a time of inherent instability in world fisheries." At present, overfishing and other human-caused stresses outweigh the impacts of climate change and are likely to continue doing so for several more decades. However, overfishing is being addressed by many countries and over the next 50 to 100 years, climate change may become the more dominant influence.

Agriculture

Warming should lengthen the growing season in many regions of the world, particularly at high latitudes. Crops could be planted earlier and they may mature faster, allowing early harvesting. In some places, it might even be possible to fit in an extra crop each season. However, these benefits could be offset by a loss of soil moisture and more frequent droughts caused by increased evaporation in a warmer climate. Some studies estimate that a 1°C rise in temperature with no increase in precipitation could reduce wheat and maize yields in many key regions by 5%, while a 2°C rise with a decrease in precipitation could cut yields by 20%. In many agricultural regions, the loss of soil moisture could exacerbate existing problems with soil erosion and infertility and cause desertification in extreme cases. Climate change is expected to degrade soil quality in several ways, including loss of organic matter, nutrient leaching, and increased salinization.

A warming climate would greatly increase demand for irrigation in many parts of the world, further straining water resources and causing conflict among multiple users. According to Peter Gleick of the Pacific Institute for Studies in Development, Environment, and Security, more than three-quarters of the water consumed by human activities in most parts of the world is used for agriculture and nearly 40% of the world's food is produced on the 16% of agricultural land that's irrigated. "Many argue that growing more food will require more irrigation water." However, irrigation is likely to become more difficult and more costly in a warmer climate, according to Cynthia Rosenzweig, an agronomist with the Goddard Institute for Space Studies, and Daniel Hillel, professor emeritus of plant and soil sciences at the University of Massachusetts.

Agriculture will also be affected by changes in the intensity of precipitation, the timing and intensity of spring runoff and flooding, and the frequency of extreme weather events. Water supplies in coastal agricultural areas could be threatened by contamination with salt water as sea levels rise. Rosenzweig and Hillel say this may make it difficult to sustain agriculture in some parts of Florida, the Netherlands, China, Egypt, and Bangladesh.

Increasing temperatures can affect crop yields and cause a shift in the types of crops that can be grown. Yields may drop steeply when temperatures exceed a crop's optimum range; some studies indicate that a 4°C rise will decrease yields of wheat, rice, corn, and soybeans in many places and that even a 2°C rise will negatively affect crop yields in some semi-arid and tropical regions. Crops that are grown close to their upper temperature limits (e.g., rice in Southeast Asia) are especially susceptible to heat waves. Increasing nighttime temperatures—an expected consequence of global warming—may also reduce yields. However, in some temperate climates, crops may benefit if higher nighttime temperatures reduce the frequency of severe frost.

A warmer climate theoretically will allow the expansion of farming to higher latitudes, but this potential may not be fully realized because these regions often don't have high quality soils. Moreover, not all plants can migrate easily; some may not adapt to the longer day length at higher latitudes. New high-latitude varieties may need to be developed to overcome these problems.

Climate change is also likely to alter the range of insect pests and diseases, allowing some to flourish where they could not previously survive. For example, the European corn borer, which attacks maize, would move 150 to 500 kilometers northward if temperature increased 1°C. A warm spell from 1986 to 1988 enabled locusts to spread to new regions in southern Europe. In 1998, after an unusually warm winter, Western Canada's canola crop was attacked by an unprecedented outbreak of lygus bugs.

Warmer weather will also allow some insects to produce more generations each year and to survive the winter, creating larger infestations the following year. However, climatic changes can also suppress insect infestations; for example, the reproduction of some worms and caterpillars is reduced when temperatures exceed their tolerances. The IPCC report notes that mild winters and warm weather can increase outbreaks of plant diseases like mildew, rust, and blight. However, very dry, hot conditions would generally reduce infestations of most fungal diseases, and many of these diseases would shift their range into currently cool regions.

Increased levels of atmospheric CO_2 are likely to increase the productivity of many food crops. Some studies indicate that a doubling of CO_2 will increase the growth of crops like wheat, rice, potatoes and soybeans by an average of about 30%, although the response can vary from −10 to +80% in difference circumstances. Most of the plant species that provide the majority of the world's food supply belong to the category most likely to benefit from carbon fertilization. However, there are several important species that do not respond to increased CO_2 as well; these include many grown in tropical regions, such as sorghum, maize, millet, and sugarcane, which provide a significant portion of the world food supply, especially in developing countries.

Many estimates of increased crop yields are based on laboratory studies done under optimal growing conditions. In the real world, the benefits of carbon fertilization could be offset by changes in temperature, precipitation, extreme weather events, and pest infestations.

Given the complexity of the climatic factors that influence agriculture, the impact of global warming is expected to vary from country to country and within different regions of the same country. For example, developing countries, which are mostly located in warm, low latitudes, are likely to experience a drop in grain yields, while mid- and high latitude countries like Canada, the United States, and Australia are likely to experience overall increases in grain production.

The many studies cited in the IPCC report project wide swings from minus to plus values depending on different assumptions about changes in temperature and precipitation, carbon fertilization, and adaptive measures (e.g., irrigation, developing new crop varieties). The data suggest some overall conclusions:

- Considering climate change alone, without the offsetting benefits of carbon fertilization or adaptive measures, there will be substantial economic losses and reductions in food production throughout the world. Estimates of annual losses range from US$115 to $248 billion.
- Economic losses can be greatly reduced if the effects of carbon fertilization and/or adaptation measures are included (ranging from annual losses of US$61 billion to gains of US$7 billion). However, most models still project net losses worldwide. These projections show that adaptive measures reduce losses more than carbon fertilization.
- Developing countries will be far more negatively affected than developed countries. Worst-case economic scenarios (no carbon fertilization, no adaptation) for OECD countries show annual losses ranging from US$13.5 to $17.6 billion, compared with losses of US$57 to $121 billion for developing countries. Even with carbon fertilization and/or adaptation factored in, developing countries will remain in the loss column, while developed countries move into positive territory. The IPCC concluded that "the population at risk of hunger...could increase despite adaptation."

The agricultural response is also expected to vary within countries. For example, an analysis done by Rosenzweig for the U.S. Environmental Protection Agency indicates that with a 2.5°C rise in temperature and a 7% increase in precipitation, wheat yields will rise between 10 and 30% in the U.S. Midwest, northeast, and along the west coast, but will drop up to 10% throughout much of the West and up to 30% in the southeast. Corn yields will rise from 10 to 30% in the eastern third of the country and in excess of 30% along the west coast, but will drop up to 20% throughout the West and Midwest. These findings suggest it's somewhat misleading to say that U.S. agriculture will benefit from climate change, according to Joel Scheraga of the EPA. "Even though the U.S. as a whole will be a winner, some regions may be net losers." He added that most studies focused on changes in average climate and do not yet fully account for the agricultural impacts of climate variability and extreme events.

A good example of the damage done by weather extremes occurred in March 1998. A warm El Niño winter caused fruit and vegetable crops in the U.S. southeast and Midwest to bloom early; then they were hit with a vicious storm that dropped torrents of rain and snow and dragged cold air down from the north. In some places, temperatures plummeted from the mid-20s to below zero in a couple of days, forcing farmers into a desperate struggle to protect their crops against a record-setting cold snap.

In Canada, which encompasses several climatic zones, the effects of climate change are likely to be quite variable. "The ranges of existing plant and animal species are projected to shift to higher latitudes and higher elevations, while southern species extend their range into Canada," the CCS notes. There could be a northward expansion of southern regions of Ontario, Quebec, and B.C where fruits and vegetables can be grown. Commercial farming could expand into northern areas of Ontario and Quebec, as well as into the Peace River region. Even the southern Yukon and lower Mackenzie River area might see some agricultural opportunities. However, lack of water and poor-quality soils are two critical factors that may prevent the benefits of such expansion from being fully realized in many cases.

In the prairies, drought is already a major problem and it's likely to intensify in a warmer climate. Studies indicate that precipitation will be a limiting factor and, while the growing season will lengthen, lower soil moisture will negatively affect crop growth. Rising temperatures will also increase the risk of insect infestations. The CCS estimates that crop yields could potentially drop by as much as 10 to 30%, with cereal crops dropping by one- to two-thirds of current levels. Dry conditions could negatively affect the production of livestock feed, but warmer winters would probably benefit the animals. While the demand for irrigation may increase, water supplies are expected to drop with climate warming.

In the Great Lakes region, where agriculture is the largest industry, the growing season will probably be longer, but the loss of soil moisture will likely decrease crop yields unless countermeasures are adopted. One study found that the growing season in parts of Quebec could increase by 33 to 40 days, the time it takes for legumes to mature could be reduced 7 to 21%, and carbon fertilization will likely increase the yields of some grains but reduce those of others. Increasing temperatures could both help and hurt the Ontario wine industry. Lisa Lepp of Stonechurch Vineyards says that a 2°C increase would help winemakers grow some "awesome reds" but it would adversely affect the grapes used for ice wine, currently Canada's "hottest" wine commodity.

Many studies indicate that grain crops may extend their range in high latitude regions such as northern regions of Canada, Scandinavia, Europe, and Russia and southern regions of New Zealand, Chile, and Argentina. However, poor soils and lack of water will limit this expansion in many cases. Irrigation demands are already depleting groundwater supplies in many regions; the Worldwatch Institute found that water tables are dropping in irrigated farmland in northern China and India's "breadbasket" Punjab region. A U.S. study found that increased irrigation would be needed in a warmer climate even if precipitation increased by 20%.

British studies indicate that drier soils would affect agriculture throughout Britain and that a 1.1°C increase in temperature by 2050 would increase irrigation demand by 25%. Warmer temperatures are expected to decrease yields of cereal crops such as wheat; yields for other crops such as potatoes and sugar beets are expected to increase, but they may be more susceptible to insects. New crops such as maize and sunflower could be grown in Britain if the average temperatures rose about 1.5°C, which is equivalent to a southward shift in latitude of about 200 to 300 kilometers.

Energy and Transportation
The impact of climate change on energy and transportation is expected to vary considerably by region and to have both positive and negative consequences. Quantifying these impacts is difficult; sometimes even figuring out the direction of change is a challenge. Energy demand, for example, can go in either direction depending on how the demand for heating and cooling evolves at local and regional levels. Studies in several countries suggest that in regions where the demand for heating declines, fossil fuel use will drop, whereas the demand for electricity could either rise or fall depending on the balance between heating and cooling.

The IPCC report summarizes several ways in which the energy and transportation sectors are sensitive to climate change. (Interestingly, it says most of these impacts are considered moderate or minor. There's a perception that these sectors aren't as sensitive as agriculture and natural ecosystems and that they can adapt as long as global warming occurs gradually. Events like the 1998 ice storm have no doubt caused some rethinking of these perceptions.)

The report said that offshore oil and gas operations will benefit from reductions in ice cover caused by rising temperatures, but will also be affected by more frequent extreme weather events and rising sea levels. It's virtually certain the demand for heating will drop and the demand for cooling will rise, but the balance will vary considerably in different places. Higher water temperatures and lower water levels will affect the availability and efficiency of cooling water used in power plants. Coastal plants and refineries may experience problems associated with sea level rise. Hydroelectric power stations and dams will be affected by increased precipitation and evaporation and lower water levels in lakes and rivers. Renewable energy sources, especially wind and solar systems, will be vulnerable to extreme weather events. Melting permafrost could threaten the stability of pipelines.

Some of these impacts are already apparent. During the bitter winter of 1996 (a three-week cold snap in January rendered many Canadian cities colder than the North Pole), Canadians paid an additional $500 million for heating compared with the year before. On the flip side, the warm winter of 1997–98 reduced U.S. heating costs by an estimated 10% in some regions, producing savings to consumers of several billion dollars. A study done by the U.S. Department of Energy found that a 1°C warming by 2010 would reduce U.S. energy costs by US$5.5 billion (1991 dollars) and a 2.5°C warming (considered unlikely by 2010) would save US$12.2 billion. It suggests that reduced winter heating costs would more than off-set increased summer cooling costs; however, it didn't consider increased humidity, which is expected to affect cooling demand. The study also said that if warming above 2.5°C occurred beyond 2010, cooling effects would start to dominate the economic calculations.

Canada is expected to benefit from reduced heating costs in winter, which will likely more than offset increased cooling costs in summer, at least in the early stages of global warming. However, southern Ontario is already undergoing a seasonal shift in energy demand, according to Don Power of Ontario Hydro. The company is already experiencing peak demand in summer, similar to utilities south of the Great Lakes, rather than the winter peaking that was typical until a few years ago. He attributed this to higher summer temperatures and higher demand for air conditioning.

The repercussions of climate change for energy production and generation will vary across Canada. Changes in water levels and flows will have a major impact on both hydroelectric and nuclear generation. According to the CCS, water availability may increase the prospects for hydro generation in Labrador and northern Quebec by perhaps 15%, but reduce it in Ontario, the prairies and southeastern B.C. During the 1988 drought, Manitoba's hydro generation dropped 4%, causing a 73% decline in export sales and a $24.6 million loss. Saskatchewan had to import energy and electricity costs increased by 29%. In the Mackenzie Basin region, surcharges have been added to electric bills because lower water levels forced a switch from hydro to more expensive thermal generation (which also puts more CO_2 into the atmosphere). In Ontario, the projected decline in Great Lakes water levels could significantly affect electrical generation, the biggest single use of water from the lakes. Ontario Hydro's nuclear plants use about 50 times Toronto's daily water use, and the fossil fuel plants use about five times the daily municipal use.

Both power generation and transmission are likely to be affected because the efficiency of power lines decreases with higher temperatures and the efficiency of power plants drops if cooling water warms up. The IPCC notes that the cost of interrupting electrical service is rising because of the increasing economic importance of electronic information and communications systems.

In the Arctic, offshore oil and gas operations and tanker shipping should benefit from a longer summer season and less severe ice conditions, but the reduced ice cover may increase the risk of more frequent and intense storms and larger storm surges.

Energy systems are particularly vulnerable to extreme weather events. Distribution and transmission networks are, of necessity, exposed to the elements and many energy installations are located either offshore or in high-risk coastal regions. A British report points out that all of the U.K.'s oil refineries and half of its power station capacity are located in coastal or estuarine regions that are vulnerable to stronger storm surges resulting from rising sea levels. It cost U.S. utilities US$64.7 million and US$17.5 million to restore power after Hurricane Hugo in 1989 and Hurricane Fran in 1996, respectively. Severe weather is particularly dangerous and costly for offshore exploration, according to David Epp of BP Exploration in Texas. Evacuations cost millions of dollars per day and weather-related delays in drilling can cost US$250,000 a day or more.

The transportation sector will experience mixed impacts from climate change. Warmer weather and reduced ice cover should lengthen the shipping season in northern regions; it's estimated the Great Lakes ice season could be reduced by as much as 5 to 12 weeks and the St. Lawrence Seaway would also have a longer season. The melting of Arctic sea ice should extend the shipping season in northern waters; transit through the Northwest Passage may become much easier at times. The barge season on the Mackenzie River, a major northern supply route, could be extended by 40%. However, lower water levels could offset this benefit and increase transportation costs in the region. In fact, shipping on many inland lakes and rivers throughout North America may encounter problems caused by low water levels, especially during droughts. Barges either would have to wait for higher water or would have to carry lighter loads, or channels would have to be dredged to provide a deeper draft. During the 1988 drought, more than 800 Mississippi barges were tied up for months.

Reductions in snowfall or shifts from snow to rain should reduce snow removal costs. One Ontario study compared a cold winter (1993–94) with a warm one (1994–95) and found large savings in snow removal costs, salt, and equipment use. Road maintenance costs dropped 8% ($14 million) between the two winters. On the other hand, maintenance costs could rise if heat waves cause increased buckling of road surfaces.

The CCS concluded that while some parts of the country will experience increased snowfall, land-based transportation costs will likely be reduced due to shorter, milder winters, especially in southern Canada. Further north, however, warmer winters could increase transportation costs by melting ice- or permafrost-based roads, forcing communities to shift to water or air transportation or to build all-weather roads. During the warm winter of 1997–98, a town in northern Manitoba that normally trucks in its yearly supplies over ice roads was forced to resort to airlifts when the ice failed to form. "With thawing of permafrost, maintenance costs for all-weather roads and rail-beds currently overlying existing permafrost may rise," CCS said.

The Mackenzie Basin study noted that permafrost and ice conditions, which are essential to land-based transportation in the region, are already being affected by climatic changes. "Ice strips for air transport are eroding in areas affected by fires. Permanent roads and ferry harbours are being damaged by erosion caused by melting permafrost. Any changes in the rates of run-off, melting, and water flows would result in revising dates for cut-offs and closures of winter roads."

The IPCC report notes that transportation infrastructure will be affected by the migration of people and economic activity caused by climate change. For example, the northward expansion of agriculture, forestry, and mining, and the consequent buildup in populations and

industry, could open up new transportation opportunities in high latitudes like Canada's North. However, in many regions, sea level rise may
force large numbers of people to migrate from low-lying coastal regions,
which could also affect transportation systems.

Transportation infrastructure in low-lying regions also will be affected
by rising sea levels. One U.S. study found that 40 to 100% of the transportation, communications, and utilities infrastructure of Galveston,
Texas, could be underwater by 2100. Another study found that a one-
meter rise in sea level and the water table could cause extensive damage
to roads, bridges, and causeways in Miami, Florida, unless they're reconstructed. It's estimated that raising the most vulnerable roads would cost
about US$237 million. In Canada, the new Confederation Bridge that
links Prince Edward Island to the mainland was constructed with a one-
meter sea level rise factored in.

Transportation is, of course, extremely vulnerable to extreme weather
events. Although improved forecasting has reduced the risks of weather-
related disasters, these gains may be offset if climate change causes an
increase in the frequency or intensity of extreme events. The CCS suggests that air travel may be more sensitive, both directly and indirectly,
than any other mode of transportation, based on the large costs currently
associated with weather-related disruptions.

Economic trends since the late 1980s have increased the vulnerability
of the transportation industry to weather disruptions. For example,
commercial transportation is now based on "just in time" delivery—
moving parts and supplies so they arrive just when needed. This reduces
costs associated with storing large inventories but increases the risk of
loss due to bad weather. These disruptions can be far-reaching. "A major
flood that closes interstate highways, such as the 1993 Midwest flood,
can literally cause factories to shut down and layoffs to occur in businesses 1000 miles from the nearest high water," according to meteorologist Michael Smith.

Infrastructure

Extreme weather events, rising sea levels, and melting permafrost are the
major threats to the human-built environment or infrastructure. The
IPCC report notes that much of this infrastructure is already at risk
because of the growth in human settlements and increasing wealth concentrated in regions vulnerable to flooding, storms, fires, landslides, and
other extremes. It's been estimated that about US$2 trillion in insured
property sits within 30 kilometers of areas exposed to Atlantic hurricanes.
Other estimates of property exposed to climatic hazards include US$807

billion in Japan, US$25 billion in Australia, and US$186 billion in the Netherlands. According to one estimate cited in the IPCC report, the global value of exposed assets is US$22 trillion, but "this value may have doubled due to economic growth since the 1970s."

Given projections that global warming will increase the frequency and intensity of extreme events in many regions, protecting infrastructure is likely to become more problematic in the future and may require a retreat from the most vulnerable zones. According to one estimate, it could cost the world roughly US$1 billion a year to protect against a 50 centimeter rise in sea level by the year 2100.

The CCS said that coastal facilities (e.g., docks, homes, ports) in low-lying southern regions of British Columbia could be flooded more frequently during severe storms if sea level rises a few tens of centimeters. Upgrading existing dikes protecting the city of Richmond could cost hundreds of millions of dollars. Beaches will be more costly to maintain and "increased winter precipitation will put greater stress on water and sewage systems, and increase the danger to environmental and human health." In the prairies, dams and other water control structures may be vulnerable if there's an increase in the magnitude and frequency of extreme events that cause flooding.

In the Great Lakes region, owners of shoreline property have already spent a lot of money building retaining walls and barricades to protect their homes against damage from fluctuating lake levels. Flooding and erosion along the U.S. and Canadian shorelines since 1985 have caused damages estimated in the hundreds of millions of dollars.

In the North, the fate of permafrost in a warming climate is a big issue. It provides the foundation not only for roads and airstrips, but for many other structures—bridges, dams, houses, pipelines. A report by the Environmental Adaptation Research Group points out that the mechanical and physical properties that make permafrost good for engineering purposes are dependent on temperature, particularly temperatures within one to two degrees Celsius of thawing. Tens of thousands of square kilometers of Arctic permafrost are within this temperature range, so the potential impact of global warming could be profound. Rising temperatures will weaken the frozen soil, threatening the stability of slopes, structures, and foundations.

The resulting landslides and damage to or collapse of structures could have serious consequences, and northern communities could experience increased socioeconomic disruption if there's a substantial increase in the failure of permafrost-based structures. According to the CCS, "buildings could become structurally unsafe or even collapse; utility lines and pipelines may rupture. Mining operations may become easier, but waste dumps, tailings dams, and water diversion channels would require costly maintenance."

It can be seen from these few examples of environmental and socioeconomic activities affected by global warming that we'll be confronted with complex and evolving challenges over the next century. Although further research is needed to better understand the economic and social costs and benefits of global warming, these impacts studies at least give us an idea of what may be in store.

seven

DISEASE AND DEATH:
Human Health and Climate Change

Coping with climate change will take more than money and technology. It will also require people with the physical health and emotional and psychological strength needed to confront the challenges that face us. Unfortunately, global warming has the potential to undermine these personal qualities as surely as it does our technological and economic resources. The spread of infectious diseases and the debilitating health impacts of heat waves and extreme weather may sap our ability to adapt to a changing climate.

Around the world there is a "deep, universal concern" about the health impacts of environmental problems, according to a 1997 survey by the Toronto polling firm Environics International Ltd. When they questioned 27,000 people in 24 countries representing 60% of the earth's population, they found that most respondents felt their own health had been affected and more than 80% believed their children's health will be affected by future environmental problems. "The results would seem to contradict the belief among many policymakers that the environment is fading as a concern for their citizens," Environics noted in a news release.

The polls showed that health concerns had greatly increased since the 1992 Rio Earth Summit, said Environics International President Doug Miller, who directed the international study. "We were surprised by the sharp rise of public health concerns in both developed and developing countries over a five-year period in which their leaders have moved in the opposite direction on the environment issue. It is as if the survival instinct of our species has been activated. Government and industry leaders ignore these deepening environmental and health concerns at their peril."

Social and demographic factors, technology, and wealth greatly influence how vulnerable different populations will be and how they will actually be affected by increased risks to human health caused by climate change. People in developing countries—the majority of the world's population—are more vulnerable than those in wealthier industrialized countries, but in all cases, the poor, the young, the elderly, and the ill will suffer disproportionately. The IPCC report notes that the vulnerability of populations is increasing even in developed countries, because of the aging of their populations as well as increasing levels of disability, chronic illness, and the fact that many people choose to retire in coastal regions vulnerable to extreme weather.

It projects that most health impacts associated with climate change would be negative and would fall into one of two categories: direct and indirect. Direct effects are those related to 1) increases in high temperature extremes and heat waves; and 2) increases in extreme weather events. Indirect effects are those caused by 1) changes in the range or activity of organisms that cause or spread infectious diseases; 2) changes in food production; 3) sea level rise; 4) increased air pollution; and 5) dislocations and conflict caused by the effect of climate change on the economy, infrastructure, and resources.

"If long-term climate change ensues, indirect impacts would predominate," the report notes, adding that different populations would differ in their vulnerability, depending on population density and crowding, the security of the food supply, local pollution, and distressed ecosystems.

The report predicted the following effects with "high confidence": 1) increased deaths and illness—mainly heart and respiratory problems—because of increases in the frequency and severity of heat waves; 2) increased allergies and cardiorespiratory illnesses and death because higher temperatures will make urban air pollution worse; 3) increased deaths, injuries, infectious diseases, and psychological distress caused by extreme weather events; and 4) increases in infectious diseases spread or caused by organisms whose geographic range and life cycles will be altered by climate change.

The following effects were predicted with "medium confidence": 1) a severalfold increase in heat-related deaths that would be partially offset by fewer cold-related deaths. The exact balance can't yet be quantified and will vary from place to place; 2) an increase in the percentage of the world's population potentially exposed to malaria and an increase in the incidence of malaria in some areas; 3) an increase in food- and water-borne infectious diseases such as cholera and salmonella, particularly in tropical and subtropical regions, because of the effect of climate change on temperatures, water, and the growth of microorganisms; 4) an increase in health problems associated with malnutrition and hunger in regions where climate change adversely affects agriculture, farming, and fisheries; and 5) an increase in physical and psychological health problems related to social upheaval and conflict resulting from shortages of natural resources, rising sea levels, and disruptive weather events.

These projections were based on studies done in different countries, which examined current worldwide public health trends and possible future health trends using climate model projections. The uncertainties in the models also create uncertainties in the projections of health impacts, particularly local or regional effects. Nevertheless, these studies identify areas with the potential for future public health problems as a first step toward more in-depth research and the development of adaptive strategies.

Infectious Diseases

Infectious diseases are currently the leading killer around the world and increasing the transmission of these diseases is the most significant indirect mechanism by which climate change could affect human health. Infectious organisms, including bacteria, viruses, and parasites, as well as their "vectors" (e.g., insects or rodents that carry disease-causing organisms), are influenced by climatic conditions such as temperature, humidity, soil moisture, surface water, and changes in ecosystems where they thrive. Climate change can potentially affect the distribution and life cycles of these organisms and thus the transmission of disease. However, it's difficult to figure out exactly what the effects will because so many factors associated with the vectors, the pathogens, and their habitats have to come together at the "right" time to facilitate the spread of these diseases.

Two major categories of infectious diseases could be affected by climatic conditions: *vector-borne diseases* (e.g., malaria, dengue, yellow fever) and *water- or food-borne diseases* (e.g., cholera and salmonellosis). These diseases already affect billions of people worldwide, mostly in tropical and subtropical regions and mostly in developing countries. Anything that amplifies their threat or expands their territory could potentially cause a substantial increase in human health problems.

Whether such problems will actually materialize in any given location is difficult to predict because so much depends on social and health programs and technologies such as vaccines, pesticides, sanitation, and water treatment facilities (some of which themselves may be affected by climatic factors such as extreme weather events and rising sea level). There's some controversy within the scientific community about the relative importance of climatic versus socioeconomic factors in determining the future spread of infectious diseases.

Climate impact studies have focused mainly on the question of how global warming might shift and expand the potential range (both latitude and altitude) of disease-causing organisms and their vectors, and how warming might influence their reproductive life cycles. "In general, increased warmth and moisture would enhance transmission of these diseases," the IPCC report says. A report by the World Health Organization says that projected increases in average temperature will likely create more favorable conditions for vectors, "which may then breed in larger numbers and invade formerly inhospitable areas....Major shifts in vector densities and distribution can be expected." Climate change may also cause diseases that now occur only during certain seasons to break out all year round.

Among the vector-borne diseases, malaria is of greatest concern because it already affects so many people; there are an estimated 300 to 500 million cases worldwide annually and an estimated 2.4 billion people are potentially at risk. A fever disease caused by a parasite spread by mosquitoes, it's most prevalent in tropical and subtropical regions. In the past, it has occurred in some parts of the U.S., southern Canada, southern Europe, and northern Australia, but today in developed countries, it's largely controlled by public health measures. In northerly regions of temperate zones, the relatively cool climate plays an important role in limiting the potential for spreading the disease.

Several species of parasites and mosquitoes transmit malaria, and the life cycle of each is sensitive to certain thresholds and ranges of temperature and humidity. For example, temperatures between 20 and 30°C and humidity above 60% are optimal for mosquitoes to grow quickly and survive long enough to incubate the parasites. Cooler, drier conditions inhibit their growth. But excessively hot conditions are bad for them, too; temperatures above 35°C drastically reduce the lifetime of mosquitoes and thus their ability to transmit the disease. The parasites are temperature-sensitive, too; for example, at 25°C they will take only one-third to one-fifth as long to develop inside the mosquito as they do at 18°C.

There have been recent outbreaks of malaria associated with heat waves in regions where the disease is not common. Record-breaking heat in June and July of 1995 caused an explosion in mosquito populations in parts of Russia, followed by the region's first outbreak of malaria in four decades.

In September 1995, a Michigan man was hospitalized with a case of malaria caused by a locally infected mosquito, the first such case to be reported that far north in 23 years, according to the U.S. Centers for Disease Control and Prevention (CDC). Indigenous malaria in the U.S. was deemed by the CDC to have been eradicated in the 1950s and since then nearly all reported cases involved people infected outside the country. In the Michigan case, the CDC noted that the average evening temperature during August that year was 3.3°C above normal and this may have facilitated the transmission of malaria by shortening the reproductive cycle of the parasite and prolonging the life of the mosquitoes. Malaria is not readily transmitted in the U.S. at present because mosquitoes often do not live long enough to incubate the parasite.

Reporting on U.S. malaria outbreaks between 1986 to 1994, Jane Zucker of the CDC wrote: "A common feature of all recent outbreaks has been weather that is hotter and more humid than usual." She suggested that the combination of environmental change, increasing drug resistance, and increased air travel could cause malaria to reemerge as a serious public health problem in the U.S. Unlike people who live in regions where malaria is common, most U.S. residents would have little resistance to the disease.

Other studies have suggested that parts of southern Canada also face an increased risk of malaria. Research done by Kirsty Duncan of the University of Windsor's geography department examined whether global warming would increase the potential for malaria in the Toronto area. She studied the effects of increased mean daily temperatures on mosquitoes, the development of parasites inside mosquitoes, and transmission of the disease.

Duncan concluded that between 1951 and 1988, malaria could not have occurred in the Toronto region, even if mosquitoes were present, because temperatures were too low for the parasite to develop. However, the increased temperatures associated with a doubling of atmospheric CO_2 would be high enough permit parasites to develop and transmit the disease. Warmer temperatures would be "more hospitable" to both mosquitoes and parasites, she writes. "Moreover, malaria parasites would have more vigorous life cycles and would likely become more virulent." She doesn't argue, however, that climate change alone will cause malaria to emerge in the Toronto area. Social, economic, behavioral, and environmental factors are also important and "climate is merely one factor."

The worldwide distribution of malaria is "highly likely" to be altered by climate change, according to the IPCC report. It cites a study that estimates that if the earth's average temperature increased 3 to 5°C by 2100 (which is higher than recent projections of 1 to 3.5°C), the percentage of the world's population living within the potential malaria transmission zone would rise from 45 to 60%. With a 3°C warming by 2100, the annual incidence of malaria could increase by 50 to 80 million cases over the 500 million that would occur without global warming—a rise of between 10 and 16%.

However, the report emphasizes that "although climate change would increase the potential transmission of malaria in some temperate areas, the existing public-health resources in those countries…would make reemergent malaria unlikely." What's more likely is that it will spread out and up from areas where it's now endemic, mostly in the tropics. Initially, there would be high death rates among newly exposed populations that had not acquired natural immunity. However, in some hot regions that get even hotter because of global warming, malaria might actually be reduced as very high temperatures shorten the life span of mosquitoes.

Another disease spread by mosquitoes is a viral infection called dengue fever. It causes severe flu-like symptoms and, in some cases, fatal hemorrhagic symptoms. It's considered the most serious viral infection transmitted in humans by insects, and there are no available vaccines or drug treatments.

Dengue currently affects about 10 to 30 million people a year, and an estimated 2.5 billion people are potentially at risk. The disease is most common in tropical regions but there's concern about its potential to spread to temperate zones. The survival of mosquitoes and the transmission of the virus are temperature-dependent and there's evidence that, with rising temperatures, the mosquitoes have already spread to higher latitudes and altitudes bordering their current habitats. For example, they've been reported above 2200 meters in Columbia where once they were restricted by temperature to regions below 1000 meters.

A group of New Zealand researchers found that since 1970, several dengue epidemics have occurred throughout South Pacific islands that are "on the fringe of the current endemic zone" and the serious hemorrhagic form of the disease emerged there for the first time. Non-climatic factors in the region haven't changed much since the 1970s so the dengue outbreaks "may have been mainly climate driven," the scientists write in the journal *Lancet*. Their research indicates there was a correlation with warm temperatures caused by El Niño. "These findings suggest that dengue will be an increasing problem if the global climate continues to warm."

In fact, a study by researchers at the Johns Hopkins School of Public

Health projected that the "epidemic potential" of dengue would increase with rising temperatures. The higher this potential, the fewer mosquitoes would be needed to spread the disease. Computer simulations indicated that global warming not only would increase the range of the mosquitoes, but would also boost their biting rate, thus increasing the chances of transmitting disease. Warmer temperatures would also reduce the time the virus requires to develop inside the mosquitoes. For example, the incubation period for one dengue virus would drop by seven days if the temperature rises from 30°C to between 32 and 35°C, a change that could potentially triple the transmission rate of the disease. The study indicated that the disease would most likely spread into temperate zones bordering areas where dengue is already entrenched. The authors conclude that climate change would have a "substantial impact" on the spread of the disease, but they do not suggest it's the only factor, nor necessarily even the most important one.

Higher temperatures and increased precipitation associated with global warming could increase the potential for spreading several other types of vector-borne diseases, including African sleeping sickness, spread by the tsetse fly; schistosomiasis, a disease spread by water snails that afflicts some 200 million people a year; and river blindness, a parasitic disease spread by blackflies.

In some cases, however, global warming may reduce the spread of infectious diseases if conditions become too hot, too wet, or too dry for certain pathogens or their vectors to survive. For example, excessive rainfall and flooding could decimate mosquito populations rather than allowing them to proliferate. Salination of coastal freshwater areas by rising sea levels could also reduce mosquito populations. Drought and the spreading of deserts would also reduce vectors that breed in water. Some disease-carrying ticks, like those that spread Rocky Mountain spotted fever, prefer cooler temperatures, so their numbers might drop in their current habitats—though they might adapt by migrating northward to infect new areas.

In addition to insects, water and food are major vehicles for transmitting infectious diseases. Several diseases caused by bacteria, viruses, and protozoa (e.g., cholera, salmonella poisoning, and cryptosporidium) are spread this way. "Many of these organisms can survive in water for months, especially at warmer temperatures, and increased rainfall could enhance their transport between groups of people," the IPCC report notes. These diseases, most of which cause severe diarrhea and dehydration that can be fatal if not treated, are contracted by drinking contaminated water or by eating raw or undercooked fish, fruits, or vegetables.

Poor communities that lack adequate sanitation facilities are chronically vulnerable to diarrheal diseases, but outbreaks can also occur in the after-math of hurricanes, floods, or storm surges that destroy sanitation facilities and displace large numbers of people. In January 1998, heavy rains and flooding destroyed the sewage system in a normally arid region of Peru, causing 200 reported cases of cholera. Cholera also broke out in Acapulco, Mexico, after Hurricane Pauline in October 1997 and in Russia in 1995, after a hot, dry summer raised water temperatures and reduced water flow in streams, concentrating contaminants.

The bacterium that causes cholera prefers warm, brackish water, but it can remain dormant in colder conditions waiting for an opportunity to emerge. It can also piggyback on other organisms, such as algae or small sea animals, and be carried long distances by ocean currents and tides. According to the IPCC, the transmission of cholera could be amplified by the explosive growth ("blooms") of coastal algae that occurs in response to rising ocean temperatures and the runoff of nutrients from fertilizers and waste water into the ocean. An increase in the incidence of algal blooms is expected to increase the contamination of seafood with toxins.

The report notes that warmer temperatures could also increase the incidence of an Australian encephalitis that occurs in summer, caused by an amoeba that proliferates in above-ground water pipes warmed by the summer sun. "Higher temperatures would also increase the problem of food poisoning by enhancing the survival and proliferation of bacteria, flies, cockroaches and so forth in foodstuffs," the report says.

Although there are still uncertainties attached to forecasts of the effects of global warming on infectious diseases, recent outbreaks—particularly in regions where they've never occurred before or where they were thought to be eradicated—as well as the emergence of new diseases, are seen by some public health experts as warning signs. Many of these outbreaks have been associated with specific weather patterns—"exactly the kind of weather patterns you'd expect to increase in frequency with climate change," says Joel Scheraga of EPA.

For example, in 1994, temperatures in India remained as much as two times higher than normal during the autumn monsoon season, creating conditions that led to outbreaks of malaria, dengue, and pneumonic plague that killed an estimated 4000 people. In the U.S. southwest, a 1993 outbreak of an emerging often-fatal pulmonary disease caused by a pathogen known as hantavirus was attributed to climatic and ecological factors that caused an explosion in the population of the vector, deer mice. Two hantavirus species have been found in Canada as far north as the Yukon.

It's often difficult to link a specific disease outbreak to changing climatic conditions; in most cases, other factors are at least as important. But taken together, these events suggest a trend that bears watching. As the IPCC comments, assessing *patterns* of change in human health will help to develop predictive models. Scientists have also argued for increased global monitoring of infectious disease outbreaks, saying this would not only allow a quicker response to contain the diseases, but also provide early warning signs of climate change.

Some scientists, however, are concerned about what they view as an undue emphasis on the role of climate in spreading infectious diseases. An article in *Science* quotes several infectious disease experts making the case that recent outbreaks are more the result of a breakdown in public health programs than climatic factors, and that public health measures will be far more important than climate in determining the pattern of future outbreaks. Duane Gubler of the CDC is quoted as dismissing concerns about outbreaks of disease in the U.S. because it has "good housing, air conditioning and screens that keep mosquitoes outside...." He said the dengue epidemic that affected at least 2000 people in Mexico in 1995 stopped dead at the border, causing only seven cases in Texas.

Gubler and others argued that overfocusing on global warming would distract from the real problem: the deterioration of public health measures and the increasing resistance of parasites and vectors to drugs and pesticides. He said efforts should be made to reduce global warming, but the most cost-effective way of dealing with infectious diseases is to "rebuild our public health infrastructure and implement better disease-prevention strategies."

Science later printed a letter written by several scientists who've raised concerns about the link between climate change and infectious diseases (and also signed by Gubler). It agreed that public health improvements are needed worldwide and that the spread of disease is influenced by multiple factors. While they weren't claiming climate was the only, or even the most important, factor, they said it will alter the risks associated with infectious diseases and extreme weather will place added burdens on health care facilities. "At issue is not which is more important...rather, it is important to assess how health risks might change in both industrialized and more vulnerable developing countries."

Global warming skeptics have vigorously attacked suggestions of a link between climate change and infectious diseases. The *World Climate Report* claims that advocates of greenhouse gas cuts are trying to scare the public by alleging that a warmer climate would cause death and disease.

Some skeptics argue that *if* any health problems arise, health care and social support systems will cope with them.

Leaving aside the fact that billions of people don't have access to even the most basic health care and social support systems (some of whom, it should be noted, live in developed countries like Canada and the U.S.), there is the question of financing. As Scheraga said, "effective health care comes at a cost. There are those who argue we can adapt, but even if that's true, adaptation costs something and those resources must be diverted from other activities."

In any event, there are doubts about how well the much-vaunted health care systems of industrialized countries would respond to added pressures caused by an increase in climate- and weather-related health problems. Many are hard-pressed to deal with existing demands. In the U.S., with a mostly private health system, more than 40 million people are without basic medical coverage. Nor is the U.S. health care system well prepared to respond to natural disasters, according to emergency medicine specialists who spoke at a 1996 conference at the University of Colorado. The report summarizing their remarks said the U.S. will face "significant problems in providing sufficient emergency medical resources at the local level following catastrophic disasters." They attributed the lack of medical preparedness to several factors, including fragmentation and downsizing of hospitals and health support systems, increased costs, and confusion in emergency planning.

In Canada, where the public health care system until recently has been a source of national pride, there's also mounting concern about the erosion caused by years of government cutbacks. There are ominous signs that the system is starting to crumble from the stress of too much demand and too few resources. With the baby boom generation just entering its senior years, the situation will continue to deteriorate unless these trends are reversed. It's hard to be sanguine about the ability of this system to handle additional pressures stemming from climate change and weather extremes.

Some infectious disease experts have also expressed concern about failures of public health infrastructure all over the world. There's worrying evidence that disease outbreaks are increasing everywhere, including in developed countries. A 1995 study by Ann Platt of the World Watch Institute found that mortality from infectious diseases was rising worldwide and that these diseases accounted for a third of all deaths, more than those caused by cancer and heart disease combined. The CDC found that U.S. deaths with infectious disease as an underlying cause increased by 58% between 1980 and 1992 (39% when adjusted for population aging).

Contrary to previous predictions that infection diseases would wane in the U.S., "these data show that infectious disease mortality has actually been increasing in recent years," said Robert Pinner of the CDC.

The crisis results from both the emergence of new diseases and the reemergence of old diseases like tuberculosis, once thought beaten in developed countries. The growing drug resistance of many disease organisms is an added problem that could be exacerbated by climate change. A warming climate is likely to accelerate the reproduction of parasites, facilitating genetic adaptations that help them fend off drugs and other control methods. Climate change may also reduce the effectiveness of programs to control disease-carrying vectors.

Platt blamed the global increase in infectious diseases on a "deadly mix of exploding populations, rampant poverty, inadequate health care, misuse of antibiotics, and severe environmental degradation." She noted that 80% of all disease in developing countries stems from unsafe drinking water and poor sanitation. Even in the U.S., with its advanced sanitation facilities, water-borne diseases cost an estimated $20 billion a year. A 1993 outbreak of cryptosporidium, which affected more than 400,000 people, was partly caused by a nonfunctioning water filtration plant; similar deficiencies have been found in other U.S. cities. Even with all its resources, the U.S. public health system hasn't been able to prevent either the resurgence of old diseases like tuberculosis or outbreaks of emerging diseases like hantavirus and Lyme disease.

When skeptics argue that the best way to cope with infectious diseases is to improve sanitation and other public health measures, there's often an assumption that this is not a big problem. The *World Climate Report,* for example, states that dealing with cholera is "simple"—merely a matter of filtering and chlorinating the water supply. "A warmer climate, if it were to occur, would not reduce the effectiveness of these water purification measures." While this is true, it's hardly the point. The effectiveness of purification measures is irrelevant if they're not implemented, and the fact is, they're not being implemented nearly enough as it is. If this were really that simple a matter, it would long since have been accomplished. However, many countries are stymied by budget cuts, population growth, and grinding poverty (which has been described as the deadliest disease of all). Eradicating world poverty is hardly more trivial a challenge than preventing global warming.

We're not dealing particularly well today with the threat of infectious diseases. Indeed, there's evidence that we're losing ground, even in developed countries. There's little reason to assume our already stressed public health systems can readily handle an increased threat caused by global warming. At a conference sponsored by the U.S. National Academy of Sciences, scientists concluded that, while more research is needed to reduce uncertainties in linking climate change and infectious diseases, "the lack of complete data should not be used as an excuse for inaction. Instead, the precautionary principle should apply: If the risk to public health is great, even if there is uncertainty, both policy and action should be biased in favor of precaution."

A report by the World Health Organization notes that humans can probably deal with even fairly major climatic changes—after all, they live in nearly all climatic environments right now—but the adjustments needed to do this could be substantial, expensive, and "may require many sacrifices in life-style and well-being to re-establish and maintain the basic needs." All aspects of life—housing, clothing, nutrition, mobility, education, health services, industrial production, and much of the established infrastructure—would be affected.

This reality highlights one of the oddest aspects of the argument that improved social and health care programs can handle any health problems global warming may throw our way—which is that it's so often proffered by people who generally are not enthusiastic supporters of government spending on social programs or foreign aid. It seems rather disingenuous to urge delay in cutting greenhouse gas emissions on the grounds that social programs are a better way of coping with health problems caused by global warming. One wonders if this can be taken as an endorsement of spending whatever's needed to eradicate world poverty and provide everyone with adequate medical care, clean water, and air conditioning.

Heat Stress

We've seen that global warming is expected to increase the frequency of very hot days and heat waves while reducing the number of very cold days and cold waves. These changes would be expected to occur with an increase in the *average* temperature, even if the climate does not become more variable, and this is what climate models currently forecast will happen over the next century. However, the situation would get worse if climate variability (the range between extremes) increases as well. Research indicates that if variability increases, both hot and cold extremes would become more frequent.

It's been estimated that a 2 to 3°C rise in average summer temperature would roughly double the number of very hot days in temperate climates (e.g., northern Europe, southern Canada, most of the U.S., and Australia). One study found that the number of days over 35°C in Victoria, Australia, would increase by 50 to 100 percent if the earth's average temperature increased by 1.5°C.

Studies in many countries show that death rates rise during excessively hot weather. One U.S. study found that between 1979 and 1982, heat caused nearly 5400 deaths—more than those attributed to all other weather-related disasters combined. Heat waves are most deadly when they persist for several days and exceed the threshold temperature to which the local population is adapted. Deaths can as much as double or triple after two to three days of excessive heat. Research has shown that the effects of heat are cumulative and a break of just two to three hours of lower temperatures or in an air conditioned environment can interrupt the processes that cause physiological damage.

The threshold of heat adaptation varies from place to place and is influenced by socioeconomic and technological factors (e.g., health care, housing, air conditioning) as well as physiological adaptation. According to one study, the threshold for significant increases in heat-related deaths in the southern U.S. is 36°C, for the northern U.S., 32°C, for Shanghai, China, 34°C, for Toronto, 33°C, and for Montreal, 29°C. In Australia, summer heat-related deaths rise steeply when the temperature exceeds 35°C.

The effects of high temperatures are exacerbated by other factors, such as high humidity, low wind, and intense sunlight. The combination of heat and humidity (often referred to as the "humidex" or *heat index*) can be especially deadly. The increased moisture-carrying capacity of a warmer atmosphere may increase the heat index in locations where relative humidity rises. The heat-island effect that makes large cities a few degrees warmer than surrounding rural areas can also amplify the effects of heat waves, as does the fact that city buildings tend to retain heat at night.

Healthy people can cope with higher temperatures up to a point, but they can be affected if oppressive heat persists too long. The risk of illness and death rises significantly if heat waves last more than a couple of days. The poor, the young, the elderly, the disabled, and the ill are disproportionately vulnerable. In the July 1995 Chicago heat wave, 85% of those who died were elderly, with an average age of 71. Darlene Tavares of EARG, who conducted a study that examined heat-related emergency hospital admissions in Toronto, found that for people over 65, the threshold above which these admissions increased significantly was 28°C, while for people younger than 65, it was 31°C. Among people under 65, weather accounted for 14% of emergency admissions, with maximum temperature being the most important element in weather-related admissions.

Most hot-weather deaths are not caused by heat itself, but by the impact of heat stress on preexisting medical conditions, such as heart, circulatory, or respiratory disorders. Some people are ill enough that they would die soon even without the added stress; U.S. data indicate that about 20 to 40% of those who die during a heat wave probably would have died anyway within a few weeks.

The vulnerability of people with preexisting medical conditions is increased by the fact that increased air pollution often accompanies heat waves; oppressive atmospheric conditions, characterized by high temperature, low wind, and high humidity, often prevents rapid dispersal of pollutants. Since burning fossil fuels is a major source of these pollutants, in addition to greenhouse gases, "climate change can be expected to entail more frequent occasions that combine very hot weather with increases in air pollution," the IPCC report notes. Epidemiological studies indicate that in urban areas, the combination of temperature and pollution has a greater impact on mortality than either taken separately.

World Climate Report points out that air quality has been improving since the 1970s because of the U.S. Clean Air Act and that reports of increased pollution are "a convenient fallacy." But the Clean Air Act only applies to the U.S. and even though air quality there has generally improved, temperature inversions and oppressive atmospheric conditions that intensify the health effects of pollution can still occur. The fact is that urban areas in developed countries are experiencing increasing health problems associated with air pollution.

For example, a study of 11 Canadian cities found that death rates increased as much as 11% when air pollution levels were high. The researchers concluded that exposure to air pollutants generated from burning fossil fuels poses a public health risk to Canadians. The study found increased risks of premature mortality attributable to air pollutants in all 11 cities. The report noted that even though air quality has improved because of pollution control policies, there's concern that energy-use trends, including increasing gasoline consumption, could offset these improvements. One researcher, Richard Burnett of the federal government's Environmental Health Directorate, added that in cities all over North and South America and Europe "when air pollution is high, more people die."[*]

The ability of a population to ride out heat waves depends to a great extent on socioeconomic factors such as housing and air conditioning.

---

[*]Global air pollution is expected to increase because of growing industrialization, according to the World Health Organization. The World Resources Institute estimates that by 2020, pollutants from burning fossil fuels could cause more than 700,000 deaths annually that could be avoided if greenhouse gas emissions are reduced. Roughly 8 million "avoidable deaths" would occur between 2000 and 2020, about 85% in developing countries.

This is why the poor are more vulnerable to extreme heat; in the Chicago heat wave, poverty was a "major underlying risk factor for heat-related mortality" according to Jan Semenza of the CDC. Social isolation of the elderly is another major risk factor; most vulnerable were people with preexisting medical conditions who were confined to bed or living alone, people who didn't go out every day and had no access to air conditioning or to transportation. Other social factors played a role as well; for example, many of those who fell victim to the heat kept their windows closed for fear their homes would be broken into.

Laurence Kalkstein of the University of Delaware and his colleagues assessed the potential impact of global warming on heat-related deaths in cities around the world. Two sets of projections were made: the first assumed that populations do not adapt to the hotter conditions and the second assumed that populations do adapt physiologically but there's no improvement in socioeconomic support to help them cope with higher temperatures. (Heat adaptation or *acclimatization* is a phenomenon in which an organism undergoes physiological changes over a period of days or weeks that enables it to cope with warmer temperatures. Other forms of adaptation include behavioral—e.g., avoiding strenuous activities during hot weather—and environmental—e.g., housing and infrastructure.)

The researchers found that in Montreal, which currently averages about 69 heat-related summer deaths a year, the numbers would increase to a range of 61 to 460 deaths in 2020 and 124 to 725 deaths in 2050, depending on the degree of acclimatization. In Toronto, which averages about 19 heat-related summer deaths annually, the equivalent range is 0 to 289 deaths in 2020 and 1 to 563 deaths in 2050. A similar study done for Atlanta indicates its death toll would rise from the current average of 78 deaths a year to 96 to 191 deaths in 2020 and 147 to 293 deaths in 2050.

Kalkstein said the figures that take acclimatization into account are more realistic. He added, however, that "adaptation will not be complete" and the magnitude of increases in heat-related mortality will depend on some extent on whether global warming causes an increase in climate variability. If variability does not increase, heat-related deaths may not rise much; however, if "hot episodes become hotter but average summer conditions remain the same, I believe heat related mortality will increase more dramatically."

The IPCC report says the results of these and similar studies for other North American cities, plus Shanghai, China, and Cairo, Egypt, suggest that "the annual number of heat-related deaths would, very approximately, double by 2020 and would increase several-fold by 2050. Thus, in very large cities with populations displaying this type of sensitivity to heat stress, climate change would cause several thousand extra heat-related deaths annually."

These data show that if populations can adapt physiologically to higher temperatures, the impact of global warming would be reduced, but there would still be additional deaths. For example, the 2050 projection for Shanghai indicated there could be as many as 2950 deaths without acclimatization and 1033 deaths with acclimatization.

Skeptics often use acclimatization to dismiss concerns about the health impacts of heat waves. So what if Chicago or Toronto become more like Houston or New Orleans, the reasoning goes; millions of people live in Houston and New Orleans without dying of the heat. This is certainly true. Studies show there are more heat-related deaths in the northeastern and midwestern U.S. than in the South. "The impact of heat on mortality is more profound where high temperatures occur irregularly, while in the southern United States, where heat is a relative constant, a much smaller impact is noted," Kalkstein writes.

But this is not the whole story. For one thing, people in the South would likely experience problems if temperatures *there* exceeded the levels to which they're adapted. India is certainly no stranger to hot weather, but that didn't prevent nearly 3000 people from dying in June 1998 during an extreme heat wave with temperatures up to 50°C, the worst to hit the country in 50 years. The UN report said that while humans can adapt well to "prolonged and gradual warming, extreme variations and rapid changes in thermal conditions, especially in low latitudes with existing high heat stress, and in densely populated urban areas, will carry an increased risk of heat-related disorders."

Moreover, in any given location, some people adapt to higher temperatures more easily than others. "Older people, people with cardiovascular disease, and those taking certain types of drugs may acclimate poorly or not at all," said Tee Guidotti, professor of occupational and environmental medicine at the University of Alberta. "The longer, more consistent and more gradual the warming, the greater the percentage of vulnerable people in the population who will adapt. But not everyone can." In any event, model projections do not suggest merely a long, gradual increase in average temperatures. "If the only relevant change were that the average temperature would be pushed up a degree or two, most people would adapt. However, it's much more likely that we'll see a slightly warmer background temperature against which are superimposed an increasing number of much hotter events, mostly in cities and even in Canada, where we already see heat-related mortality in Toronto and Montreal during heat waves."

Under these conditions, populations remain vulnerable to heat waves because they're unlikely to undergo widespread heat adaptation unless the average temperature increases more than anticipated. Instead, the heat waves "will hit more often and worse than they do now, but the weather between them might not feel much different than it does now."

Karen Smoyer, a professor of earth and atmospheric sciences at the University of Alberta, points out another important consideration that will influence heat adaptation in places more accustomed to cool weather: the "built environment" is unlikely to respond as quickly to climate change as people do. She points out that heat wave mortality is rare in hot and humid New Orleans in part because houses there are built to provide comfort in hot weather and many are air conditioned as well. "In northern cities, housing tends to be built for cold rather than hot weather. The housing stock of Toronto and Montreal may exacerbate hot weather conditions during heat waves, making it more difficult for the population to avoid the impacts of hot weather."[*]

According to Jerry Mahlman, there will also be an amplifying effect on human health of higher moisture levels in a greenhouse-warmed atmosphere. In subtropical regions close to the ocean—which provides large amounts of water for evaporation—climate warming could greatly increase the heat index. Mahlman's research indicates there could be as much as a 50% amplification of the warming effect in subtropical areas like the southeastern U.S. While Americans may have the resources to deal with this added heat stress, many subtropical regions are poor and thus will be more vulnerable.

Skeptics have challenged projections of increased heat-related mortality by citing data showing that nights are warming more than days and that winters are warming more than summers. In testimony before the U.S. Congress, a prominent skeptic, Patrick Michaels of the University of Virginia, stated that almost all the warming in the U.S. has taken place at night and described this as a "benign climate change."

It's true that in many places, minimum (nighttime) temperatures are increasing faster than maximum (daytime) temperatures. It's also true that

---

[*] *World Climate Report* argues that replacement of housing stock in northern regions will prevent heat-related mortality from rising sharply. "Should the climate warm, builders will move toward structures that protect the inhabitants from extreme heat, as housing in the South supposedly does now." This argument would have some merit if not for the fact that making such major changes in infrastructure requires society to take the threat of global warming seriously—something skeptics are relentlessly working to prevent.

daily maximum temperatures generally have increased more in winter and spring than in summer and fall. However, just because there's a difference in the *rates* of warming between day and night and between summer and winter, it does not necessarily follow that daytime temperatures in summer are not rising, nor does it preclude the occurrence of very hot daytime temperatures that may exceed local thresholds of heat adaptation. In fact, maximum daily temperatures have been increasing in summer on a global scale (although there are some regions where this is not the case) and hot extremes are also increasing in most parts of the world.

In any event, the fact that nighttime temperatures are rising faster than daytime temperatures is not the consolation skeptics think it is. Quite the contrary—higher nighttime temperatures can actually lengthen the duration of heat waves and thus increase their human impact. Moreover, ill people are more likely to succumb to their medical problems at night, a time when their biological resources are at the lowest ebb. A study of heat-related deaths in New York City found that high minimum temperatures and the *duration* of hot conditions above the local threshold significantly influenced mortality. Minimum temperatures also played a significant role in Chicago's killer heat wave in 1995. An analysis by Tom Karl and Richard Knight of NCDC showed that this event was extremely unusual, not just because the daytime heat index reached 49°C and nearly 48°C on two consecutive days, but because the nighttime heat index did not fall below 31.5°C and 34°C on the two nights. "Clearly, there was little relief from the heat during any time of the two-day period, even during nighttime hours," the scientists commented. This amplified the health impact of the event. It was unprecedented for such high nighttime values to persist for forty-eight hours. "No heat wave during the 20th Century was comparable in this respect," the researchers said.

Karen Smoyer added that cities tend to cool slowly at night because building materials trap outgoing heat. "The interior of dwellings may become particularly hot at night. During heat waves, hot night-time temperatures do not allow a respite from the heat, putting further stress on the body." Therefore, increases in *minimum* daily temperatures in summer may cause as great a problem with heat-related mortality as increases in *maximum* daytime temperatures.

Karl and Knight did not attribute the Chicago heat wave to global warming, though they said urban heating may have played a role. Their analysis suggests that events like the Chicago heat wave would increase in frequency with global warming, but it's uncertain how much. Such extreme events would likely remain "rather uncommon," they noted.

It's sometimes suggested that the best way to deal with heat waves is social support programs, not cutting greenhouse gases. Barbara Rippel, a policy analyst with an organization called Consumer Alert, comments that "the problem of people dying during extreme heat waves is more related to the social structure of big cites." Many people cannot afford air conditioning and many elderly people live alone, do not leave their houses every day, and are not regularly visited by family members. An international treaty limiting CO_2 emissions would not prevent these people from dying, she argues. "Researchers suggest that the best way to tackle these problems is to provide more social care...and start information campaigns about the dangers of extreme heat....These approaches have already been tested with success in some bigger cities."

For example, after a 1980 heat wave killed 88 people in Memphis, the city initiated a plan to check on vulnerable people and transport them to air conditioned shelters if necessary. Everyone, from family members and visiting nurses to Meals on Wheels volunteers and letter carriers, is urged to check up on people at risk.

There's no doubt that improving social support would save many lives in heat waves, but it would be foolish to ignore the potential for increased risks associated with such events if we do nothing to slow global warming. "Physicians and public health officials need to be aware that heat waves, typically seen as local and unavoidable, may actually be part of a preventable global process," commented David Shumway Jones of Harvard Medical School in a letter to the *New England Journal of Medicine*.

As for improving social services, the issue is not whether this would combat heat-related mortality—clearly it would. The real question is how likely it is that social programs will be expanded and improved. Even in developed countries, it seems optimistic to assume that government programs or private industry will suddenly undertake to eradicate the poverty and social inequities that make some people more vulnerable to heat waves. Nor would it be possible to eliminate the increased risks associated with aging without increased spending on public health and social programs. Such efforts are even less likely to occur in poor developing countries.

If we haven't accomplished this feat so far, why is it logical to assume such efforts would become more commonplace in the future? As Scheraga commented: "The public health system in the U.S. does not deal well with heat stress. If we don't even deal well with it today, what makes you think we'll deal well with it in the future?"

It's often suggested that increases in heat-related deaths and illnesses would be offset by reductions in cold-related deaths and illnesses, which, in temperate countries, are currently more numerous. This seems logical, given that current global warming projections suggest a decrease in the number of very cold days.

On this question, the IPCC notes that, at present, there are no studies that directly compare projected winter deaths and summer deaths in a warmer world. However, studies of the relationship between daily death rates and temperature show there's a "comfort zone" where deaths are lowest; death rates climb much more sharply when temperatures rise higher than this comfort zone than they do when temperatures drop below it. In short, summer deaths are more closely tied to temperature extremes than winter deaths. The implication is that increases in heat-related deaths during hotter summers are unlikely to be completely offset by reductions in cold-related deaths during milder winters.

Other studies of winter deaths indicate that a significant proportion are due to the increased risk to people with cardiovascular disease or respiratory disorders. A British study estimated there would be 9000 fewer winter deaths annually by the year 2050 under typical global warming scenarios, about half attributed to people with heart disease and another 10 to 20% to people with bronchitis and pneumonia. However, while the risk associated with some respiratory disorders may be reduced with milder winters, this may not be the case for influenza outbreaks, which don't appear to be correlated with temperature.

"Overall, the sensitivity of death rates to hotter summers is likely to be greater than to the accompanying increase in average winter temperature," the IPCC report notes. While the balance is difficult to quantify and would be influenced by adaptive responses, "research to date suggests that global warming would, via an increased frequency of heat waves, cause a net increase in mortality...."

Extreme Weather Events

Extreme weather events have taken an enormous human toll in recent years. Data compiled by a United Nations program known as the International Decade for Natural Disaster Reduction (IDNDR) indicate that during the 1980s and 1990s, the number of people affected by natural disasters has risen on average by 6% annually, triple the rate of global population growth. This figure includes people affected by earthquakes as well as climate-related disasters, but analysis indicates that climate-related disasters are primarily responsible for the human impacts. For example, the Brussels-based Centre for Research on the Epidemiology of Disasters (CRED) estimated that about 75% of the 760,000 people who died from natural disasters between 1986 and 1995 were killed during climate-related disasters such as floods, droughts, and storms. A further 11% died in epidemics and famines (which can be exacerbated by climatic factors). Earthquakes accounted for only 14% of the deaths.

CRED also estimated that about 1.9 billion people were affected by climate-related disasters during this period, roughly a third of the world's population (though this figure may include people counted more than once because they were affected by multiple disasters). Among survivors of natural disasters, 96% were affected by climate-related events; flooding alone affected 69% of survivors.

As bad as these numbers are, they could have been worse; during the 1990s, the human impact of weather-related disasters has increased at a slower rate than the economic losses because of preventive measures undertaken as part of the IDNDR program, such as better building codes and improvements in warning systems and disaster preparedness. Most of the human losses—88% of deaths and 92% of people affected between 1985 and 1992—occurred in poorer countries that are less able to pursue loss prevention programs.

In 1994, IDNDR commissioned a study to examine trends in major natural disasters between 1963 and 1992. "Major" events were defined as those causing economic damage in excess of 1% of GDP, affecting more than 1% of the population, or causing more than 100 deaths. James Bruce, former chair of the scientific and technical committee for IDNDR, said these data show "an inexorable rise" in the number of disasters by all three criteria. In terms of deaths and the numbers of people affected, climate-related disasters dominated; nearly 85% of the disasters affecting more than 1% of the population were storms, floods, and droughts, compared with only 10% related to earthquakes. More than half of the disasters that claimed 100 or more lives were climate-related, compared with 20% related to earthquakes and another 17% related to epidemics.

These data suggest that if global warming increases the frequency or severity of extreme weather events, it has the potential to increase the deaths and injuries associated with such events. However, it's impossible to predict what will actually happen because of the uncertainties in climate projections and because so much depends on socioeconomic factors and loss prevention programs.

Poor countries with large populations in low-lying coastal areas are likely to be most severely affected. However, trends in developed countries are increasing their vulnerability as well; growing urbanization is crowding more people together and the migration to the coasts is increasing exposure to natural disasters and sea level rise. A paper by Rodney White of the University of Toronto and David Etkin of EARG notes that "the coastal movement" began in the U.S. early in this century and "the same trend is now gathering momentum in China. Thus, it is a phenomenon which is prevalent in both the richest countries and the poorest."

Extreme weather events can affect human health in several other ways, by contributing to outbreaks of infectious diseases and to the contamination or destruction of water and food supplies. A 1970 cyclone in Bangladesh destroyed two-thirds of the country's coastal fisheries and 125,000 animals. In 1983, flooding linked to El Niño increased salmonella infections in Bolivia by an estimated 70%. Heavy rains in British Columbia in 1995 were partly responsible for an outbreak of a parasitic disease called toxoplasmosis. In California, public health officials expressed concern that standing water left by the heavy rains and flooding caused by El Niño would cause an explosion in populations of disease-carrying mosquitoes and rodents.

Warm, wet conditions like those produced by El Niño and global warming can also increase the incidence of allergies and asthma. Warm temperatures cause premature budding of plants, which puts more pollen into the air earlier in the season. Wet conditions increase the output of mold spores. By February 1998, with El Nino holding sway, parts of the U.S. were already reporting high pollen counts and a 20 to 30% increase in allergy complaints. By mid-March, after an unusually warm and snowless winter, Canadian medical authorities reported an early increase in asthma cases. The 1998 allergy season was likely to be "a lulu," as Toronto allergy expert Norman Epstein put it.

The damage caused to infrastructure and food production by natural disasters like droughts, floods, hurricanes, storm surges, and rising sea levels can also create health problems. Malnutrition resulting from prolonged drought can increase susceptibility to infections. Crowding among dislocated people provides fertile ground for the spread of infectious diseases; during the ice storm, one long-term care facility faced increased risks of contamination because of the influx of ill and elderly evacuees suffering from diarrhea and vomiting. With only one generator, washing clothes and keeping the place clean was a major challenge.

Damage to water treatment and sanitation facilities can lead to contamination of water supplies with fecal matter or toxic chemicals. There were reports of meningitis and hepatitis among people rebuilding their homes after the Mississippi flood in 1993; the U.S. Geological Survey also reported that huge amounts of herbicides had been washed into the river. In March 1997, Kentucky vaccinated thousands of people against tetanus and distributed bleach cleaning kits to protect residents exposed to floodwaters contaminated by human and animal waste from damaged sewage systems and farm runoff. Residents affected by the Saguenay floods were urged to boil their water before drinking it. Flooding can also create conditions that allow toxic molds to thrive in water-damaged walls and floors, presenting an ongoing health hazard, especially to children.

Extreme events pose another threat to health. Though less obvious than broken bones or heart attacks, the emotional and psychological trauma of surviving a natural disaster takes as great a human toll.

eight

TRAUMA:
Psychological
Impacts of Extreme
Weather

The man sat in his truck, surveying the wreckage of his home. He'd had it half built when the tornado blasted it to bits. He'd start over, he vowed. What else could he do? His voice wavered. Then he started to cry.

There's no doubt that extreme weather takes a devastating emotional toll, but this is harder to quantify than the economic losses—nightmares, flashbacks, and suicides don't show up on spreadsheets. The most dramatic impacts occur in the wake of sudden events, but more gradual changes that affect water supplies, food production, and other resources also cause stress. According to the IPCC report, several factors determine the psychological impact of natural disasters, including the unexpectedness of the event, its intensity, the extent of personal and community disruption, and long-term visible reminders of the disaster.

Extreme events vary considerably in the rapidity of onset and the degree to which affected populations can prepare themselves. For example, during the 1997 Red River flood, residents of southern Manitoba had several weeks to prepare for the crest but they also had several weeks to worry about their homes and farms, especially after seeing the unprecedented devastation the flooding caused in North Dakota. The 1998 ice storm came on more quickly, but it still took several days before the extent of the disaster began to sink in. On the other hand, residents of central Florida had no warning at all when they were struck in the middle of the night by an unusual and deadly outbreak of multiple tornadoes in February 1998.

Much of the stress caused by natural disasters occurs after the event. While the crisis is underway, most people rise to the challenge pretty well, according to Joe Scanlon, director of the Emergency Communications Research Unit at Carleton University in Ottawa. It's largely a myth that victims of extreme events are panic-stricken, dazed, or confused. "The fact is, victims cope very well. In destructive types of events, where search and rescue is required, the bulk of the initial rescue and transport of the injured is done by the survivors. After the Edmonton tornado, it was 21 minutes before anyone got there and by that time, the moveable injured were gone." As for panic, it's "so non-existent you can't even measure it," he said. During floods and ice storms, for example, people resist leaving their homes even when warned of life-threatening conditions. "By and large people treat it as if they can handle it. In fact, non-panic is a problem because people don't take the danger seriously," Scanlon said. And when people do evacuate, it's usually orderly. "Large scale evacuations are not difficult," he said, noting that some 6 million people were evacuated in Florida before a hurricane without accidents or injuries.

No matter how much people try to make the best of it, however, being caught in a weather disaster is stressful. Fatigue and exhaustion are common, which increases anxiety, anger, and conflict. During the ice storm, sleep deprivation was a problem in the shelters and fatigue made many people irritable and angry. One man told CBC Newsworld that it was difficult to sleep because of distressed elderly and ill people who needed care. "Every night there's another catastrophe. Somebody yells, somebody falls out of bed." Small children, too, were alternately frightened, confused, bored, and cranky from being uprooted and forced to live in unfamiliar and crowded shelters.

Disaster workers face particularly severe emotional pressure and can easily become "secondary victims" due to working long hours in hazardous conditions, according to a report by the American Red Cross. Most are dedicated people, perfectionists who push themselves too hard and find it hard to say no without feeling guilty. "With so much yet to do, they often fail to take credit for the amount of work completed and the effort contributed to the operation," the report notes. Frustration, exhaustion, and a loss of humor puts them on edge, and it's often difficult for them to deal with the anger of others, which "may be seen as a personal attack on the worker rather than as a normal response to exhaustion. Survivor guilt may emerge as workers see the losses of others when they have suffered none themselves."

During the ice storm, many people on the front lines—hospital workers, shelter volunteers, and repair crews—were faced with incredibly demanding tasks. Hydro employees worked 16-hour shifts, battling freezing temperatures and high winds to repair power lines that wouldn't stay fixed. Doctors, nurses, and other health care workers burned out while trying to cope with storm-related injuries and stress, as well as their regular caseloads, with diminishing resources.

The *Toronto Star* documented the harrowing problems faced by the hospital in St. Jean-sur-Richelieu, one of the hardest-hit towns. Emergency cases doubled during the blackout, not only because of carbon monoxide poisoning but because elderly people experienced medical complications from the freezing temperatures. The long-term care center at the hospital took in 120 evacuated elderly people whose home care services were suspended, almost doubling the normal caseload. The center's director, Flore Barrière, was reduced to tears by the condition they were in. Some of the elderly hadn't eaten for two days and many were so confused they didn't know their own names. "They were dirty—some of them can't go to the washroom by themselves—they hadn't taken their medication, they were scared," Barrière told the *Star*. "It wasn't a pretty sight."

Among people who remained at home, an emotional divide sometimes occurred between those who had power and those who didn't. One woman told the CBC of her annoyance that neighbors whose power had been restored weren't offering to help those who were freezing in the dark across the street. If her power had been restored, she said, she would have invited neighbors in to "boil their kettles. I don't know…people react in different ways." However, other residents in similar situations strung extension cords across the street to share electricity with their neighbors. And, of course, many people who didn't lose their power took in family, friends, and complete strangers. In fact, an Alberta town literally airlifted a group of Montrealers out of the mess by inviting them into their homes halfway across the country. While such generosity is appreciated, the crowding can still be stressful.

People coping with a crisis are quick to turn on anyone perceived to be taking advantage of their vulnerability. Within the first week of the ice storm, Quebec consumer organizations received complaints about price gouging and they publicly embarrassed offenders by calling attention to these abuses. Global TV reported that some people were charging $8 for flashlights or $3.50 to fill a thermos with hot water. There were reports of prices on wood, generators, and camping fuel being doubled. Gas prices rose that week in Montreal, which provoked outrage, although oil companies insisted it was coincidental. Federal industry minister John Manley later said oil companies should have been giving out free gas.

As the crisis wore on, stress levels increased not only because of the continuing disruption of daily life but because of mounting financial losses. Although the Quebec government quickly offered a small relief payment to tide people over, many found it insufficient to offset the loss of income. In Maine, some of the anger was directed toward the power company, which was forced to hire off-duty police officers for protection after receiving threats against its workers. A stabbing in Augusta, Maine, was attributed to the storm; the police chief commented that people were getting confrontational about things that normally wouldn't cause trouble.

Looting is often a concern during events that force people to evacuate their homes, but Scanlon said the idea that many people suddenly turn into criminals is a media-inspired myth. "In fact, crime rates go down."

However, extreme events often precipitate increases in domestic violence. Montreal police reported an "alarming" jump in domestic incidents during the ice storm. Reports of increased abuse followed Hurricane Andrew in Florida and the Red River flood in North Dakota, among others. One survey of 77 Canadian and U.S. domestic violence programs found that some experienced increased demand for services as long as six months to a year after a disaster.

This study, done by Elaine Enarson of Florida International University and a visiting scholar with the Disaster Preparedness Resources Centre at the University of British Columbia, notes that the special vulnerability of women during disasters has only recently attracted academic attention. "[P]ets, tourists, and cultural artifacts receive more attention than battered women," she said.

It's difficult to obtain reliable data on the increase in domestic violence caused by natural disasters because many agencies normally involved— police, the courts, social service agencies, hospitals, etc.—are hard-pressed to maintain even basic services while coping with the disaster. In Hawaii, for example, abusers who violated protection orders were not arrested because the jail was closed by a hurricane. A woman in Grand Forks, North Dakota, who'd been within days of filing for divorce before the flood was still waiting a year later. Staff at the women's shelter reported that obtaining protection orders was a "huge struggle" because they had to drive long distances to a different courthouse to get the papers signed.

Post-disaster domestic violence is often not even defined by authorities as a disaster-related issue. For example, a survey of medical examiners done by Gib Parish of the CDC found that 60% of those questioned did not define as "disaster-related" an incident in which a man shot his wife and himself six months after a hurricane because the stress of living in temporary housing had destabilized their marriage. Only 16% felt the deaths were even indirectly related to the disaster, while another 24% said they possibly were. None felt there was any direct relationship. In contrast, more than 90% of the examiners defined as directly or indirectly disaster-related an incident in which a worker was asphyxiated while trapped under a tree that had been toppled in a tornado. Only 3% of those surveyed felt this incident was not disaster-related.

It's not surprising therefore that post-disaster increases in the incidence of domestic abuse may go largely unnoticed by organizations managing the recovery. The problem is more likely to be dealt with by organizations already involved in domestic abuse, such as battered women's shelters, according to Jennifer Wilson of Florida International University. Domestic violence programs in Grand Forks and Fargo, North Dakota, reported substantial increases in crisis calls, protection orders, referrals to hospital emergency rooms, and the demand for counseling after the Red River flood. A year later, the crisis center in Grand Forks was still fielding calls at a rate 47% higher than before the flood and they'd had to increase staff from 7 to 17 people. Interestingly, however, similar programs in southern Manitoba reported no flood-related increases in service demand in the six months after the event.

Wilson notes that many women living in abusive situations before the disaster are more vulnerable afterward and it's even harder for them to leave the relationship. They're often isolated and lack a social support network, so they end up viewing their abuser as their only option in a crisis. Enarson reported that one woman who'd received a protection order against her partner just before a flood felt compelled to go back to him when an evacuation was ordered because she had no one else to turn to.

Jane Ollenburger of Boise State University found that women experienced more anxiety, depression, and overall stress than men after a flood. One reason is that older women are the largest group of people living alone and many often live in rental accommodation. (Research has shown that renters exhibit higher levels of post-disaster stress than people who own their homes.) Women are usually also responsible for children and elderly parents, two vulnerable groups that themselves suffer great stress during disasters.

Robert Bolin of New Mexico State University noted that until recently disaster researchers have largely ignored how gender inequality creates differences in the way men and women experience natural disasters. They tend to treat families as cooperative units, ignoring inequalities of power and violence problems that may exist within the family. Consequently, policies designed to help families may not necessarily help individuals within the family.

Elaine Enarson and Betty Morrow of Florida State University found examples of this when they studied the impact of Hurricane Andrew on women, particularly those from low-income households, which generally experience the most damage during natural disasters. They found that women bore the brunt of the work involved in relocating and resettling the family in temporary housing, often repeatedly. Women took on added domestic chores (e.g., finding supplies for infants) and spent a lot of time in difficult circumstances seeking post-disaster assistance. "Women spent interminable hours in stifling heat, often with children in tow, walking or riding buses...and waiting in line at various relief agencies, often returning again and again," the authors commented in a report. "Facilities to provide day care for their children, handicapped adult or elderly family members were rarely available."

They noted that this was one reason why male family members were sometimes able to get relief payments first and divert them to their own needs, leaving their families destitute. In some cases men actually deserted their families after obtaining insurance or government funds. On the other hand, some women were able to use the relief money to relocate and leave abusive relationships.

Like other researchers, Enarson and Morrow found that incidence of abuse against women and children increased after a disaster. One woman said that before the hurricane, she would "get beat up maybe once a month if I was lucky. Afterward, it was like every other day." More than a year after Hurricane Andrew, many low-income families were still living in temporary housing where overcrowding led to increased violence against women, including incest and rape, as well as increases in teen pregnancies and conflicts between parents and children.

Enarson said disaster planners give little thought to the special problems faced by women's shelters and other social service agencies that deal with domestic violence after a disaster. Women evacuated from a protected shelter usually end up in unprotected temporary housing where their abusers can find them. The men just "put two and two together and say, OK, where is she going to go," one shelter worker told Enarson.

The domestic violence programs and women's shelters surveyed by Enarson appeared to be almost totally ignored by disaster management agencies. Few reported receiving official information on preparing for a disaster or being part of disaster planning groups. Less than half of the responding programs had taken steps to prepare for a disaster, saying they were too strapped by inadequate resources that left them struggling to provide basic services and meet existing demand. These programs "cannot move toward disaster readiness," Enarson concluded.

British researcher Maureen Fordham of Anglia Polytechnic University agreed that women's vulnerability is amplified by their generally lower socioeconomic status and their child-care responsibilities. But she said that some women gain a more positive self-image and feel empowered when they find they can cope with the crisis. One important way in which they differ from men is that they quickly form social networks— for example, by meeting at the shelter laundromat—and talk about their experiences. Men are less likely to mix and discuss their experiences; they're often unable to acknowledge their feelings of helplessness and have difficulty asking for help, preferring to tough it out. Fordham found that the stress sometimes causes couples to separate.

Even when families stick together, they face many psychological challenges in attempting to rebuild their lives. Focusing on all the practical tasks that need doing may cause people to neglect the emotional needs of the family. Meals are eaten hastily, sleep is sacrificed, there's little time for relaxation or conversation. Sudden mood changes may occur and family members may alternate between bickering and withdrawal. Being able to talk openly about the experience is important, but there may be psychological barriers inhibiting communication. A brochure published on the Internet by North Dakota State University in the aftermath of the Red River flood points out that "when people stop talking with others who have suffered loss or who are facing financial trouble, they send the message that they don't care." Friends and neighbors may not be indifferent but rather, are "caught up in their own losses, uncertainties and problems. Those who were not hurt directly by the floods may feel guilty and not know what to say."

Children are especially vulnerable to anxiety and stress when their world is suddenly turned upside down. According to David Hart, a Calgary psychologist and trauma specialist, they're prone to sleep disturbances, flashbacks, persistent fears about a recurrence of the event, and a loss of interest in school. They may exhibit many symptoms, including psychosomatic illnesses and regressive behavior (e.g., thumbsucking and bedwetting).

Children of different ages react differently to trauma, Hart says. Preschoolers won't verbalize their anxieties without prompting, but clues can be found in their play activities. Children this young may be strongly affected by an awareness of their parents' reactions and vulnerability. Regressive behavior, sleep disturbances, and increased clinginess are common. Children between five and eleven commonly exhibit regressive behavior; they become irritable and whiney, get into fights with their friends and siblings, and compete for attention. They may get into trouble at school or refuse to attend. Sleep disturbances are common and contribute to their irritability.

Adolescents are likely to have played a direct role in the disaster; they may have been involved in rescuing people and helping survivors and may have witnessed death, injuries, and the physical destruction of their world. Psychosomatic complaints and social withdrawal are not uncommon, and this group should be watched for evidence of suicidal impulses, Hart said. With this group, it's important not to ignore the event; Hart recommends activities like helping to rebuild the community, visits to emergency measures organizations and weather offices and providing opportunities to talk about their experience (e.g. school newspaper, counselors, call-in talk shows, etc.).

It's a tall order for parents to respond to their childrens' need for increased reassurance when they themselves are feeling emotionally and physically overwhelmed. Children are sensitive to increased tension and changes in their parents' behavior, and some may worry that they're somehow to blame, so it's especially important for parents to reassure children that what's happened isn't their fault. Allowing children to help the family recover—for example, asking them to think of ways to economize if money is tight—enhances their feelings of having some control over the situation.

The stress associated with a natural disaster can persist for months and years. After the initial shock has passed and people are confronted with an uncertain future, it's common for them to experience a gamut of emotions: anger, fear, cynicism, depression, and exhaustion, according to DonnaRae Jacobson, a family science specialist with North Dakota State University. She pointed out that victims of natural disasters like the Red River flood should watch for symptoms of prolonged stress, including sleep disturbances; irritability, outbursts of anger, and loss of self-control; excessive drinking or drug use; excessive worry; withdrawal and suspicion; feelings of guilt and self-doubt; and apathy, denial, avoidance of emotions, and feelings of detachment from other people or from life.

The disaster can mark a sudden turning point that causes people to lose their dreams and goals in life. "Some people may go through the rest of their lives angry with the unfairness of the flood," Jacobson said. One of the most important keys to recovery is to fight the feelings of isolation and detachment by connecting with others, by helping others as repayment for the help received, by saying thanks and celebrating survival, by being willing to seek and accept support, and by "reframing the meaning of life and setting priorities."

For people who fail in this effort, delayed stress can literally be a killer.

Researchers from the CDC surveyed U.S. counties that had experienced a severe natural disaster between 1982 and 1989 and found that suicides increased nearly 14% during the four years after the event, compared with an increase of only 1% nationwide. Suicide victims included people who were injured or who lost family, friends, property, or jobs in the disaster. The increases in suicides affected both sexes and all age groups.

There were differences in the suicides associated with different kinds of disasters. With earthquakes and hurricanes, the biggest increase in deaths occurred right after the event and then dropped off. After hurricanes, the rate increased 31% during the first two years, then returned to normal. Earthquakes had the biggest immediate effect: a 63% increase in the first year. Floods caused a 14% increase in suicides; they had the most sustained impact, rising in each of the years following the flood, with the biggest increase (nearly 24%) occurring in the fourth year. This might be due to the fact that flood victims suffer three times the financial losses and four times the injuries experienced by victims of hurricanes or earthquakes. This, plus the destruction of much of their social support network, may account for the lingering psychological distress. Interestingly, the researchers did not find significant increases in suicides after tornadoes or severe storms.

In a paper published in the *New England Journal of Medicine*, the CDC researchers concluded that victims of floods, earthquakes, and hurricanes experience an "increased prevalence of post-traumatic stress disorder and depression, which are risk factors for suicidal thinking," a finding that reinforces the need for mental health support after disasters.

There is one thing most mental health experts agree on: victims of natural disasters, no matter what age or gender, must talk about their experiences. Repressing or denying the feelings of helplessness, fear, and anxiety only causes lingering physical and emotional problems.

Social Disruption and Conflict

When a natural disaster affects large numbers of people, causes tremendous damage to infrastructure, and disrupts the daily functioning of the social order, it makes people feel edgy and vulnerable, even those who experience the event vicariously. They need reassurance from the authorities that things are under control. They expect their political leaders to show strength, decisiveness, and empathy. But, most of all, they expect their leaders to show up. In February 1998, when both Florida and California were relentlessly pummeled by El Nino–related storms, President Bill Clinton made personal visits to the sites of the Florida tornadoes and flood-ravaged areas of California, promising federal disaster aid. After Hurricane Pauline devastated Acapulco in October 1997, President Ernesto Zedillo of Mexico cut short a visit to Germany to return home, only to meet with the wrath of victims complaining about the inadequacies of the government relief effort.

During the ice storm, Prime Minister Jean Chrétien, Ontario Premier Mike Harris, and Quebec Premier Lucien Bouchard personally surveyed the damage and tried to reassure people that help was on the way. Both Chrétien and Harris delayed joining a Canadian trade mission to Mexico and South America, and Bouchard never made the trip at all.

How effectively political leaders handle a crisis has a tremendous influence on how threatened people feel, according to Allan Bonner, a Toronto-based crisis-management consultant. "If there's low trust in high-risk situations, people will feel threatened. And if they feel the leadership is not helping them, they'll punish that person," he told Montreal's *The Gazette.* "People won't blame a political leader because there was an ice storm or whatever, but they will judge the political leadership on how it was handled."

Scanlon argues that Quebec authorities made a major mistake when they delayed telling people how precariously close they came to losing the entire power grid serving Montreal. It was nearly two weeks after the height of the crisis before they revealed they'd purposely blacked out the downtown core to keep pumping the city's water, which had no backup power system. Telling the truth is crucial in disasters, even if it's not very palatable, Scanlon said. But those managing a crisis often conceal important facts in the mistaken belief that people will panic or that they won't be able to cope. "Research originally done in the 1940s showed that people cope best with the truth and nothing has contradicted it since," Scanlon said. "The reality is that people behave pretty rationally."

We've seen that temporary displacements of large groups of people during a disaster can cause psychological problems and social conflict. But what happens when disasters are so large that they force people from their homes and lands permanently? Climate-related changes, particularly rising sea levels and weather extremes, have the potential to create such environmental "refugees."

Many populations around the world are vulnerable to displacement by climate change. Nearly half the world's people are already living within a few dozen kilometers of the ocean, and these numbers are expected to rise as more people migrate to the coast. Many of the world's largest cities are located in low-lying coastal regions. As rising sea levels encroach on these places, their residents will be concentrated into smaller areas, increasing the risks of social conflict associated with urban crowding. This will only add to the problems already caused by rapid urbanization. The number of people living in cities increased 50% in the last half of this century and is expected to double by 2025.

Rising sea levels and weather extremes can also precipitate temporary or permanent migrations to higher ground or to less threatened areas. A report by the World Health Organization warns that catastrophes caused by climate change "can generate large refugee and population movements, with a need for resettlement in what may already be densely populated areas." Moreover, these circumstances can exacerbate the spread of disease.

A large influx of people can also have an environmental impact on the receiving region and "may, in some situations, result in conflict," according to a report by the Canadian Global Change Program, which began a study on the links between environmental change and human security in 1996. It notes that, although researchers only recently began to explore this subject, there's a growing recognition that environmental degradation of all kinds, including climate change, presents a potential threat to peace that must be considered along with military, political, and economic factors that cause conflict. Environmental threats to human security transcend national boundaries; Gro Harlem Brundtland, the Norwegian president who headed the World Commission on Environment and Development, predicted that "the environmental problems of the poor will affect the rich ...transmitted through political instability and turmoil."

Tad Homer-Dixon, director of the Peace and Conflict Studies Program at the University of Toronto, is a leading expert on the links between environmental degradation and social conflict. His research indicates that environmental scarcity does cause violent conflict and "its frequency will probably jump sharply in the next decades as scarcities rapidly worsen in many parts of the world." At present, these conflicts are primarily associated with scarcities of resources such as water, agricultural land, forests, and fisheries; many of these environmental problems (e.g., deforestation and poor water) have already caused "acute hardship" in many parts of the world and are causing serious social disruption. Global warming, he says, "will probably not have a major effect for several decades, and then mainly by interacting with already existing scarcities."

Homer-Dixon argues that conflicts arising from environmental stresses are likely to occur in poor countries first. Reductions in food production, economic decline, social disruption, and the displacement of populations could lead to conflicts such as civil strife and insurrections, disputes between ethnic groups, and cross-border altercations. Countries that experience extreme environmental stress will either fragment or move toward authoritarianism. Fragmentation will prompt large migrations of refugees, he suggests. "As different ethnic and cultural groups are propelled together under circumstances of deprivation and stress, we should expect intergroup hostility." The fact that newcomers will stretch already-depleted resources will only intensify such conflicts.

Homer-Dixon predicts that many migrants from developing countries will seek to move to the developed world—from Latin America to the U.S. and Canada, from Africa and the Middle East to Europe, and from Asia to Australia. This has already occurred to some extent and has resulted in a "xenophobic backlash" in some cases, a problem that "will undoubtedly become much worse." However, he notes that some countries, including Canada, have demonstrated an "astonishing capacity" to accept immigrants without serious conflict.

The question is whether that will remain true if countries are confronted with a major influx of environmental refugees. An Environment Canada report said that Canada will likely become even more attractive than it is today and the Canada Country Study (CCS) notes that increased immigration would have "implications for health and social systems, as well as potential impacts on the labour market."

Migration is not the only source of environment-related conflict. Disputes over natural resources such as water, agricultural land, forests, and fisheries have occurred within and between countries. The CCS notes that climate change could lead to "increased diplomatic turmoil" over issues like managing transboundary waters and fisheries. Disputes in other parts of the world could also increase the demand for Canadian peacekeepers.

It's often argued that human ingenuity will find a way to cope with social conflicts arising from environmental problems. Homer-Dixon argues, however, that environmental degradation and population growth will undermine the adaptive capabilities of many societies, particularly those in developing countries. "As environmental degradation proceeds, the size of the potential social disruption will increase, while our capacity to intervene to prevent this disruption decreases." It's not reasonable to "assume we can intervene at a late stage, when the crisis is upon us."

Efforts to stabilize the global climate, as well as to stop population growth and other environmental losses, are key to human security in the future, according to Michael Renner, a senior researcher with the Worldwatch Institute. "In the struggle for environmental and social security, tanks and automatic weapons are at best irrelevant and often obstacles," he argues. "Security in the nineties has less to do with how many tanks or soldiers a country can field and more with how well it protects its arable lands and watersheds, and whether it is able to reduce social pressures." One policy he recommends is strengthening the conventions on climate change and biodiversity. "Unless the climate is stabilized and natural ecosystems protected, other efforts to improve human security may be overwhelmed."

Faced with the prospect of myriad environmental, social, psychological, and economic consequences from global warming, we have just two choices: either we take steps to avoid further warming or we find ways to adapt to it. They're not mutually exclusive, except in the sense that both involve costs and social adjustments and we don't have unlimited stores of either. Finding the right mix is the issue that confronts us now.

nine

TOO LITTLE, TOO LATE:
Avoiding Climate Change

For most of its brief history, the global warming debate has been dominated by the issue of *avoiding* future climate change by reducing greenhouse gas emissions. The IPCC report defines avoidance (or what most climate scientists refer to as *mitigation*[*]) as "actions that prevent or retard the increase of atmospheric greenhouse gas concentrations by limiting current and future emissions from sources of greenhouse gases and enhancing potential sinks."

Most environmentalists and many scientists regard avoidance as the best approach to dealing with climate change, arguing that the uncertainties associated with global warming, far from being a reason to delay cutting greenhouse gas emissions, are actually the best argument in favor of it. If we're unsure of what awaits us, the prudent response is to slow down, not speed up.

It's also been argued that avoidance is the most ethical approach, particularly regarding future generations. The costs and benefits of policies to deal with climate change will be borne by different generations; if aggressive action is taken today to reduce greenhouse gas emissions, the

---

[*] Unfortunately, *mitigation* is also used, with an almost exactly opposite meaning, by other researchers who study how we cope with natural disasters. To them, mitigation doesn't mean stopping global warming but rather *adapting* to it with measures that reduce vulnerability to the consequences of climate change. (See Chapter Eleven.)

costs will largely be borne by the present and perhaps one or two subsequent generations while the benefits will accrue primarily to generations further down the line. A "wait-and-see" approach shifts the costs of a warmer climate to future generations.

This issue of intergenerational equity is central to the concept of sustainable development, defined as meeting the needs of the present without compromising the ability of future generations to meet their own needs. However, the IPCC report notes that there's disagreement on whether this necessarily implies the present generation is ethically bound to leave future generations an unchanged climate. "There are different views...on the extent to which infrastructure and knowledge, on the one hand, and natural resources, such as a healthy environment, on the other hand, are substitutes....Some analysts stress that there are exhaustible resources that are unique and cannot be substituted for. Others believe that current generations can compensate future generations for decreases in the quality or quantity of environmental resources by increases in other resources."

However, even if one accepts that avoidance is the best and the right thing to do, it's clearly not the simplest or easiest thing to do. Emitting greenhouse gases is what makes the world go round, economically speaking, and it's proving hard for us to stop. The IPCC report estimates that stabilizing atmospheric *concentrations* of greenhouse gases at 1990 levels would require reducing global *emissions* about 50 to 70%—a goal beyond anything contemplated by even the most aggressively pro-environmental governments at present.

Given the large uncertainties associated with both the science and the economics of climate change, governments face a problem in weighing the up-front costs and benefits of immediate, aggressive action to cut emissions versus the longer-term costs and benefits of delaying or rejecting such action. Disputes over these equations, which involve much guesswork, have kept governments, business leaders, and environmentalists in a perpetual state of disagreement. It's no surprise that the FCCC, and its offspring, the Kyoto Protocol, have had to travel such a long and rocky road toward implementation. (See Chapter Ten.)

As the new millennium approaches, it's unclear where this road will lead. The greenhouse gas cuts now being contemplated, even if fully implemented, are unlikely to prevent a doubling of CO_2 levels over the next century. How much worse it will get depends on two things: whether the Kyoto Protocol is ratified and whether world governments, including those of developing nations not bound by the Protocol, agree to work together and move beyond the Kyoto agreement during the next few decades. Unfortunately, the fractious history of international climate negotiations does not inspire optimism.

Before examining why the FCCC has encountered such a rough ride, it's worth examining what avoidance of climate change entails and what measures may help to achieve reductions in greenhouse gas emissions.

Avoiding Greenhouse Gas Emissions

Measures to limit greenhouse gas emissions fall into two broad categories:
- *Reducing the amount of greenhouse gases emitted by various activities and processes.* Examples include increasing the energy efficiency of power plants, equipment, and vehicles; adopting energy conservation measures; designing buildings to use less energy; and reducing automobile use.
- *Using alternative methods and technologies that provide the same benefits without emitting greenhouse gases at all.* Examples include switching to renewable or nuclear energy and developing new energy technologies such as fuel cells and advanced solar and wind devices.

These approaches are not mutually exclusive and economic instruments and incentives, such as "carbon taxes," tax credits, and subsidies, can be used to encourage increased efficiency, the development of new technologies, and the implementation of programs to encourage energy production and lifestyle changes that reduce or eliminate greenhouse gas emissions. Some controversial economic mechanisms have also been proposed by developed countries to give them "flexibility" in meeting their emission targets. One example is *emissions trading*, which allows companies or countries that cannot meet their targets to "buy" the unused portions of another company's or country's emissions allowance. Another is *joint implementation* (a.k.a. *activities implemented jointly),* which allows one country (usually an industrialized country) to receive credit for funding emissions-reducing activities in another country (usually a developing country), where achieving such reductions may be easier and cheaper than they would be domestically. In the Kyoto Protocol, this approach was given a new name: *clean development mechanism.*

Most avoidance efforts focus on reducing CO_2 emissions, which account for more than 80% of greenhouse gases released by developed countries. However, some programs focus on methane, the second most important greenhouse gas, which is released by oil and gas production, agricultural activities, and landfill sites. (The Kyoto Protocol calls for reductions of six greenhouse gases in aggregate: CO_2, methane, nitrogen oxides, and three halocarbons used as substitutes for CFCs.)

Since the burning of fossil fuels—oil, gas, and coal—is the major source of CO_2 emissions, energy use is the major target of reduction programs. But it's not going to be easy to wean the world from fossil fuels. According to a report by the UN's World Bank, by the middle of the 21st century, fossil fuels will still provide nearly two-thirds of primary energy, with coal—one of the worst CO_2 emitters—still accounting for about a third of that. "Coal is too abundant…and too low in cost to be replaced easily as a major fuel." Three sectors will use fossil fuels in roughly equal amounts: electrical generation, transportation, and all other sectors, including residential and industrial uses.

Developed countries have been and remain the largest emitters of greenhouse gases; they not only release the largest amounts in absolute terms, but also have the highest per capita emission rates. However, developing countries, particularly those that are industrializing rapidly, are projected to become the major emitters within the next 25 to 30 years, although their per capita emissions are expected to remain below those in the developed world. Climate treaty negotiations have repeatedly bogged down because of political disputes between developed and developing countries over who should reduce emissions and when.

At the Rio Earth Summit in 1992, developed countries agreed to a nonbinding target of aiming to stabilize their emissions at 1990 levels by 2000. Both Canada and the U.S. established voluntary programs to achieve this goal. The U.S. program, called the Climate Change Action Plan, was adopted in 1993, while Canada established its National Action Program on Climate Change (NAPCC) in 1995. NAPCC outlined strategies for both governments and private industry to reduce greenhouse gas emissions and improve adaptation to climate change. The plan endorsed the *precautionary principle* set forth in the FCCC by stating: "Where there are threats of serious or irreversible damage to our health and livelihood, lack of scientific certainty should not be used as a reason for postponing mitigative actions that are cost-effective or justified for other reasons." It also called for *shared responsibility*, meaning that action must be taken by all sectors of Canadian society—governments, private industry and the public—and "no one region or economic sector should be unduly disadvantaged." Priority should be given to actions that also solve other environmental problems (i.e., "no regrets" measures).

A key element of the plan was the Voluntary Climate Challenge and Registry Program, a signal that Canadian governments intended to rely on voluntary rather than mandatory measures to achieve the cuts. The registry is to be a repository of information about the commitments, action plans, and achievements of the participants in the voluntary program; it would be used to publicize examples of their actions and to determine the progress being made.

The Canadian government claimed some success with this program. In its second report to the UN outlining progress since Rio, it noted that between 1990 and 1995, CO_2 emissions from secondary energy use (i.e., energy used by the final consumer) were 3.5% lower than they would have been without the program. But by the last half of the 1990s, it was clear that voluntary measures weren't getting the job done. There's no possibility that Canada—nor, indeed, most other developed countries—will meet their commitment to stabilize emissions at 1990 levels by 2000. Most countries aren't even headed in the right direction. By the end of 1995, both U.S. and Canadian emissions were up about 8% from 1990 and were on course for an increase of roughly 13% above 1990 levels by 2000.

Canada's second report to the UN acknowledged that it was unlikely to meet its Rio commitment, but said that progress is being made in lowering the projected "gap" from an increase of 13% to one of 8% by 2000. This was a far cry from the goal set by Prime Minister Jean Chrétien in the 1993 federal election campaign, when he said that Canada would aim for emission reductions of 20% below 1988 levels by 2005.

The Canadian report to the UN noted that Canada is "a northern country, with climate extremes, vast distances to cover and a heavy dependence on energy-intensive natural resource development." A statement in the report articulated the dilemma that most developed countries faced in trying to meet its Rio targets: "Canada's environment and economic interests both need to be protected." However, as events unfolded over the next several years, it's clear the scales were tipping toward the economic side of that equation.

Unlike the FCCC, whose provisions are voluntary, the Kyoto Protocol commits developed countries to legally binding targets and timetables for reducing emissions (although as yet there's no mechanism to punish offenders). Under the treaty, 38 countries agreed to reduce emissions of CO_2, methane, nitrous oxide, and three halocarbons to below 1990 levels between 2008 and 2012. Overall, a 5.2% reduction must be achieved, but individual countries have different targets: Canada and Japan agreed to 6%, the U.S. to 7%, and the European Union to 8%. (Since emissions in most countries have *increased* since 1990, the effective reductions required to get below 1990s levels are actually much larger, in the range of 15% or more for Canada and the U.S., for example.) However, none of these obligations become binding until the Protocol comes into force, which requires ratification by 55 countries. Even then, it doesn't bind any individual country until ratified by its national government. In the U.S., the likelihood of that happening appears to have the proverbial snowball's chance in a greenhouse-warmed world. It will be a while before we see any measurable impact on atmospheric greenhouse gas concentrations as a result of the Kyoto Protocol.

The Cost of Avoiding Greenhouse Gas Emissions

Countries that agreed to reduce their emissions are faced with the dilemma of achieving even the modest objectives of the Kyoto Protocol without stifling economic growth or forcing politically unpalatable price or tax increases or lifestyle changes on their populations. There are also concerns that jobs, industrial production, and capital will flee to developing countries that remain exempt from emissions targets.

The cost of limiting emissions is an issue as fraught with controversy, uncertainty, and dueling experts as climate change science itself. The fossil fuel industry, much of the business community, and other groups opposed to the climate treaty predict dire consequences, including rising prices for energy and consumer goods, job losses, and general economic catastrophe. Environmentalists and other supporters of emissions limits argue that energy efficiency and new technologies not only can achieve emissions reductions without great economic hardship but also, in fact, produce economic benefits. They claim that projections of economic doom and gloom don't account for the benefits of technological innovation and the energy savings that will result from increased efficiency.

Several economic studies have tried to quantify the cost of reducing emissions, but their results are all over the map. The most negative estimates come from research organizations or lobby groups that are avowedly conservative in their political and economic outlook and/or openly opposed to measures that would limit fossil fuel use. For example, the Global Climate Coalition (GCC), an industry lobby group with a cleverly ambiguous name but a very clear agenda, released a study that estimated a loss of US$227 billion in the U.S. gross domestic product in 2010 alone if emissions were cut to 1990 levels. William O'Keefe, executive vice president of the American Petroleum Institute and until recently GCC chairman, has argued that stabilizing CO_2 emissions at 1990 levels by 2010 would cause a 25% reduction in energy use in the U.S., cost the country US$250 to $300 billion a year (4% of GDP), eliminate half a million jobs for a decade, increase gasoline taxes by 60 cents a gallon, and cost the average family US$4000 extra a year. U.S.

automakers suggest that the rising gas prices will cost U.S. motorists an extra US$1600 a year. Jack Kemp, a former Republican Congressman, has argued that emissions limits would increase energy costs by as much as 50%.

During the Kyoto conference, the Organization of Petroleum Exporting Countries (OPEC) argued that cutting emissions would cost its member countries as much as US$20 billion a year and thus would threaten their "legitimate right to economic development."

A report by the Conference Board of Canada, which summarized several forecasting studies, estimated that stabilizing emissions at 1990 levels could reduce Canada's projected 30% growth to the year 2010 by 0.5 to 2.3 percentage points. Estimates of the dollar loss by 2010 ranged from $5.4 billion to $30 billion.

Labor and farm groups have expressed concern about rising prices and job losses. The executive council of the AFL-CIO argued the climate treaty will create an "uneven playing field" for U.S. workers. Carbon taxes or emissions trading programs would raise domestic energy prices. "These taxes are highly regressive and will be most harmful to citizens who live on fixed incomes or work at poverty-level wages." Higher prices would lead to the closing of factories, mines, and mills and the transfer of these operations, along with high-paying jobs, to countries without emissions restrictions. "Carbon emissions, therefore, will be transferred to the developing world along with the jobs, thus providing no real benefit to the environment." The American Farm Bureau Federation claims that if a 25 cent per gallon fuel tax were introduced in the U.S., net farm income would drop by 24%.

The conviction that cutting emissions will cause economic catastrophe is the foundation for the argument that we must wait for "proof" that climate change is happening and that it will have severe negative consequences. (The question of what the skeptics would accept as proof is another matter, which will be explored in Chapter 12.) The *Globe and Mail,* which has consistently espoused the skeptical agenda, published two editorials before the Kyoto conference that provide examples of this viewpoint. The first argued that global warming is so gradual a process that waiting 25 years to cut emissions would cause little additional warming by 2100. "Governments need not be hasty as they grapple with the global warming issue. Indeed they must not." The second editorial cited a report by Daniel Schwanen of the C.D. Howe Institute, a conservative think tank, to support its argument that "most scientists who believe in global warming do not expect it to have serious effects for many decades....Mr. Schwanen says that the world could easily wait 20 years before imposing mandatory emissions controls. If, by that time, science proves a link between human activity and global warming, there will still be plenty of time to act." The rationale for not acting right away was the high cost of cutting emissions by shifting away from fossil fuels. Waiting would be "far less costly," the editorial claimed, because this would give industry time to adapt, to retool, and to switch to new technologies.

There's little doubt that industries would find a gradual adjustment more palatable than a sudden one, but it doesn't follow that there's little or no price to be paid for waiting. Even accepting the questionable claim that "most scientists" don't expect serious impacts for "many decades," it misses the point. The issue is not when the serious impacts will first show up, but *how long they will continue to worsen afterward.* The idea that there will "still be plenty of time to act" ignores the fact that it takes a long time for the atmosphere and oceans to respond to changes in greenhouse gas emissions. Even if it does take decades for "serious impacts" to appear, it will take many more decades after that, perhaps a century or more, for those impacts to reverse themselves in response to cuts in greenhouse gas emissions—assuming, of course, that the climate system doesn't trip over an irreversible threshold into a significantly different state.

We've seen that, with business-as-usual emission scenarios, it will be virtually impossible to avoid a doubling of CO_2 by the end of the 21st century, and a tripling is not impossible. Moreover, with such scenarios, atmospheric greenhouse gas concentrations will continue rising after 2100, in some cases substantially. If we do nothing for another few decades and then find proof that global warming really is happening, we can't stop it on a dime—not even in the unlikely event that we manage to stop our *emissions* on a dime. If we continue pumping greenhouse gases into the atmosphere at an accelerating rate until the day scientists prove to the satisfaction of the fossil fuel industry and the skeptics that global warming is real, then we will very likely have major trouble on our hands until well into the 22nd century.

The IPCC report points out, moreover, that "decisions taken during the next few years may limit the range of possible policy options in the future, because high near-term emissions would require deeper reductions in the future." While delay might reduce the cost of avoidance measures by allowing time for technological advances, it would likely increase both the rate and the eventual magnitude of climate change and, in turn, increase the costs associated with the damage it causes and measures to adapt to it.

Danny Harvey of the University of Toronto argues that delay could result in higher, rather than lower, costs because of "lost windows of opportunity." For example, it's less expensive to design cities at the outset to require less vehicle use and encourage energy efficiency than to redesign them later. Delay might force an acceleration of capital turnover, which could be very expensive, especially if climatic impacts are worse than expected.

Some have argued, nevertheless, that it's better strategy to continue building economic wealth and developing technology so we'll be better equipped to deal with global warming if and when we have to. As the *Economist* put it: "The world economy could, if recent growth continues, be more than 300% bigger in 2095 than today, and that much better able to bear the costs of coping with climate change."

Cornelius van Kooten, a professor of agricultural economics at the University of British Columbia, has argued that avoidance measures will take time to implement and probably won't work as well as their proponents hope. If such measures don't succeed, future generations will not only inherit a warmer climate but will also be less well-off than if we hadn't tried avoidance. Avoidance measures "will make current generations poorer, and poorer people pass on less wealth to future generations." It might be wiser *not* to invest now in costly avoidance measures, but to concentrate on increasing human wealth and capital—including knowledge and technology—that will enable future generations to cope with climate warming.

There's some merit to this argument, particularly given the foot-dragging that has characterized international climate negotiations to date. The half-hearted targets currently being contemplated may not even be fully implemented, and they're unlikely to be sufficient even if they are. Unless the world stiffens its resolve to achieve further reductions, it's legitimate to question whether it makes sense to invest in measures that won't get the job done as opposed to investing in measures to deal with the consequences.

This strategy is, however, based on two assumptions that aren't universally accepted: first, that avoidance measures implemented now will be so costly that they'll devastate the economy and, second, that money and technology can handle any future climate changes (i.e., that future costs associated with *not* avoiding climate change will be more easily borne than the costs of trying to stop it today). This last assumption is biased by the conviction of many skeptics that future climatic changes won't be very serious, if they occur at all. Both assumptions are open to challenge.

The predictions of price hikes, job losses, and economic slowdowns made by global warming skeptics are ironic in a couple of ways. Many of the same people strongly favored the war on inflation and national deficits in the early 1990s that resulted in massive corporate downsizing and government cutbacks, which caused job losses and severe economic hardship. Statistics Canada figures show that the income of Canadians dropped each year between 1990 and 1995, the first time that has happened since the 1930s. This wiped out all income gains from the last half of the 1980s. Over the period, governments and the private sector achieved with paychecks what they failed to achieve with greenhouse gas emissions—a drop of 6% below 1990 levels.

The rationale for the downsizing was this: the deficit/debt problem is too significant to be ignored; if we don't do something now, it will get worse; we'll leave a terrible legacy for future generations; we must pay the price to deal with the reality of rising debt; once we adjust and get the situation under control, things will get better again.

This kind of reasoning can also be applied to global warming. There's a high probability that if greenhouse gas emissions continue rising on their present course, the world will face serious socioeconomic problems associated with climate change. In other words, the problem is too significant to be ignored; if we don't do something about it, it will get progressively worse; we will leave a terrible legacy for future generations; we must pay the price to deal with the reality of global warming; once we adjust and get the situation under control, things will get better again.

There's yet another irony associated with the predictions of economic disaster: while global warming is actually *more* certain to happen than the skeptics claim, the economic cost of fighting global warming is *less* certain than they claim. While they're happy to use scientific uncertainty as an excuse to delay avoidance measures, they rarely mention the uncertainties associated with their own predictions. Rather, they assert unequivocally that avoidance measures will inevitably cause economic disaster.

For example, the American Petroleum Institute has said that stabilizing emissions at 1990 levels "is impossible to achieve without enormous economic pain." Fred Palmer of the Western Fuels Association, a coal industry group, commented that a treaty limiting CO_2 emissions would have "terrible impacts on all of us, our quality of life, the future of our children, our health, and our wealth." Jonathan Adler of the Competitive Enterprise Institute, an organization "dedicated to the principles of free enterprise and limited government," has said, "While global warming is highly uncertain, the impacts of global warming policies are not. Dramatic restrictions on energy use would have severe economic effects...." And Alberta Premier Ralph Klein stated, "I don't have any hard figures right now but I'm told the economic impact [on Alberta's oil and gas industry] would be almost catastrophic." The *Economist* advised that it's "far too early to be panicked into Draconian actions to avert global warming, especially when most actions would pose a bigger threat to human well-being than does global warming.... *One of the few certainties* about global warming is that the costs of severely curbing emissions of greenhouse gases now would be huge." [Emphasis added.]

Given that skeptics accuse scientists and environmentalists of scaremongering, it's hard to see why they should not be similarly charged for predictions of economic doom if we try to fight global warming. After all, it's not as though economics is any more of an exact science than climatology. Nor does it have a better crystal ball. Economists use computer models too, and it's hard to see how their projections can lay claim to any greater certainty than those of climate models. As an article in the *Globe and Mail* pointed out (in contradiction of the paper's editorial stance), "the costs are hard to predict because so many other factors affect the course of the economy. The mix of energy sources in each country is crucial, as are urban planning, transportation and energy-pricing policies."

David Runnalls of the International Institute for Sustainable Development says that "nobody has poked enough holes" in assumptions about the cost of switching to new energy systems. Economic models based on present technologies and costs produce "horrendous figures that have everybody living in caves. It's all been diabolically wrong. When somebody increases prices and puts in efficiency standards, the private sector invents their way into something that's more efficient and doesn't cost as much."

With regard to climate models, William O'Keefe of the Global Climate Coalition advises governments not to "bet your economy and society on computer projections. These are nothing more than a set of highly uncertain best guesses." This is sound advice as long as it's applied equally to climate *and* economic models, since both depend on underlying assumptions (and climate models at least have the advantage of being partly based on natural laws of physics and chemistry that come as close to a sure thing as we get in this world). This fact was highlighted in a revealing study done by the Washington-based think tank World Resources Institute entitled "The Costs of Climate Protection: A Guide for the Perplexed." It analyzed 16 commonly used economic models that have produced a wide range of predictions about the economic impact of emissions reductions and identified several key assumptions that account for most of the differences in these projections. For example: Does the model assume that non–fossil fuel alternative energy will be available at competitive prices? Does it assume that revenues raised from energy taxes and emissions permits will be used to reduce other taxes? Does it include the impacts of joint implementation and emissions trading programs? Does it consider the impact of emissions cuts on reducing air pollution? Does it account for the cost of environmental damages from climate change that would be avoided by emissions reductions?

The study examined how these assumptions affected the outcome of a particular policy—stabilizing emissions at 1990 levels by the year 2010 and keeping them at that level afterward. Models that used the most unfavorable assumptions predicted a reduction of 2.4% in the growth of the U.S. GDP by 2020. This would mean the U.S. economy would be *only* 75% larger by 2020, instead of 77.4% larger. Models that used more favorable assumptions predicted an *increase* in GDP of the same amount—2.4%. Their advice to the perplexed: when you hear predictions about the economic effects of climate protection policies, ask about the underlying assumptions. The study concludes that "people need not accept all the best-case or all the worst-case assumptions. It is more reasonable to predict that with sensible public policies and international cooperation, carbon dioxide emissions can be reduced with minimal impacts on the economy."

In late 1997, the U.S. Department of Energy (DOE) released a massive study, which found that domestic emissions trading, energy efficiency programs, and new energy technologies could both reduce U.S. emissions and produce energy savings that would roughly equal or exceed the costs of these measures. In short, the DOE concluded, emissions cuts could be achieved "without increasing the nation's energy bills," and could even save people money. In March 1998, economist Janet Yellen, head of President Clinton's Council of Economic Advisers, told a Congressional committee that implementing the Kyoto Protocol would add only a "modest" US$70 to $100 a year to the average household's yearly energy bill. Their analysis suggested a 3 to 5% increase in energy prices and a 3 to 4% increase in gasoline prices.

The World Bank, which funds environmental programs in developing countries, also argues that efforts to reduce emissions need not spell economic doom. One of its papers concludes that countries can pursue lower-carbon energy options and that these choices will "ultimately help them achieve stronger more sustainable energy growth. The Bank believes that its clients stand to benefit from global efforts to address climate change."

A panel on Canadian Options for Greenhouse Gas Emission Reduction (COGGER), established by the Canadian Global Change Program and the Canadian Climate Program Board, examined the potential for reducing Canada's emissions through increased energy efficiency or switching to fuels that emit less CO_2. In 1993 (when the Rio commitment was still fresh), the panel concluded that it appeared "feasible and cost-effective" for Canada to achieve the Rio target and, moreover, to achieve a cut of about 20% by 2010. Improved energy efficiency was the key to reaching these goals, but this wouldn't happen without strong government policies, including setting targets and timetables, some regulation, and economic incentives. (Little of this happened and it soon became apparent that Canada wouldn't meet the stabilization goal, much less a 20% reduction by 2010.)

Several groups of economists have issued statements supporting the view that emissions cuts can have positive economic benefits. In 1997, in an effort to counter the Australian government's resistance to the Kyoto Protocol, 131 Australian economists stated that cost-effective economic instruments such as carbon taxes and emissions trading could reduce Australia's emissions without damaging its economy, employment, or living standards. With its large land mass and climatic conditions, Australia also had great potential for developing renewable energy sources, the statement said.

Another group of some 2000 economists, including half a dozen Nobel laureates, issued a statement saying they believed that climate change "carries significant environmental, economic, social and geopolitical risks and that preventative steps are justified. Economic studies have found that there are many potential policies to reduce greenhouse gas emissions for which the total benefits outweigh the total costs." They said that for the U.S. particularly, "sound economic analysis shows that there are policy options that would slow climate change without harming American living standards and these measures may in fact improve U.S. productivity in the longer run." They advocated the use of market-based mechanisms like carbon taxes and emissions trading, which would generate revenues that could be used to reduce deficits or other taxes.

Not surprisingly, environmental groups focus strongly on the potential savings resulting from increased energy efficiency and new technologies. In 1997, five U.S. environmental organizations released a study that found the U.S. could reduce emissions to 10% below 1990 levels by 2010 while reducing energy costs by US$530 per household per year and creating nearly 800,000 jobs. According to the report, these benefits could be attained by stimulating the introduction of innovative energy-efficient and renewable energy technologies, including fuel cells, advanced gas turbines, improved building designs, biomass fuels, advanced wind turbines, and advanced solar/electric technologies. This "innovation path" would increase the proportion of U.S. energy needs supplied by renewables to 14% by 2010 and 32% by 2030, while the contribution of oil would drop to 36% and 34%, respectively. Carbon emissions would drop 10% between 1990 and 2010 and 46% by 2030. A significant benefit of this program would be to reduce U.S. dependence on imported oil, which accounted for most of the US$56 billion bill for energy imports in 1996.

John Adams of the Natural Resources Defense Council argues that "the cost of solving environmental problems has routinely been overestimated." For example, neither phasing out ozone-destroying CFCs nor reducing acid rain cost as much as originally projected. "If we make firm commitments to solve our environmental problems, we can spur innovations that let us meet these goals more quickly and more cheaply than just about anyone expects ahead of time. The same is true for global warming."

Christopher Flavin of the Worldwatch Institute is optimistic that the same human ingenuity that caused the climate problems we now face can help to solve them and create economic opportunities at the same time. Building an energy system based on solar and wind energy would create "hundreds of billions of dollars of business and millions of jobs."

U.S. President Bill Clinton picked up this theme in his effort to promote the Kyoto Protocol in the U.S. "Environmental initiatives, if sensibly designed, flexibly implemented, cost less than expected and provide unforeseen economic opportunities." He also said that "every single action the United States has taken since 1970 to clean up our own environment has led to more jobs [and] a diversifying economy."

These arguments do not fly very well with the skeptics. William O'Keefe, of the American Petroleum Institute, criticized the Clinton administration for perpetuating "the myth that 'free-lunch' technologies are readily available." The *Globe and Mail* opined that "Clinton was talking nonsense when he contended...that emissions control would be painless and even economically beneficial." The *Globe* also took Prime Minister Chrétien to task for similar suggestions; an article by Marcus Gee accused Chrétien of "trying to persuade us that it will be easy, cheap and maybe even profitable. That is both dishonest and irresponsible. The campaign against global warming is the most ambitious and costly ever undertaken by a Canadian government...."

What's interesting about arguments like these is that skeptics have so little confidence in our ability to develop cost-effective technologies to cut emissions and slow down climate change, yet they argue we can readily develop the technologies needed to cope with future climate change.

There's a certain inconsistency in rejecting innovation as the answer to the first problem while relying on it for the solution to the second.

Environmentalists also accuse the skeptics of ignoring two other aspects of the cost/benefit equation. The first is that measures to reduce greenhouse gases produce multiple benefits besides reducing the threat of global warming, some of which are not easily quantified in economic terms. The second is that there are long-term costs associated with *not* cutting greenhouse gas emissions, which many skeptics don't factor into their economic analyses.

Multiple Benefits and "No Regrets" Measures

Because of the uncertainties associated with predicting climate change and its consequences, many who advocate immediate action have focused on "no regrets" measures that provide economic and social benefits in *addition* to reducing the threat of global warming. The IPCC defines these as measures "whose benefits, such as reduced energy costs and reduced emissions of local/regional pollutants, equal or exceed their cost to society, excluding the benefits of climate change mitigation. They are sometimes known as 'measures worth doing anyway.'...Significant 'no regrets' opportunities are available in most countries."

The reasoning is simple: such measures will produce a net benefit *even if the projected impacts of global warming turn out not to be as bad as anticipated.* No regrets provides a cautious middle path between incurring large upfront costs that might prove to have been unnecessary and doing nothing at all until it's too late to avoid serious environmental damage. It's an attractive option given the political realities that make it hard to muster support for expensive actions with an uncertain payoff. As a report by the Canadian Global Change Program points out, "In many cases, the cost of action is both immediate and very large, while the potential consequences of inaction are not felt until ten, fifty or one hundred years in the future. Society tends to focus more on the immediate, and leave future problems to future generations. As a result, it will be difficult to convince either the Canadian public or Canadian politicians to choose the second option and act immediately, even if we do agree that the consequences of inaction are potentially catastrophic."

Environmentalists argue that any measures that reduce emissions will generate benefits aside from their impact on global warming. "There are so many no regrets options that make a lot of economic sense, that have little or no economic cost, it's crazy not to do these on the grounds that maybe the [climate] models aren't accurate," says David Runnalls.

For example, reduced burning of fossil fuels or switching to cleaner fuels will cut emissions of deadly pollutants like sulfur dioxide, nitrogen oxides, particulates, and heavy metals such as mercury and lead. This would reduce acid rain and smog, which not only damage trees, lakes, crops, and buildings, but also create a public health hazard. In 1997, a panel of climate and public health experts convened by the World Resources Institute published a study in the journal *Lancet,* which concluded that by 2020, more than 700,000 deaths will occur worldwide each year from exposure to particulates released by burning fossil fuels that could have been avoided if policies to limit greenhouse gas emissions were adopted by both developed and developing countries. The total number of "avoidable deaths" between 2000 and 2020 was estimated at around 8 million—1 million in developed countries and nearly seven million in developing countries. Although these figures were characterized as indicators of the magnitude of the problem rather than precise predictions, the scientists concluded that the "short-term public-health impacts of reduced [particulate matter] exposures associated with greenhouse-gas reductions are likely to be substantial even under the most conservative set of assumptions....Regardless of how or when greenhouse gases alter climate, reducing them now will save lives worldwide by lessening particulate air pollution."

In May 1998, the Ontario Medical Association called for stronger, mandatory measures to reduce emissions from automobiles and coal-fired power plants. "Ontario doctors have made a diagnosis. There is a health crisis. It needs major surgery, not cosmetic treatment," said OMA president Dr. John Gray. The pollutants cause throat, eye, nose, and lung irritation in healthy people and can exacerbate heart and lung disorders. An estimated 1400 people are hospitalized and about 1800 die each year in Ontario due to respiratory problems aggravated by air pollution. A few days after the OMA announcement, Toronto issued its first smog alert of the year, the first time this had happened as early as May since the air-quality index was established in 1988. Weather experts said May temperatures were averaging 5°C above normal.

Measures to cut transportation emissions—such as reducing automobile usage by encouraging efficient public transportation and redesigning communities to be less dependent on vehicles—could also help cut costs associated with traffic congestion, noise, and accidents, as well as infrastructure costs such as road maintenance and repair. According to the Worldwatch Institute, the U.S. paid a US$100 billion "traffic congestion price tag" in 1996 for wasted fuel, health care costs from air pollution, and lost productivity. It estimated that in Bangkok, one of the world's most congested cities, "the typical motorist spends the equivalent of

44 working days each year sitting in traffic jams." Australian researchers have found that cities that invest the most in roads and automobiles do not gain an economic advantage; in fact, they have less efficient transportation systems and are less competitive than cities with more balanced transportation infrastructures.

Increasing the energy efficiency of homes, appliances, and vehicles would reduce energy bills as well as greenhouse gas emissions. Although the initial cost of these technologies may be higher than conventional alternatives, they recoup these up-front costs over time by using less energy. Reduced energy demand also decreases the need for capital investment to expand the power generation and distribution system and, if it reduces the need to transport energy over long distances, it could also decrease the vulnerability of the power system to extreme weather events like the 1998 ice storm.

All of these are "no regrets" benefits independent of the impact on global warming—and they're easier to sell to politicians and taxpayers because they produce more visible near-term results at the local and regional level than the climatic benefits, which may, in any event, not occur for decades.

Increased employment is also a bonus of investing in energy efficiency, according to Philip Jessup, director of the Cities for Climate Protection Program of the International Council for Local Environmental Initiatives (ICLEI). "When energy efficiency substitutes for energy supply, it creates jobs. Generating electricity is capital intensive; efficiency is labour intensive." If, instead of importing fossil fuels, money is spent on "an electrician installing efficient lighting, that money is staying in this economy."

By 1995, there were an estimated 4500 firms in Canada specializing in goods and services related to energy efficiency, conservation, and environmental protection. They employed around 150,000 people and had annual sales of more than $11 billion. This business was expected to more than double by the turn of the century, creating many more skilled jobs.

Jessup argued there's a need to reassess the "payback" criteria used to determine whether efficiency measures are economic. Businesses generally want to recoup their initial investment in one to two years, and longer periods are considered uneconomic. It usually takes longer for efficiency measures to pay off in terms of energy savings and "unless there are incentives for longer-term paybacks, five, six or seven year paybacks wouldn't be considered economic. So there's an argument over what's economic, what's the rate of return, what can be quantified and how do you weigh benefits that can't be quantified." Many analyses of measures to reduce greenhouse gas emissions fail to consider benefits to the local community and, in particular, ignore those that cannot be quantified in economic terms. Social or aesthetic benefits, such as clean air and more livable communities, are not factored into many economic studies simply because they're not considered to *be* economic benefits, Jessup said.

Many of those involved in the climate debate—on both sides of the issue—can agree, at least in principle, on the wisdom of implementing no regrets measures that generate economic returns aside from their effect on global warming. Even William O'Keefe recommends that countries "look for actions that will produce benefits under any set of circumstances."

However, once you get beyond statements of principle, agreement is harder to come by. The fossil fuel industry, for example, would certainly have regrets about measures that cut back on fossil fuel use. Many environmentalists, on the other hand, feel that no regrets measures aren't enough—that they should be considered only a foot in the door, not the ultimate solution. The IPCC report notes that the risk of damage from climate change provides a rationale for "action beyond no regrets." The World Bank also said that while it's in a country's interest to pursue "win-win" actions that offer both economic and climate benefits, there's a need to move beyond such measures. "Countries must identify environmentally friendly alternatives and put them into place early in the policy and investment cycle, and aggressively mobilize new resources to support them....The Bank itself will substantially increase its own resources allocated to these areas, and strengthen its own capacity to assess the risks and rewards of market introduction of new technologies."

The Cost of Not Preventing Global Warming

Many environmentalists and climate scientists criticize economic analyses of climate change that fail to consider the long-term costs of failing to prevent greenhouse warming. For example, Ed Ayers, editor of the Worldwatch Institute's magazine, commented in an editorial that the fossil fuel industry's rationale for delaying avoidance measures "pays due attention to the alleged risk of damage to the economy but it is oblivious to the vastly greater risk of undermining the planetary systems (including climate) on which that economy utterly depends."

In some cases, this failure reflects the bias of skeptics who simply don't accept there will be any serious long-term costs, because they don't believe global warming is really happening or that it will have bad consequences even if it is. Mostly, however, this failure stems from the fact that it's easier to quantify the short-term costs of avoidance measures than to estimate accurately the cost of future climatic events whose timing, location, and severity cannot be predicted with any precision. It's not surprising, therefore, that many economic projections related to climate change are skewed toward emphasis on the up-front costs.

In fact, the method economists use to assess future costs and benefits in terms of today's dollars, a procedure called "discounting," always rates future events at a lower value than present events. The IPCC report says that the choice of a discount rate is crucial when analyzing long-term issues like climate change, where the up-front costs of reducing greenhouse gases occur much earlier than the benefits from avoiding environmental damage due to global warming. "The higher the discount rate, the less future benefits and the more current costs matter in the analysis. For example, in today's dollars, $1000 of damage 100 years from now would be valued at $370 using a 1% discount rate...but would be valued at $7.60 using a 5% discount rate." The IPCC says there's little agreement on how to choose a discount rate and, in fact, many different figures are used.

The IPCC reviewed economic studies that attempt to estimate the costs associated with an atmosphere containing twice the current levels of CO_2. One of the difficulties is finding a way to assign a dollar value to the wide range of social and environmental consequences of climate change, many of which—like human health and mortality or the viability of natural ecosystems—are hard to quantify in economic terms. The IPCC says there's little agreement and major uncertainties involved in valuing such "non-market" impacts, yet these are important in assessing the social cost of climate change. "While some regard monetary valuation of such impacts as essential to sound decision-making, others reject monetary valuation of some impacts, such as risk of human mortality, on ethical grounds."

Not surprisingly, efforts to value human life in financial terms are controversial. The IPCC notes that "the value of life has meaning beyond monetary valuation." Nevertheless, many economic studies do place a value on "statistical lives" and, in fact, place a lesser value on statistical lives in developing countries than those in developed countries. Other social damages are not valued equally in developed and developing countries either. This approach elicited a good deal of reproach from developing countries, which were not pleased with the concept that deaths and environmental damage in their countries were worth less that those in industrialized countries. However, James Bruce, co-chairman of the volume of the IPCC report that dealt with these issues, said economic theorists weren't making moral or ethical judgments that people in developing countries are worth less; rather, their studies were based on the assumption that higher prices would be paid in developed countries to reduce the impacts of global warming than in developing countries.

The IPCC report notes that the few studies available indicate that developing countries will suffer more monetary damage from global warming, as a percentage of GDP, than developed countries. If the value of a statistical life were equalized at the level typical of developed countries, the estimates of global damage would increase several times and, moreover, the gap between damages suffered in developing and developed countries would be even larger.

Estimates of global damages from a warming of 2 to 3°C are in the range of a few percent of world GDP, "with, in general, considerably higher estimates of damage to developing countries as a share of their GDP," the report says. More specifically, countries with "a diversified industrial economy and an educated and flexible labour force" would suffer damages in the range of one to a few percent of GDP. For countries with a "specialized and natural resource–based economy (e.g. heavily emphasizing agriculture or forestry) and a poorly-developed and land-tied labour force," the damage estimates were several times higher. The report notes that these estimates refer to *net* damages; in other words, the findings take account of the beneficial impacts of global warming but are still dominated by negative damages.

There are very few, if any, countries in the world that are totally free of dependence on agriculture, forestry, fisheries, water resources, and other natural ecosystems for their economic well-being. There's no question that these systems are influenced by weather and climate. While we don't know exactly how, when, and to what extent these resources will be affected by global warming, scientists can give us a reasonable idea of the direction in which those impacts will proceed. Even with all the uncertainties that plague our understanding of this problem, it defies common sense to blunder ahead on the misguided assumption that there will be no impact at all—or that all the impacts that do occur will be beneficial.

The skeptics' confident assertions to the contrary, we have no reason to assume that waiting would be "far less costly." In fact, the IPCC computer projections indicate that the longer the delay in reducing emissions, the steeper the cuts needed in later years to achieve a given level of CO_2 concentrations in the atmosphere. Steep cuts over a short period are bound to be more expensive than gradual cuts, whether they happen sooner or later—but they're far more likely to be necessary the later we start on them. An early start provides a less taxing path to an equivalent result. If you want to reach the top of a mountain by a certain time, starting early means you can hike along the sloping road to the summit rather than having to clamber up a sheer face.

Avoidance Measures

There are many ways to reduce greenhouse gas emissions, but most methods fall into five broad categories:
* using energy more efficiently
* changing lifestyles and infrastructure
* switching to lower carbon fuels and renewable energy
* increasing greenhouse gas "sinks"
* economic policies and instruments

Developments in these areas are too myriad to discuss in detail, but a few representative examples provide some insight into the potential for reducing greenhouse gas emissions over the next few decades.

Using Energy More Efficiently

We waste a great deal of energy, so increasing energy efficiency is viewed by nearly everyone as a good thing. It falls into the "no regrets" category because increased efficiency provides benefits independent of climatic effects—it can save money on energy bills. Increased energy efficiency could benefit all economic sectors, including power generation, transportation, infrastructure, and agriculture. The IPCC report says the efficiency of power generation around the world can be doubled from about 30% at present to about 60%. Switching from coal to natural gas is one way to achieve such improvements; not only would this reduce CO_2 emissions by 40% (because natural gas contains less carbon than coal), another 20% would be cut because of the higher efficiency of natural gas.

Improving energy efficiency in commercial and residential buildings is another major area of research. The Canadian government has started a "federal buildings initiative" that aims to reduce energy use by 15 to 20% or more. About $200 million has already been invested in more than 90 projects to increase the efficiency of 2500 federal buildings. The government projects this program ultimately will involve capital expenditures of more than $1 billion, create 20,000 person-years of employment, and save more than $160 million a year in reduced energy costs.

A number of provincial, territorial, and municipal governments have started similar programs. In Toronto, a pilot project known as the Better Buildings Partnership has been retrofitting more than 100 buildings, including offices, schools, and homes, with new lighting, windows, pumps, pipes, and insulation—all designed to conserve energy. The largest office building in Canada, the Bank of Montreal tower in downtown

Toronto, is part of the program. Toronto councillor Jack Layton, who was involved in setting up the program, said that energy costs in the retrofitted buildings are dropping dramatically, hundreds of people are employed in doing the retrofits, and CO_2 emissions are significantly reduced. Layton says the retrofitting can be done at no cost to the building owner because the cost is recovered in energy savings.

The pilot project was successfully concluded in the fall of 1998 and Toronto's city council voted to continue and expand it. Layton said the goal is to retrofit all buildings in the city on a continuous cycle. If 6 to 8% of the building stock is done annually, the entire city can be retrofitted in a decade.

Layton was also involved in another project to design an apartment building that would use only 15% of the energy consumed by a comparable building but would cost no more to build. This building was given an award by the United Nations for being the most sustainable building in the world. "If we can get it down to 15%, we can get it down to zero," said Layton, who participated in the design of another building that would achieve 90% energy savings through efficiency measures and the remaining 10% from a combination of solar, wind, and biomass energy. The building would take no electricity, oil, or gas from the utility companies; in fact, it could actually become a net exporter of energy. "It would become a little power plant that looks like just another building," Layton said. Unfortunately, the building was never built because of cutbacks to social housing by the Ontario government.

Improvements in heating, cooling, lighting, and appliances are also important in reducing energy demand. A report by the U.S. Department of Energy estimated that consumer appliances such as televisions and computers consume the energy output of about eight to ten large power plants when they're not even switched on. (These devices draw power as long as they're plugged in.) Even lightbulbs can be a source of big savings. For example, one study found a factory manufacturing energy-efficient fluorescent lightbulbs would cost around US$7.5 million to build, compared with US$135 million to build the power plants needed to generate the electricity the lightbulbs would save. Substantial reductions in energy use and CO_2 emissions, up to around 40%, can be achieved by switching from oil to natural gas for home heating or by using high-efficiency oil or gas furnaces.

Efforts are also being made to improve the fuel efficiency of vehicles, which are almost completely dependent on petroleum products. Transportation consumes about a quarter of all the energy used globally and is responsible for about 22% of CO_2 emissions from fossil fuels. (In Canada, transportation is responsible for about 26% of the country's emissions.) The IPCC report estimates that emissions could rise between 40 and 100% by 2025. A significant proportion of the transportation industry is devoted to transporting energy. In the U.S., for example, coal and petroleum products comprise about 30% of shipments by rail and 60% of shipments by water.

According to the IPCC, improved vehicle fuel efficiency could, by 2025, reduce greenhouse gas emissions 20 to 50% below 1990 levels without requiring changes in vehicle performance or size. It also said that if users would accept some changes in performance or size, the "energy intensity" (energy use per passenger-kilometer or tonne-kilometer) of the transportation sector could be reduced by 60 to 80% in 2025.

Technological advances in car design provide opportunities for reducing fuel consumption and greenhouse gas emissions. Using new lightweight materials and reducing vehicle and engine size help to reduce the mass of vehicles. The IPCC says concept cars have been developed with masses 30 to 40% below vehicles of similar size and performance. "The technical potential in 2010 for mass reductions without compromising comfort, safety and performance is probably in the range of 30% to 50% for most vehicle types." This could reduce the energy intensity of the vehicles about 15 to 30%.

A U.S. program called Partnership for a New Generation of Vehicles is aimed at making improvements that will increase fuel efficiency by a factor of three while preserving safety and comfort. According to R. J. Eaton, head of Chrysler, this "supercar" project is expected to produce vehicles early in the next century that will get 80 miles to the gallon and emit less than half the CO_2 of today's cars. Low emission vehicles were unveiled at the North American International Auto Show in early 1998, but it's not clear when they will come to market. Japanese automakers, on the other hand, have made their move. In the U.S., two-thirds of the 1998 Honda Accords were designed to meet California's strict emission standards. In Japan, Toyota has started selling its new Prius sedan, a hybrid vehicle that uses both gasoline and electricity and gets 66 miles to the gallon in stop-and-go traffic. It plans to introduce the vehicle in the U.S. in 2001.

These new vehicles aren't cheap; the Prius costs the equivalent of US$39,000 to make, but Toyota is selling it for just over US$17,000 in an apparent effort to penetrate the market quickly. To encourage consumers to buy energy-efficient cars, President Clinton has proposed to offer tax credits for vehicles that get two to three times the fuel economy of equivalent vehicles in their class.

Research is also being done on alternate fuels, such as ethanol produced from biomass (sugarcane, corn, or wood). The use of ethanol in vehicles would reduce greenhouse gas emissions an estimated 30 to 50% compared with gasoline.

Interest is also growing rapidly in the potential for using fuel cells in vehicles. Fuel cells use an electrochemical process to convert hydrogen and oxygen into electricity; when pure hydrogen is used, the only exhaust product is water vapor. Fuel cells can also run on hydrogen-rich fuels like methanol, natural gas, or gasoline, but they do produce some CO_2 emissions as a result. The U.S. Department of Energy says fuel cells could double the fuel economy of automobiles and cut their CO_2 emissions in half.

Most of the world's automakers are now doing research on fuel cells. Germany's Daimler-Benz is the most committed; it's putting US$450 million into the effort and says it wants to produce 100,000 vehicles by 2005. Daimler recently bought a 25% share in a Vancouver-based company, Ballard Power Systems, which has become a trendsetter in fuel-cell development. Ballard has already produced a prototype of a fuel cell-powered bus.

There are still many impediments to the introduction of fuels cells. One is money: a fuel cell engine costs 10 times more than a conventional one—but that's down from 1000 times more just five years ago. Another problem is developing a network for storing, transporting, and distributing hydrogen fuel, which according to some estimates could cost up to US$1 trillion to build in the U.S. alone. The IPCC says that fuel cells are unlikely to be widely used in cars before 2025, but if the impediments can be overcome, they could dominate motorized transportation by 2050.

Reductions in greenhouse gas emissions can also be achieved by switching the mode of transportation to less energy-intensive options. The IPCC notes that switching from car to bus or rail could potentially reduce energy use by 30 to 70%, but it adds that confidence in such measures is low because "human behaviour and choice play a central role and these factors are poorly understood."

Changing Lifestyles and Infrastructure

Weaning people from their cars will be a major challenge. Worldwide, cars and light trucks account for nearly three-quarters of the energy consumed by vehicles. The IPCC says that the global fleet of road vehicles rose 140% between 1970 and 1990 and is expected to continue rising dramatically throughout the next century. Estimates of the increase in vehicle population range from 60 to 120% by 2025 and 140 to 600% by 2100. Cars and light trucks are expected to make up three-quarters of these increases. The largest growth in light vehicle traffic has occurred in

Southeast Asia, Africa, Latin America, and central and eastern Europe.

Some countries see bicycles as part of the solution to these trends. Although motor vehicle use is growing rapidly in developing countries, bicycles still hold sway in many of them. According to the Worldwatch Institute, "in Asia, where more than half the world's people live, the use of bicycles dwarfs that of automobiles." China is the world leader with 41 million bicycles, while India has more than 12 million. However, the institute notes that both countries are "in the midst of a car boom" and warns that if developing countries repeat the mistakes of industrialized countries by building transportation systems around the private automobile, it could have "tragic consequences," including the paving over of cropland needed for food production.

Some developed countries are turning to bicycles, the WWI says. It estimated that 20 to 30% of all trips in Denmark and Holland involve bicycles, and bicycle use in Germany has increased 50% in the last 20 years. "Copenhagen provides free bicycles for use in the city. In the European Union, bicycles have been included for the first time in the comprehensive transportation plan. The United Kingdom has developed a plan to quadruple bicycle use by the year 2012."

North America might be more of a challenge. Consumers have a strong emotional attachment to their cars and it's questionable how willingly they'd give up or reduce their vehicle use to protect the environment, particularly when the environmental problem is not perceived as imminent. In recent years, the most popular and fastest-selling vehicles are gas-guzzlers like minivans, small trucks, and sport utility vehicles; many people buy SUVs for their image, with little thought to gas prices or fuel efficiency.* However, in Toronto, new bicycle lanes are being created and "we think we can increase the bicycle commuting population by 40,000," Jack Layton said. Even if Torontonians only use bikes for nine months of the year (few people hazard the winter on bikes— although with global warming, Toronto's streets might become less treacherous), this could have a significant impact in reducing greenhouse gas emissions.

One of the most important ways that greenhouse gas emissions from transportation could be reduced would be to redesign cities so that people don't have to travel long distances to work, shop, or play. Jessup says it would be possible to cut emissions in half by designing more compact

---

* In late 1998, California became the first state to adopt regulations requiring most of these larger vehicles to meet the same strict emissions standards that apply to cars, starting in 2004.

communities where residential and commercial activities are better integrated, making it feasible for people to walk or ride bicycles more frequently.

Switching to Renewable Energy

Renewable technologies—solar, wind, biomass, hydro, and geothermal systems—are already widely used around the world; in 1990, they made up about a fifth of primary energy consumption globally, most of it from wood burning and hydroelectric power. The IPCC report says that technological advances and declining costs are improving the prospects for these systems and renewables could satisfy a major part of the world's energy needs in the future.

The Worldwatch Institute says that energy production from renewable sources is "expanding at a breakneck pace." Wind generation grew by more than 25% annually during the 1990s. According to a U.S. Department of Energy study, three states—North and South Dakota and Texas—could provide enough wind power to supply the electrical needs of the entire country. China could double its current supply of electricity by using wind power. The WWI says that Germany and India have recently become "wind superpowers" and a Japanese firm is investing more than US$1 billion to install 1000 wind turbines throughout Europe in the next five years.

In September, 1998, Calgary's electrical utility became the first in Canada to offer residential customers the option of buying wind power. Demand has been high, even though users must pay an extra $7.50 a month. Some business clients are already receiving all their power from wind sources.

Solar power is growing rapidly too—by about 16% annually during the 1990s, ten times the growth rate of the oil industry, according to the WWI. Sales of photovoltaic cells jumped 40% in 1997 and now provide electricity to about 500,000 homes, mostly in developing countries. Programs to install solar systems on more than a million homes by 2010 have been started in the U.S., Europe, and Japan.

Even some of the world's major fossil fuel companies are getting into renewables. A U.S. natural gas company, Enron, has invested in wind and solar power companies. British Petroleum and Royal Dutch Shell also plan to invest hundreds of millions of dollars in renewable technology. BP plans to build a US$20 million plant in California to manufacture solar cells.

John Browne, CEO of British Petroleum of America, has said that solar energy and other renewables could provide up to half of the world's energy needs by the middle of the next century. Solar energy is likely to make the most significant contribution; while it's not yet commercially viable, in

1997 Browne expressed confidence that solar energy would be "competitive in supplying peak electricity demand within the next 10 years."

Advocates of renewable energy sources say they would be economically competitive a lot sooner if they were given the same tax breaks and incentives enjoyed by the fossil fuel industry. Jack Layton says the "carbon industry" has been sustained by "massive tax subsidies to their operations. If the alternative energy sources were given a level playing field—if either the subsidy is taken away from oil or the same subsidy was provided to alternatives—the carbon industry will collapse of its own weight."

A 1995 study done by the U.S. Environmental Protection Agency projected the emissions reductions that would occur as a result of reducing or eliminating subsidies to different economic sectors. It found that the greatest benefit would be obtained by reducing subsides in the energy and transportation sectors. Energy subsidies, including tax breaks and low-interest loans, amounted to between US\$9.4 and \$15.4 billion and their elimination could produce substantial reductions in CO_2 emissions. According to one model, subsidy elimination could contribute almost a third of the reductions needed to stabilize CO_2 emissions by 2050. This strategy would be even more effective in reducing emissions from the transportation sector, the study found. Cutting highway and parking subsidies could reduce emissions by more than 50 million tonnes below what they'd otherwise be in 2010.

The study concludes that the benefits of eliminating subsidies would not derive solely from reducing greenhouse gas emissions; the revenues freed up could be used to help the economy grow faster and in a more environmentally sustainable fashion. But the report acknowledged that subsidies "are notoriously persistent and difficult to reduce or eliminate. They often enjoy popular support and promote widely-accepted objectives."

In Canada, a 1997 report issued by the House of Commons committee on environment and sustainable development recommended that the government gradually phase out subsidies to the fossil fuel industry in favor of less-polluting energy sources. In a 1996 speech, committee chairman Charles Caccia noted that more than 80% of government support for energy goes to fossil fuels and nuclear energy, while only 7.2% goes to renewable energy. Moreover, renewables get only half the tax advantages of fossil fuels. To achieve a level playing field, he said, "we may have to offer to the renewable sector the same advantages or greater advantages than those...enjoyed by the non-renewable sector."

Nuclear energy is another alternative that doesn't produce greenhouse gases, and its proponents argue that switching to nuclear energy would be one way to achieve emission reductions. However, nuclear energy has its own environmental problems and is not accepted as a good alternative by most environmentalists.

Increasing Greenhouse Gas "Sinks"

The most obvious way to increase greenhouse gas sinks is to stop the galloping destruction of forests around the world and to implement reforestation programs. Industrialized countries have argued that they should receive credits for funding reforestation projects in developing countries to apply against their emissions targets under the Kyoto Protocol. Although such proposals are controversial, some reforestation projects are underway. For example, British Petroleum has invested in reforestation and conservation projects in Turkey and Bolivia. The World Bank and the World Wildlife Fund have started a program to increase the number of forests around the world protected from development from 6 to 10% by the year 2000. This would protect an additional 129 million hectares, including 50 million hectares in developing countries.

While reforestation is generally regarded as a good way to reduce atmospheric CO_2 levels, it must be remembered that forests can fairly easily be converted to a net *source* of CO_2 by other environmental factors. A report by the Woods Hole Research Centre notes that unless global warming is "gradual enough to avoid widespread mortality of forests, the additional releases of carbon caused by the warming itself, through increased respiration, decay, and fires, may cancel the intended effects of forest management."

Economic Policies and Instruments

One of the more contentious issues in the global warming debate concerns economic "instruments"—taxes, credits, subsidies, emissions trading, and joint implementation—that may be used by governments to influence changes in energy use and pricing. These mechanisms interact with the economy in complex ways and the debate over their net economic and environmental impacts is as fierce and divided as that over global warming itself.

Carbon or energy taxes start off with a disadvantage, since it's hard to find a word that inspires more dread than "taxes." There's much debate about the economic impact of carbon taxes, but most studies that have examined the issue suggest that the best way to minimize the impact is to use revenues raised by these taxes to reduce other taxes, such as income and payroll taxes. This "recycling" of tax revenue could stimulate economic growth as well as reduce emissions. According to the Worldwatch Institute, several European countries have already started reducing taxes on personal income and wages and increasing taxes on carbon emissions and vehicle ownership. Runnalls suggests Canada should try a tax trade-off: increasing gas prices and reducing the Goods and Services Tax, which is "the most unpopular tax in our history." While this would affect the energy industry, it would stimulate other industries by reducing sales taxes.

An article in *New Scientist* argues that carbon taxes are the only thing that would stop people from using more energy. Energy efficiency measures will not solve the global warming problem because they will have a paradoxical effect: by reducing the cost of energy, they will make energy-consuming technologies more affordable for more people and thereby will increase the overall demand for energy.

It doesn't matter how much sense carbon taxes make, however. In the current political environment, this strategy is a nonstarter in both Canada and the U.S. Powerful business and industry groups, particularly those associated with energy-intensive industries, and their political supporters fiercely oppose new taxes, fearing job losses and the migration of industries to countries without carbon taxes. To date, both Prime Minister Chrétien and President Clinton have shied away from any suggestion of implementing carbon taxes. In a speech to the Petroleum Club in Calgary, Chrétien pledged there would be no carbon tax. Before the Kyoto conference, economic advisers to President Clinton emphasized that energy taxes were not among the options that had been presented to him.

Instead, both governments intend to rely heavily on emissions trading and joint implementation or "clean development" projects, mechanisms permitted under the Kyoto treaty that countries may use to meet their emissions targets. An emissions trading program, which can operate domestically or internationally, is based on the premise that a regulatory body sets a limit on the total permitted emissions of a pollutant and then distributes "allowances," or emissions permits, among the sources of the pollutant (e.g., factories, power plants, industrial operations, etc.). Countries as a whole can also be considered sources.

Those who emit greenhouse gases can either use up their allowance or they can reduce their emissions and sell the unused portion of their permit to another source that cannot or does not wish to reduce their own emissions. Giving these permits a sale value provides an economic incentive for reducing emissions. Permits can be bought and sold domestically or internationally. The cost of reducing greenhouse gases varies from place to place, but the benefit—reducing total global emissions—does not depend on where the reductions take place; some operations may find it cheaper to pay for emissions reductions elsewhere rather than invest in pollution abatement themselves.

Emissions trading for other pollutants has already been tried with some success; for example, the U.S. set up a domestic trading system to reduce sulfur dioxide emissions from power plants. A few projects aimed at reducing greenhouse gases are already underway. Suncor Energy Inc. of Canada has agreed to purchase 100,000 tonnes of greenhouse gas reductions from Niagara Mohawk Power Corp. of New York; Niagara has agreed to use at least 70% of the net proceeds from the sale on other projects to reduce emissions. Ontario Hydro has an agreement to buy CO_2 emissions reductions from Southern California Edison. British Petroleum, one of the world's largest oil companies, is working with the U.S. Environmental Defense Fund to develop a system of voluntary emissions trading within BP's diverse operations.

It's anticipated that considerable emissions trading will take place between developed and developing countries because it's generally cheaper to improve older plants and factories in Third World countries than to upgrade more efficient operations in industrialized countries. Jonathan Wiener, a Duke University professor of law and environment and a former U.S. government economic adviser, has suggested that emissions trading makes a climate treaty more palatable for both industrialized and developing countries. For the former, it reduces the cost of participation by allowing them to make emissions reductions in a cost-efficient way; for the latter, it allows them to profit from selling emissions permits to developed countries.

According to Wiener, the transfer of money from rich to poor countries—which could exceed amounts currently spent on foreign aid—would not only encourage developed countries to participate in the treaty as their contribution to total global emissions rises, but also help them to "shift to a more prosperous but lower-emissions development path." The participation of developing countries is also important to allay concerns in developed countries about the competitive advantage of regions not subject to emissions controls.

Emissions trading has its critics, however. Among environmental groups and developing countries, there's concern that emissions trading will be a loophole for industrialized countries to do virtually nothing about limiting greenhouse gases at home. This is not the intent of the Kyoto Protocol, which specifies that trading should be "supplemental to domestic actions."

University of Toronto geography professor Danny Harvey debated this issue with Daniel Schwanen of the C.D. Howe Institute on CBC Newsworld. Schwanen argued that emissions cuts should occur "where they are least costly for the world as a whole....If you're having trouble cutting emissions, if it's very costly for you, you might as well compensate someone for whom it's easier to cut emissions." However, Harvey said there are many things industrialized countries can do to reduce emissions at little or no net cost. "Flexibility options" like emissions trading can "take the pressure off industrialized countries to do things that may be politically difficult but in fact are technically feasible and economically attractive in their own right...." To prevent this, it has been proposed that emissions trading be limited to a certain percentage of the developed countries' emissions targets.

Regardless, not all business leaders are enthusiastic about emissions trading with developing countries, raising concerns about the amount of money that would be transferred. Indeed, some skeptics depict emissions trading as little more than a disguised foreign aid grab and suggest that a purely domestic system is the best way to avoid this.

According to Ian Parry of Resources for the Future, governments must sell emissions permits, rather than giving them away, for an emissions trading program to pay off economically. "If the permits were auctioned off, the government could use the revenues to reduce other taxes in the economy. But if the permits were given out free...no revenue would be collected and there would be no potential for a revenue-recycling effect....Recent research warns that even modest emissions reductions might be especially costly if the policies used do not raise revenues for the government that are returned to the economy in other tax reductions."

Joint implementation is a concept introduced by the U.S. and supported by many industrialized countries because it would reduce the cost of achieving their emissions reductions targets under the Kyoto Protocol. It's generally cheaper to fund projects to reduce emissions in developing countries than to undertake such projects domestically.

Canada's National Action Plan on Climate Change endorses "sponsoring actions to reduce greenhouse gas emissions in other countries" and sets up a program called the Canadian Joint Implementation Pilot Initiative to encourage projects that are "voluntary, enhance opportunities to transfer energy efficiency technologies to other countries, produce results in a cost-effective manner, are sustainable over the longer term, and have other environmental, social, and economic benefits." Among other things, the program is intended to encourage the private sector to get involved in developing and disseminating new technologies to developing countries.

However, environmentalists and developing countries are less enthusiastic. As with emissions trading, the concern is that developed countries might use joint implementation to avoid their commitments under the

Kyoto Protocol. The World Wildlife Fund argues that joint implementation is "wide open to abuse" and that it "could allow rich nations to focus on reducing emissions abroad and evade their ecological responsibilities at home." Some developing countries have also expressed fears that if developed countries only invest in projects like reforestation, they'll get credit for reducing emissions but they won't have done much to improve the economies of developing countries. James Bruce notes that there's also concern that industrialized countries will fund the cheapest and easiest emissions reduction projects and, when developing countries have to start meeting their own targets after the turn of the century, only the most expensive projects will remain.

Success Stories

The greatest proof that reducing greenhouse gas emissions is not a prescription for economic disaster is that many cities around the world have achieved substantial reductions without falling into utter ruin. Some have done so well that they've put their national governments to shame. The cities are members of the International Council for Local Environmental Initiatives (ICLEI), an international environmental agency with a membership of more than 275 municipalities worldwide. ICLEI was founded in 1990 with the goal of building a worldwide network of municipal governments committed to improving the environment through actions taken at the local level.

In 1993, ICLEI focused its attention on global warming and urban emissions of greenhouse gases when it hosted the first Local Government Summit on Climate Change and the Urban Environment, where an international project called Cities for Climate Protection (CCP) was born. Its goal is to slow global warming and improve urban air quality by reducing urban greenhouse gas emissions. By the end of 1997, 182 cities were involved, representing 100 million people and nearly 5% of global greenhouse gas emissions. By the year 2000, the campaign hopes to involve enough cities to represent 10% of global emissions.

In November 1997, ICLEI announced that Toronto had beaten 149 other cities by reducing its emissions by 7.85 million tonnes between 1990 and 1996. The core of the city achieved a 6% reduction during a period when Canadian emissions as a whole increased about 10%. Berlin, Germany, was a close second with a reduction of 7.49 million tonnes. The reductions achieved by other cities in the top 10 were considerably lower, ranging from 2.62 to 1.12 million tonnes. Germany had four cities in the top group, Canada had two (Edmonton, Alberta, was 7th), the U.S. had two, and the Czech Republic and Finland had one each.

In a report prepared for the Kyoto conference, Philip Jessup, director of the CCP program, noted that participating cities were taking their commitments seriously and making progress in achieving their targets and timetables. About half of the 48 cities that had set emissions targets were striving to meet the so-called "Toronto Target"—reducing emissions by 20% over a period of about 15 years. In 1989, Toronto became the first municipality in the world to commit to greenhouse gas reductions, setting a goal of cutting its 1988 emissions level by 20% by the year 2005 and many other cities have followed suit. Others have less ambitious targets, but one in five of the CCP participants, including most of the Canadian and U.S. cities, are aiming for cuts of 20% or more — as high as 50% in some cases.

At first glance, it might appear that fighting a global problem like climate change at the local level is a contradiction in terms, if not an exercise in futility, but this is far from the case. Cities are significant players in the effort to slow global warming and, in many cases, are demonstrably more committed and successful in reducing emissions than their national governments have been. Jessup's report notes that this is particularly true for the Canadian and U.S. cities in the ICLEI program.

Cities have a tremendous stake in preventing global warming. Nearly half the world's population now lives in cities and the United Nations estimates that will rise to 65% by 2025. More than three-quarters of Canadians live in cities. The density of urban populations exposes large numbers of people and a great deal of valuable property to extreme weather events. Many cities are located near and dependent on rivers, lakes, or the ocean, so rising or falling water levels are of great concern. Many cities are also plagued by air pollution, and there's a growing recognition that reducing greenhouse gas emissions has the added benefit of improving urban air quality and public health.

Cities are also uniquely positioned to encourage significant changes in infrastructure and human behavior that can have a tremendous impact in reducing emissions. Local governments are responsible for managing and regulating many activities that produce emissions, such as public transportation, energy utilities, and industrial production. They also control urban development and land use patterns; decisions about the density, layout, and mixture of services and activities in residential, commercial, and industrial areas can significantly affect both energy use and emissions levels.

"Local governments must and can play a major role in implementing the [FCCC], primarily because urban emissions of greenhouse gases are significant worldwide," states the ICLEI report. One of the major incentives for local governments to undertake emissions reduction programs is the ancillary benefit of improving urban air quality, reducing traffic congestion, creating jobs, and improving economic competitiveness. "They especially realize that implementation of energy efficiency measures, which can be largely financed with private capital, is a win-win situation for taxpayers, local businesses and labor. This insight has spurred them into action."

Toronto's experience provides several excellent success stories. A program of recovering methane gas from landfill sites had a dramatic impact, accounting for 79% of the reduction in the city's emissions. The methane is used to fire an electricity generating plant that would otherwise have burned coal. Recycling and retrofitting municipal buildings to make them more energy efficient accounted for another 13%. Other measures included water conservation and increasing the energy efficiency of street lighting.

One innovative project that got underway in 1998 is a district cooling system that will pump cold water drawn from the depths of Lake Ontario through pipes that circulate around buildings in the downtown core, eliminating the need for individual fossil fuel–powered air conditioning systems in each building. "Most of the cooling is [currently] done by old electric chillers," said Toronto councillor Jack Layton. "They use a lot of electricity, marginal electricity that's provided by burning coal." The district cooling system would replace this equipment with efficient, centrally located, gas-fired chillers to cool water. "The plan is to connect the buildings with insulated pipes and run very cold water through it, just above freezing. Then you have centralized cooling plants that operate at peak efficiencies, managed by a team of specialists working to keep it at the most efficient level."

This cooling system is expected to use only about one-tenth of the electricity used for conventional air conditioning and will reduce Toronto's annual CO_2 emissions by some 30,000 tonnes. The estimated $80 to $100 million cost of the project will be paid for by customers who use the cooling system. Layton said that about $10 million will be spent to extend the city's water intake pipe deeper into the lake so it will withdraw colder water.

There were concerns that warm water discharged from the system into Lake Ontario would affect the ecology of the lake. But analysis found the discharged water would not be much warmer than it is at present and, in fact, would not be as hot as water discharged by the coal-fired plant that would have provided electricity to power conventional air conditioning systems used by downtown buildings. "The result is that you actually end up putting less heat into the lake than you did by using electrical chillers," said Layton.

He described this project as a "triple win situation." It's a winner in terms of reducing CO_2 emissions and reducing the warming of Lake Ontario, and it's an economic winner, too. Layton says building owners will find the system cost competitive with their current cooling systems and their costs will drop over time because they don't have to maintain their own equipment and staff. Climate change, he says, "doesn't have to be the reason they sign on the dotted line."

Toronto is now trying to persuade the Canadian government to encourage expansion of local programs to other cities. Layton describes this as the "Lilliputianism" strategy for taking down the "carbon giant. It's the towns and cities and individuals with photovoltaics on their roofs that will do it."

Demonstrations of economic payoffs from reducing local greenhouse gas emissions are extremely important in light of the controversy over cost/benefit trade-offs of avoidance measures. These grassroot successes are the only really encouraging development in the long, torturous saga of humanity's reluctant efforts to come to grips with global warming. Certainly what's happened at the national and international level in the past decade gives little cause for optimism.

It's clear there are many ways we can cut greenhouse gases—if we so choose. It won't be easy or cheap, but emissions reductions *are* achievable, both technically and economically. What's lacking is the political will. As the next chapter shows, for most of the 1990s, world governments have engaged in an elaborate political gavotte whose purported objective is the avoidance of human-induced climate changes that pose a danger to human society and the natural environment. The word "purported" is apropos, because to nonparticipating observers, the spectacle of international climate negotiations is at once comical and dispiriting—an effort that appears to be characterized mostly by an intense and remarkably inventive struggle to avoid avoidance at all costs.

ten

THE ROCKY
ROAD FROM RIO:
The Struggle for a
Climate Treaty

The international climate treaty signed at Kyoto in December 1997 was the culmination of a long and difficult political process that was four decades in the making. Warnings that greenhouse gases could warm the earth's climate, with potentially dangerous impacts on human society and natural ecosystems, started as early as 1957, when scientists at the Scripps Institute of Oceanography found that the oceans were not absorbing much of the CO_2 being emitted into the atmosphere. They warned that warming caused by the emissions would constitute a "large-scale geophysical experiment" on the earth's climate.

It was not until two decades later, in 1979, that momentum to do something about global warming picked up. At the First World Climate Conference held that year, scientists in attendance called for efforts to prevent human-induced climate changes from having adverse impacts. The UN World Climate Program was born. A panel of the U.S. National Academy of Sciences (NAS) stated that adopting a "wait and see" approach could mean "waiting until it is too late." In 1983, an NAS report suggested that doubling of CO_2 would raise the earth's average temperature by 1.5 to 4.5°C and the Environmental Protection Agency warned of disruptions to environmental, agricultural, and economic systems. In 1988, the IPCC was formed with a mandate to aid policymakers by assessing the science of climate change. In 1990, it released its first report, which provided a firm scientific foundation for the theory that increases in greenhouse gas concentrations since pre-industrial times had warmed the climate, but said it was not yet possible to determine the relative contributions of human activities and natural variability.

With the release of the first IPCC report, the momentum for an international climate change treaty accelerated. By 1990, the international community of climate researchers was reaching a widespread consensus that global warming was a serious problem with potentially significant consequences and that political action was needed to deal with the threat. More than 700 scientists attending the Second World Climate Conference sponsored by the United Nations Environment Programme, the World Meteorological Organization, and the International Council of Scientific Unions, issued a statement saying there was "a clear scientific consensus" on estimates of the range of global warming to be expected during the 21st century. Calling for an international treaty, they warned that without cuts in greenhouse gas emissions, the predicted warming would stress natural and social systems to an extent "unprecedented in the past 10,000 years."

In December 1990, the UN General Assembly started the international consultation process that would ultimately result in the introduction of the Framework Convention on Climate Change (FCCC) at the Rio Earth Summit 18 months later. That year, 700 members of the NAS, including 49 Nobel Prize winners, issued a statement saying "there is broad agreement within the scientific community that amplification of the Earth's natural greenhouse effect by the buildup of various gases introduced by human activity has the potential to produce dramatic changes in climate." They said immediate action was necessary to ensure that "future generations will not be put at risk." In 1992, another NAS report was published, which stated that the scientific evidence for global warming was strong enough to justify "prompt responses" to deal with it. It argued that reducing emissions would not be too costly and would provide "insurance protection against the great uncertainties and the possibility of dramatic surprises."

Rio Earth Summit, June 1992

The FCCC was adopted by the United Nations and signed by 154 countries and the European Union during the Earth Summit in Rio. It came into force two years later, after being ratified by 50 countries; by the end of 1997, nearly 170 countries had ratified it. The treaty contained no binding targets for reducing emissions or timetables for achieving those reductions. However, developed countries and countries with economies "in transition" (designated "Annex I" countries) made nonbinding commitments to try to reduce their greenhouse gas emissions to 1990 levels by the year 2000. Annex I countries included the U.S., Canada, Japan, Australia, New Zealand, the countries of western and eastern Europe, Scandinavia, and the Russian Federation.

Developing countries, designated "Annex II" countries, included primarily countries in Asia, South America, and Africa, including two of the most populous, China and India. Under the FCCC, Annex I countries agreed to transfer funds and technology to Annex II countries to help them deal with climate change. The Global Environment Facility was set up within the World Bank to distribute funds to green projects in developing countries.

Annex II countries were not asked to make commitments to reduce their emissions. This was viewed as a matter of equity, a recognition of their lower economic status and the fact that most past and present emissions come from developed countries. The FCCC secretariat noted that developed countries should "take the lead" in the fight against global warming, noting they are "responsible for over two thirds of past emissions and some 75% of current emissions, but they are best positioned to protect themselves from damage. Developing countries tend to have low per-capita emissions, are in great need of economic development, and are more vulnerable to climate change impacts." When one considers the percentage of global CO_2 emissions produced by individual countries, compared with their percentage of the world's population, the per capita disparities between developed and developing countries become clear; emissions from Canada, the U.S., and Australia are 4 to 5.5 times their percentage of the world's population, while those from Europe or Japan are 2 to 2.5 times. Comparable figures for China and India are 0.6 and 0.2 times. The FCCC recognized that developing countries needed to improve their living standards and that they'd need technological and financial assistance from developed countries to achieve growth in ways that were less harmful to the climate.

At the time, there was little emphasis on the undeniable fact that, ultimately, developing countries would have to join the battle against global warming or the battle would certainly be lost. While per capita emissions in developing countries are low compared with those in developed countries, many have large and still-growing populations, so their total contribution to greenhouse gas emissions is too large to ignore. And as their economies have grown in recent years, their relative contribution to the global total has also increased; for example, between 1950 and 1985, China's emissions grew from 2 to 11% and India's grew from 1 to 3%, while emissions in other developing countries rose from 16 to 28%. Meanwhile, the U.S. contribution dropped from 45 to 22% and Britain's dropped from 7 to 3%. The Worldwatch Institute estimated economic growth in developing countries in 1996 at three times that of industrialized countries. Assuming these trends continue, it's clear there's little hope of reversing global warming unless developing countries can achieve emissions reductions within the next century.

Although the principle of "differentiated" responsibilities for developing and developed countries was accepted in theory by the signatories to the FCCC, it would later become a major source of dissension in the squabbling that dogged international negotiations up to and beyond the Kyoto conference.

The FCCC articulated another important principle that would also hit rough patches on the road to Kyoto. It stated that "the Parties should take precautionary measures to anticipate, prevent or minimize the causes of climate change and mitigate its adverse effects. *Where there are threats of serious or irreversible damage, lack of full scientific certainty should not be used as a reason for postponing such measures,* taking into account that policies and measures to deal with climate change should be cost-effective so as to ensure global benefits at the lowest possible cost." [Emphasis added.] The idea that scientific uncertainty should not be invoked as an excuse for inaction was destined to become one of the most contentious elements of the global warming debate, particularly in the U.S. where the most vocal and relentless opponents of the treaty explicitly rejected the precautionary principle by demanding that a very high standard of scientific proof be met before measures to reduce greenhouse gas emissions are implemented. (This debate is examined in greater detail in Chapter Twelve.)

The Rio conference ended with a great deal of self-congratulatory hoopla. It appeared to many participants that humanity was well embarked on a plan to turn back the threat of global warming. Unfortunately, as events unfolded, it became increasingly clear that this optimism was woefully premature. The heady enthusiasm of Rio would, just a few years later, give way to the sober realization that the global commitment to reducing greenhouse gases was half-hearted at best. Nations remained as divided as ever about what to do about global warming, how much should be done, who should do it, and, especially, what it would cost. Moreover, an ever-widening chasm between industrialized countries and developing countries threatened to bring the whole effort to a grinding halt.

COP-1, Berlin, March–April 1995

These problems were already apparent in early 1995 when the parties to the convention gathered in Berlin for the first of a series of meetings (known as Conferences of the Parties or COP) that ultimately produced the Kyoto Protocol. The voluntary commitments in the FCCC dealt only with emissions up to the year 2000; nothing was said about what would happen after that. Moreover, it was clear that even if the promises of the FCCC were kept, they would not achieve the objective of avoiding dangerous interference with the climate. Even if *emissions* were reduced to 1990 levels, *concentrations* (the total accumulation of greenhouse gases in the atmosphere)—and consequently warming—would continue to increase for a century or more. To stabilize atmospheric concentrations at 1990 levels, very large reductions in emissions would be required—as much as 60%, according to an IPCC estimate.

While it was politically impossible to advocate such large cuts, there were suggestions that a goal of 20% by 2005—the so-called "Toronto Target"—would be a good starting point. However, the UN Climate Change bulletin noted that even this target, without further measures after 2005, would lead to a 2°C rise in global temperature by the end of the next century. As it turned out, even a 20% target was dreaming in technicolor.

As the labyrinthine process of international treaty-making kicked into high gear, countries began wrangling not only over targets but also over emissions trading, joint implementation, and differentiation (allowing developed countries to adopt different emissions targets as long as the combined effort met an overall target). Various special interest groups with different and often conflicting agendas took interest in the proceedings, seeking to ensure that their points of view were represented. The Alliance of Small Island States (AOSIS), fearing their member countries were facing disaster from rising sea levels, pushed for strong measures to combat global warming. The Organization of Petroleum Exporting Countries (OPEC), fearing the negative impact of reduced international demand for petroleum products on their oil-dependent economies, emphasized the lack of scientific certainty and urged a go-slow approach. Both groups included countries that belong to the Annex II bloc of developing countries, yet their positions on emissions reductions could hardly have been further apart.

Environmental groups lobbied for stricter controls while business interests weighed in on both sides of the debate; energy-related industries and their lobby groups emphasized the negative economic impacts of reducing emissions, but the global insurance industry was becoming increasingly alarmed by their mounting losses to weather-related disasters. Representatives of European insurance companies pressed for limits on greenhouse gas emissions, saying that global warming could bankrupt insurers if the incidence of extreme weather events continued to increase.

Although they acknowledged the uncertainties in global warming predictions, they said they couldn't afford to wait for proof.

The Berlin meeting got off to a rocky start when oil-exporting nations demanded that consensus be reached on all decisions, which would effectively have given them a veto. Some environmental activists suspected these countries had ratified the FCCC primarily to gain access to the bargaining table so they could disrupt the discussions as much as possible. There certainly was no lack of dissension in Berlin. *The Economist* described the talks as "viscous with regional politics and national self-interest." In fact, practically every issue that supposedly had been settled in Rio had taken a turn for the worse. The Annex I countries were not reducing greenhouse gas emissions; in fact, emissions were *rising*, which rendered the talk of setting a 20% target for emission reductions almost ludicrously optimistic. Nor had developed countries made much headway in transferring money and technology to developing countries to support environmental projects. From 1994 to 1997, the World Bank's Global Environment Facility received pledges for only about US$2 billion, far short of the estimated tens of billions needed to meet demand from developing countries.

Finally and most significantly, the concept of differentiated responsibilities for developed and developing countries began to fall apart. Developing countries, especially China and India, fiercely resisted the idea of being required to set targets and timetables for reducing their emissions; India's environment minister, Kamal Nath, said developing countries should not suffer economically by being forced to address a pollution problem they had not created. The developing countries issued a statement arguing that the Berlin conference should set "no new commitments whatsoever" for Annex II countries. Some of the industrialized countries, particularly the U.S., began to grumble openly about this. Daniel Esty, a former official of the U.S. Environmental Protection Agency who'd been with the U.S. negotiating team at Rio, told the *New York Times* that the impact of emissions from developing countries "can no longer be ignored as they have been to date and were at Rio."

The two groups were also at odds over the joint implementation concept proposed by the U.S. Developing countries were dubious, believing that if developed countries were allowed to fund projects to reduce emissions in developing countries, they would use it as an excuse not to do anything at home. There was also concern that funding projects like reforestation would do nothing to advance economic growth in developing countries.

Meanwhile, developed countries were squabbling among themselves. European nations were pushing to strengthen the Annex I commitments, while the U.S., Canada, Australia, New Zealand, and Japan stymied efforts to set specific targets and timetables beyond the year 2000 and blocked a proposal to set a target of 20% below 1990 levels by 2005. (This marked Canada's loss of the environmental "good guy" reputation it had acquired at Rio.) The vision of global cooperation embodied in the FCCC was fast becoming unraveled. It appeared that Berlin might undo practically everything that had been accomplished at Rio, modest as those accomplishments were. At one point, Timothy Wirth, then the U.S. Under-Secretary of State for Global Affairs, remarked: "We will be fortunate if we can keep this treaty alive."

The political embarrassment of coming up empty-handed after nearly two weeks of international negotiations finally greased the wheels of diplomacy enough to produce what became known as the Berlin Mandate. This was an agreement to spend the next two years working on a protocol that would establish binding post-2000 targets and timetables for reducing emissions in developed countries. This protocol would be presented at the COP-3 meeting in Kyoto in December 1997.

Developing countries remained exempt from commitments to reduce their emissions, despite urging by the U.S. and Canada that they agree to making such commitments in the future. It was decided this issue would be revisited after Kyoto. (However, as events unfolded over the next 18 months, this question, far from being put on the back burner, became the central focus of efforts to shoot down the treaty, particularly in the U.S.) Developing countries relented somewhat in their opposition to joint implementation (later renamed *activities implemented jointly* or AIJ), which allowed several voluntary pilot projects to go ahead—a development the U.S. viewed as a victory.

Reactions to the outcome of the Berlin meeting were mixed. Sheila Copps, then Canada's environment minister, declared it a great success, while Louise Comeau of the Sierra Club of Canada described it as "a very big failure," though she acknowledged to *Maclean's* that it was "the best we could squeeze out of the political process at this time." The Global Climate Coalition, a U.S.-based lobby group that opposes cutting greenhouse gas emissions, criticized the conference for giving developing countries such as China and India "a free ride."

Summing up the Berlin conference, the *New York Times* made an observation that was almost prophetic, but not precisely in the way it was intended. Commenting that the Berlin Mandate had pushed the treaty process into a "vastly more difficult phase," the *Times* said that "while the coming negotiations pose a new order of difficulty, they can shift into a higher gear at any time if scientists turn up clear evidence that humans are indeed changing the climate."

Fireworks over the IPCC Report, 1995–1996

For many in the international scientific and policy community, "clear evidence" turned up in early 1996 with the release of the second IPCC report and its infamous statement about the "discernible human influence" on the climate. *Nature* commented that, without this statement, the prospects for an international agreement would have been "considerably dimmer." But it wasn't long before skeptics attacked the IPCC's work as so much "junk science." Even before the report was officially published, a statement issued by the World Energy Council (WEC), which represents energy industries in 100 countries, argued that the conclusion about discernible human influence, contained in the Summary for Policymakers, was "not adequately supported by the scientific papers…which adopt a more tentative tone" and that the summary had been subject to "political influence." It also criticized scenarios used in the study as "seriously defective" and said the treatment of aerosol cooling was "questionable." It deemed the report to be of little practical use to policymakers. Interestingly, however—and contrary to the stance taken by many in the energy industry—the WEC argued these uncertainties should not be used to excuse complacency or inaction, saying that precautionary "minimum-regrets" measures to respond to the "possible risk of climate change" were warranted.

One of the most vituperative battles erupted when the Global Climate Coalition, the Science and Environmental Policy Project, and other skeptics picked up the allegation that a chapter in the IPCC report had been revised in an unauthorized manner to underplay the scientific uncertainties and provide unwarranted support for the statement attributing climate change to human activities. An early draft of the summary, which had been leaked on the Internet in September 1995, was worded differently, saying that the observed warming "is unlikely to be entirely due to natural causes and that a pattern of climatic response to human activities is identifiable in the climatological record." This statement was hailed by environmentalists as "the smoking gun," so it was not surprising that it caused a major brouhaha in December 1995 at an IPCC conference in Spain, prior to the official release of the report. Delegates from two oil-producing nations, Saudi Arabia and Kuwait, reportedly supported by a U.S.-based fossil fuel lobby group, argued strongly against the statement; the eight-hour standoff that ensued was resolved only when the IPCC agreed to add a footnote saying that "two countries" considered the evidence linking climate change to human activities as "preliminary." In all the wrangling, the human influence morphed from "identifiable" to "discernible."

Another major battle erupted at the meeting when the authors of one chapter of the IPCC report were instructed to make revisions to the text to clarify ambiguities in the wording about the detection of climate change and its attribution to human activities. Global warming skeptics immediately pounced on this as evidence that the report was essentially being doctored to toe the IPCC's political party line. A fierce war of words over the IPCC's scientific integrity ensued. Frederick Seitz of the conservative George C. Marshall Institute later wrote in the *Wall Street Journal* that he had "never witnessed a more disturbing corruption of the peer-review process."

These charges were vehemently denied by the authors of the chapter and other IPCC participants, including Sir John Houghton, head of the group of scientists who worked on the chapter. He said the revisions were made to improve the clarity of the text and were in accordance with established IPCC procedures. They rejected the skeptics' view that the scientific integrity of the report had been compromised by political pressure, pointing out that the chapter still contained many references to the scientific uncertainties associated with climate change. They added, moreover, that the statement about the "discernible" human influence in the Summary for Policymakers had been unanimously approved by nearly 100 countries participating in the IPCC process and that, in fact, a number of delegates felt the statement was too weak. However, some scientists have expressed concern that the summary (and media reports based on it) failed to adequately address all of the caveats and uncertainties spelled out in excruciating detail in the full report. Thus, while it can certainly be argued that the IPCC report shifted climate treaty negotiations into "high gear," it also lit the fuse of a ferocious public relations war designed to forestall binding emissions targets for developed countries. A mixed legacy, to be sure.

COP-2, Geneva, July 1996

Notwithstanding the skeptics, the IPCC report was endorsed in July 1996 at the second Conference of the Parties in Geneva. A statement described it as "the most comprehensive and authoritative assessment of the science of climate change, its impacts and response options now available" and said it provided a scientific basis for "urgently strengthening action," particularly by Annex I countries, to cut greenhouse gas emissions.

The Geneva conference was also notable for the high-profile intervention of the insurance industry, which had just created a program called The Insurance Industry Initiative for the Environment in cooperation with the UN Environment Programme. Some 60 insurance companies, mostly European, had signed a "statement of environmental commitment" that required them to make "every realistic effort" to achieve a balance between economic development and environmentally sound management and to incorporate environmental considerations into their business practices. (Notably absent was major representation from North American insurers. Frank Nutter, of the Reinsurance Association of America, said U.S. insurers feared that legal liability might result from signing the statement, but European insurers had obtained legal advice that it would not.)

In an effort to influence negotiations, the insurance group issued a paper stating their belief that human activities were already affecting the climate and that human-induced climate changes would probably lead to more extreme weather in some regions, with serious implications for insurers. The statement strongly supported the precautionary principle, saying that "research is needed to reduce uncertainty but cannot eliminate it entirely." Arguing that reducing greenhouse gas emissions would be the "most efficient precautionary measure," the insurers "insisted" that negotiators forge a treaty that would "achieve early, substantial reductions." Finally, they said research was urgently needed to establish what concentrations and rate of increase in greenhouse gases would be "dangerous."

The forcefulness of this statement caused a stir, since it pitted two huge business giants—the global energy industry and the global insurance industry—against each other. Writing in the *Washington Post*, Evan Mills of the Lawrence Berkeley National Laboratory noted that the insurance industry has annual revenues four times that of the oil industry so its concerns "can hardly be ignored. The scientific uncertainties about climate change are a side issue for insurers. For this industry, the absence of certainty is not synonymous with the absence of risk."

The COP-2 conference came up with the "Geneva Declaration," which supported acceleration of the process of developing legally binding emissions targets for developed countries after the year 2000 (with dissents from OPEC, Australia, New Zealand, and the Russian Federation). There were still no specifics about what those targets would be or whether different targets would be set for different countries—issues that would soon become highly contentious—but the evolution from the nonbinding provisions of the FCCC to a legally binding protocol was an important step. A legal instrument, which would contain penalties for noncompliance, wouldn't be as easy to live with—or ignore—as the voluntary commitments undertaken at Rio. Underscoring the significance of this step was the fact that the 1996 review of Annex I countries' CO_2 emissions showed them to be rising, not falling. It was becoming painfully obvious that most Annex I countries were not only failing to meet their voluntary targets, but, in fact, were still headed in the wrong direction.

British Petroleum Breaks Ranks, May 1997

In a speech at Stanford University, John Browne, CEO of British Petroleum of America, dropped a bombshell by coming out strongly in favor of precautionary measures against global warming. He stated that there "is now an effective consensus among the world's leading scientists and serious and well-informed people outside the scientific community that there is a discernible human influence on the climate and a link between the concentration of carbon dioxide and the increase in temperature." While there were still uncertainties about the amount of warming and especially its consequences, he said it would be "unwise and potentially dangerous to ignore the mounting concern. The time to consider the policy dimensions of climate change is not when the link between greenhouse gases and climate change is conclusively proven...but when the possibility cannot be discounted and is taken seriously by the society of which we are part....We in BP have reached that point."

Browne didn't advocate immediate, radical cuts to CO_2 emissions, saying such measures would fail because of their economic impact and because they'd be viewed as discriminatory, especially by developing countries. Instead, he suggested implementing measures that would balance economic development with environmental protection, including technological innovations applied all over the world. He emphasized that developing countries must not be prevented from improving their standard of living. BP thus became the first major oil company to formally endorse the precautionary principle and to accept a link between greenhouse gases and warming temperatures.

United Nations Earth Summit+5, New York, June 1997

When the UN convened a special session of the General Assembly in the summer of 1997 to assess progress since the Rio conference five years earlier, what emerged was not a pretty sight. In contrast to the high spirits and optimism at Rio, this conference was full of pessimism, recriminations, admissions of failure, and mea culpas. Paradoxically, the embarrassment caused by failure, far from stiffening the resolve of the world's nations to do better, instead induced a palpable reluctance to make any more promises that were going to be hard to keep.

Three major sources of failure and dissension emerged at the conference:
- the failure of voluntary programs to stabilize greenhouse gas emissions in developed countries at 1990 levels
- the reluctance of many developed countries to specify binding targets and timetables for future reductions in greenhouse gas emissions (the

Berlin Mandate required them to announce their negotiating position by June 1997, but only the Europeans did so)

- the growing rancor between developed and developing countries over the failure of developed countries to provide financial assistance and the refusal of developing countries to commit to reducing their own emissions.

With few exceptions, developed countries were forced to admit openly that they weren't going to meet the goal established at Rio of stabilizing their emissions at 1990 levels by 2000. In fact, 1996 set a record for CO_2 emissions—about 6.35 billion tonnes—producing the largest one-year jump (2.4%) since 1988. The World Energy Council estimated that global emissions had risen 6.4% between 1990 and 1996. This figure was lower than it might have been because the economic and political turmoil in central and eastern Europe and among the former Soviet republics caused a 31% drop in their emissions. By 1996, the rate of this decline had slowed and, in some eastern European countries, emissions had started climbing again. Among OECD countries, CO_2 emissions over the six years climbed 7.8%. (All of the OECD countries, except for Mexico, are Annex I countries.) So much for high hopes that Rio would at least start turning the ship around.

By the end of 1995, both U.S. and Canadian emissions were up about 8% from 1990 and were on track for an increase of roughly 13% over 1990 levels by 2000. Most other developed countries were in the same boat; for example, Japan's emissions rose 8.3% between 1990 and 1995. The major exceptions were Britain and Germany, which came close to meeting the Rio targets.

European countries came to the Rio+5 conference urging acceptance of even steeper emission cuts: 15% below 1990 levels by 2010. This was the first official proposal for specific targets offered by any of the industrialized countries. Newly elected Prime Minister Tony Blair said Britain was prepared to go as high as 20%. In a chiding reference to the reluctance of the U.S., Canada, and Australia to set specific targets, he noted that "the biggest responsibility falls on those with the biggest emissions. We in Europe have put our cards on the table. It is time for the special pleading to stop and for others to follow suit." (Although Canada accounts for only 2% of global emissions, it is second only to the U.S. in per capita emissions; this is due in large part to its cold climate and resource-intensive economy, but also to its failure to use energy efficiently.)

Blair's comments did not sit well; the other developed countries quickly pointed out that the apparent success of Britain and Germany in meeting the Rio targets was largely the inadvertent consequence of unrelated economic and political factors. Germany's emissions statistics benefited from the collapse of inefficient and highly polluting industries in communist East Germany after the fall of the Soviet Union. Britain's drop in emissions came when former Conservative governments privatized the country's power and coal industries, precipitating a shift from coal to cleaner fossil fuels like oil and natural gas. The World Energy Council report noted that the factors that enabled Germany and Britain to reduce emissions during the first half of the 1990s were no longer having the same impact and emissions were rising again. The report commented that the European Union, despite its rhetoric, was on track to exceed 1990 emission levels by 8% in 2000.

The U.S., Canada, Japan, and Australia resisted the EU's efforts to pin them down on what targets they would propose at Kyoto. They used the word "realistic" a lot, however, and the European targets clearly did not qualify. In his speech, President Bill Clinton acknowledged the U.S. "must do better ...and we will" but beyond that would say only that the U.S. would bring to Kyoto "a strong American commitment to realistic and binding limits that will significantly reduce our emissions of greenhouse gases." In a similar vein, Prime Minister Jean Chrétien told the assembly that Canada would come up with "a practical, step-by-step plan with realistic interim, medium-term targets." Canada would support binding targets, he said, but evaded questions about what they should be, saying the federal government would discuss the matter with the provinces and other stakeholders before developing a negotiating position for Kyoto.

The small island states appealed for a 20% cut below 1990 levels by 2005. Abdul Gayoom, president of the Maldives, told the assembly he'd left the Rio meeting "confident that we had an agreed agenda that would save not only us but the whole world....Today I leave here with the fear that unless we all act now with a renewed commitment, my country and many like it would neither have a voice nor a seat at a future Rio conference." One observer summed up the attitude of the industrialized nations with the phrase: "Too bad, Tuvalu." (Tuvalu is a group of nine small coral atolls in the South Pacific with a population of about 10,000. Their highest point is only 5 meters above sea level.)

Trouble was brewing on another front. A growing rift between developed and developing countries burst into the open at the UN meeting for two reasons. The first concerned promises of financial aid; the goal established at Rio was for developed countries to increase their aid funding from about 0.33% of GNP to 0.7%, but by 1997, the average had actually dropped to about 0.27%—yet another example of the road from Rio heading in the wrong direction. While private investment had increased, several leaders of developing countries said this aid was going to only a few of the more prosperous nations and was not benefiting the poorest ones. They rejected the idea that such private investment should be considered a substitute for the funding promised by the governments of developed countries. "The decisions of Rio remain empty slogans," Elfatih Mohamed Ahmed Erwa of Sudan told the assembly.

However, the U.S., which had already disavowed the 0.7% target, was not about to change its mind. A senior U.S. official, Eileen Claussen, told the *San Francisco Chronicle* that industrialized countries weren't going to give that much and developing countries would have to be "more practical." However, during his speech to the assembly, President Clinton pledged that the U.S. (which by 1997 ranked fourth among the Annex I industrialized countries in development aid and last in terms of percentage of GNP) would give a billion dollars over the next five years to help developing countries with energy and resource projects.

The second issue causing rancor was the continuing exemption of developing countries from having to commit to emissions reductions. They weren't about to budge on this matter as long as developed countries were so clearly failing to meet the goals agreed to at Rio; in their view, developed countries had created the problem and had to make a serious attempt to deal with it before developing countries could be expected to get on board. This sentiment was echoed by Prime Minister Tony Blair, who said that "our targets will not be taken seriously by the poorer countries until the richer countries are meeting them."

However, support for this position among developed countries was eroding amid concerns that the exemption from emissions targets would give developing countries an economic advantage and in light of alarming statistics about how fast their emissions were growing, especially among countries that were industrializing rapidly. An analysis by the World Energy Council (WEC) in July 1997 indicated that CO_2 emissions from developing countries had increased 32% and their share of global emissions had increased from 29 to 36% between 1990 and 1996. Emissions from the Asia/Pacific region had climbed 37% while Middle East emissions had grown 40%. The WEC projected that developing countries would become the "primary emitters" of greenhouse gases by 2020 and "their share will continue to grow, though at lower per capita emission levels than in industrial countries." It said that consumption of the world's primary energy by developing countries would rise from about 35% in 1997 to 60% by 2050. It was not surprising, therefore, that many opponents of the climate treaty balked at the idea of continuing exemptions for developing countries.

This dispute has defied resolution because neither side can view the question of emissions reductions purely as a mathematical matter. They've become hopelessly deadlocked because, for both sides, the question has been cast as a matter of fairness and economic competitiveness. Developing countries don't think it's fair they should be expected to help solve a problem they didn't create, especially when developed countries aren't honoring their commitments to emissions reductions. Some developed countries, particularly the U.S., don't think it's fair that developing countries should gain a competitive advantage by being exempt from emissions limits. Many who make this argument simply ignore what's happened in the past; their philosophy appears to be: what's done is done, let's move on.

The conference was permeated with disappointment and dwindling expectations. The *Toronto Star* described it as "a litany of failure," while *Time* magazine said it was full of "empty promises, hollow rhetoric, hypocritical posturing, bickering between rich and poor, and irrelevant initiatives." Razali Ismail, president of the General Assembly, opened the conference by describing the achievements since Rio as "paltry" and warned that humanity was "teetering on the edge, living unsustainably and perpetuating inequity, and may soon pass the point of no return." In his closing remarks, he said the nations attending the conference lacked the political will to tackle critical problems and were at the mercy of special interests. "Our words have not been matched by deeds."

In what would become a typical feature of these international gabfests, the delegates struggled down to the wire to produce a concluding statement that everyone could accept. After hours of wrangling, they managed to come up with a document that expressed concern that "overall trends for sustainable development are worse today than they were in 1992" and said there was "widespread but not universal agreement" on the need for "legally binding meaningful, realistic and equitable targets" to reduce greenhouse gas emissions. All of which amounted to no real progress. The document was variously described by the media and environmentalists as "vague," "face-saving," and "pathetic." About the best anyone could say was that at least the conference had not actually lost ground from Rio.

It was a conclusion that gave an especially hollow ring to the warning issued at the start of the meeting by UN Secretary-General Kofi Annan: "At stake this week is the capacity of the international system of states to act decisively in the global interest. Failure to act now could damage our planet irreversibly."

And the Band Played On...

The summer and fall of 1997 marked the beginning of the endgame. With Kyoto looming, there was a major escalation in the political rhetoric surrounding the climate treaty and a hardening of seemingly irreconcilable positions that had been evolving since Rio. Backpedaling on the Berlin Mandate was the order of the day as objections to the exemption of developing countries from the climate treaty mounted rapidly, particularly in the U.S. Injected into this volatile mix was a high-powered and expensive media campaign launched in the U.S. by powerful opponents of the climate treaty—including automakers, labor and consumer organizations, chambers of commerce, agricultural groups, and conservative politicians—who intensified their effort to influence public opinion and dissuade the U.S. government from agreeing to binding reductions in Kyoto. Combined with the growing number of weather disasters caused by El Niño, the earth's weather and climate had never had such a high public profile.

Spooked by the highly vocal opposition of these influential groups and the growing public visibility of their predictions of economic catastrophe, the U.S. and Canadian governments waffled as long as possible on specifying the targets they would propose at Kyoto. They also had to contend with mounting political opposition to the treaty. In July, the U.S. Senate passed a bipartisan resolution, by a margin of 95 to 0, that called on the Clinton administration to refuse to sign a protocol at Kyoto that failed to impose binding emissions limits on developing countries or that would cause "serious harm" to the U.S. economy. Republican Senator Chuck Hagel, a sponsor of the resolution, said that since developing countries would become the major emitters of greenhouse gases within 25 years, it was "complete folly" to exclude them. Democratic Senator Robert Byrd said the Senate should not support a treaty that "requires only half of the world, in other words the developed countries, to endure the economic cost of reducing emissions while developing countries are free to pollute the atmosphere, and in so doing, siphon off American industries." While this resolution was not binding on Clinton, it was a strong signal that any protocol that failed to meet the Senate's criteria would fail to get the two-thirds majority required for ratification.

In early September the public relations war against the Kyoto treaty was ratcheted up to a new level in the U.S. with the launch of a US$13 million TV and print ad campaign by the Global Climate Information Project, yet another ambiguously named coalition of interest groups opposed to greenhouse gas cuts. GCIP claimed that cutting emissions would cause drastic harm to the U.S. economy by raising energy and consumer prices and causing job losses, while developing countries would not be similarly hampered. Ads run on CNN depicted big hikes in gasoline and grocery prices. One ad showed someone cutting up a map of the world while pointing out that countries like China, India, and Mexico were exempt from emission cuts. "It's not global and it won't work," the commentary said of the Kyoto treaty, while another said: "We pay the price and they're exempt." The ads were significant because they marked the first time the growing rift between the industrialized and developing countries had been put before the public in a way that so strongly linked it to direct economic ramifications for citizens in the developed world. Until this campaign, the issue had remained a rather arcane aspect of international treaty negotiations.

Predictably, the ads simply ignored the underlying "fairness" rationale for the exemption of developing countries: the fact that developed countries were still the world's major emitters of greenhouse gases and were responsible for most of the greenhouse gases already in the atmosphere. At the time, the U.S. had the highest per capita emissions in the world; with about 4% of the world's population, it was responsible for about a quarter of global greenhouse gas emissions. "While Americans have more than one vehicle for every licensed driver, most Indians and Chinese are just now getting their first refrigerators," wrote Daniel Lashof of the Natural Resources Defense Council and Howard Geller of the American Council for an Energy-Efficient Economy in a letter to the *Washington Post*. While countries like India and China should be encouraged to use energy efficiently, they said, demanding that they adopt the same emissions timetable as industrialized countries was both unfair and "certain to undermine the Kyoto talks."

The ads caused an uproar. Then-U.S. Interior Secretary Bruce Babbitt described it as a "campaign of disinformation" and environmental groups complained to CNN that the claims were inaccurate. CNN pulled the ads, saying their policy was not to run advocacy ads on issues receiving intense media coverage and announcing they would not air ads on either side of the issue. This move predictably drew charges of censorship from the ads' sponsors, who declared that "open and honest debate is not welcome at CNN." In an open letter in the *Wall Street Journal*, they pointed out that CNN had previously allowed advocacy ads on other hot political issues like health care. A week later, CNN capitulated—"wilting in the heat of furious criticism," as the *Washington Post* put it—and the ads went back on the air.

The view that the climate treaty was really a thinly disguised effort to undermine the U.S. economy was common among corporate opponents of the agreement. Writing in the *Washington Post*, R. J. Eaton, chairman of Chrysler Corp., argued that the global warming debate had become "a trade, economic and foreign-aid issue disguised as environmentalism." Rejecting the idea that developed countries should stabilize their use of fossil fuels while allowing developing countries to increase theirs, he said such measures would involve a "massive transfer of American wealth" that wouldn't prevent ice caps from melting but would "severely undermine" U.S. competitiveness. Developing countries, if not bound by emissions limits, would enjoy cheaper energy that would attract American industries and jobs; the net impact would not be a reduction in global CO_2 emissions, but rather a transfer of emissions from the U.S. to the Third World. Eaton also suggested that the U.S. would be uniquely disadvantaged because other developed countries, though theoretically bound by emissions restrictions, were unlikely to enforce them as rigorously. Characterizing the climate negotiations as promoting a "punish the U.S. first" agenda, he said that if a global warming problem exists, it must be solved by "partnerships and technology, not economically threatening targets and timetables."

Labor and farm groups also registered their disapproval. The AFL-CIO said exempting developing countries was a "fundamental error" that would provide international companies with the incentive to transfer jobs, capital, and pollution to the Third World and yet would do nothing to stabilize greenhouse gas emissions. It demanded that the U.S. "insist upon the incorporation of appropriate commitments from all nations to reduce carbon emissions; and seek a reduction schedule compatible with the urgent need to avoid unfair and unnecessary job loss in developed economies."

These views were not universally held in the corporate world; later, a group of senior executives of major companies like Mitsubishi Motor Corp., Nike Inc., and Bechtel Group Inc. ran a newspaper ad calling for the U.S. to show "strong leadership" in dealing with climate change.

Clearly alarmed by the direction the debate was taking in the U.S., two groups of scientists also weighed in. In June, 2600 scientists, including 35 members of the U.S. National Academy of Sciences and two Nobel laureates, issued the "Scientists' Statement on Global Climatic Disruption," directed primarily at Clinton. The statement not only endorsed the IPCC findings, but went a step further by saying unequivocally that "human-induced global climate change is underway." The authors said that their scientific familiarity with the causes and effects of climate change and with the "scale, severity and costs to human welfare" of climatic disruptions led them to "introduce this note of urgency and to call for early domestic action to reduce U.S. emissions via the most cost-effective means." The risks associated with climate change "justify preventive action," they said, urging the U.S., as the world's largest emitter of greenhouse gases, to fulfill its FCCC commitments and to "demonstrate leadership in a global effort."

In early October, another group of 1500 scientists from 60 countries, including 98 Nobel laureates, issued a statement urging world leaders to adopt a strong treaty at Kyoto. Nobel laureate Henry Kendall, chairman of the Union of Concerned Scientists and author of the statement, said: "Let there be no doubt about the conclusion of the scientific community: the threat of global warming is very real and action is needed immediately. It is a grave error to believe that we can continue to procrastinate. Scientists do not believe this and no one else should either."

As the fall of 1997 wore on, events—and the accompanying rhetoric—escalated rapidly. The European Union was still the only major player that had stated its negotiating position. It was obvious that Canada and the U.S. just didn't have their hearts in it; their elusive behavior suggested a largely defensive effort to get through the game with as little political damage as possible both at home and abroad. While they clearly preferred to avoid the embarrassment of a complete failure at Kyoto, they also did not want to get caught again making promises they couldn't keep. And they knew that whatever they agreed to, or failed to agree to, it was almost certain to make someone at home mad as hell. It was little wonder they were playing their cards close to the vest.

Australia, on the other hand, was making no secret of its position—it was marching firmly in the opposite direction. It dismissed the European target as "unrealistic and unachievable, even for the Europeans themselves" and argued against a single emissions target for developed countries. In fact, it rejected the idea of binding limits altogether, saying that each country should be allowed to set its own target. Like the U.S., Australia also insisted that developing countries should be part of any treaty, arguing that jobs and industrial production would migrate to countries exempt from emissions targets. "The Australian national interest does not lie in rolling over and accepting binding mandatory greenhouse targets because that will cost thousands of jobs and seriously damage the Australian economy," said Prime Minister John Howard. Australia prevailed over a group of island nations that wanted strong emissions targets because they fear being swamped by rising sea levels. At a meeting of the South Pacific Forum, the island states reluctantly endorsed Australia's position against binding, uniform targets because of their dependence on Australia's foreign aid. But they weren't happy. "Being small, we depend on them so much. We had to give in to what they wanted," said Prime Minister Bikenibeu Paeniu of Tuvalu. "There was no compromise. It was just no, no, no, no." (Too bad, Tuvalu.) Later, Howard claimed a victory in persuading other members of the British Commonwealth to endorse the principle of differentiated targets and to require the participation of developing countries.

Japan put its cards on the table in early October. Attempting to break the logjam, it proposed that developed countries reduce emissions of the three major greenhouse gases (CO_2, nitrous oxide, and methane) by an average of 5% below 1990 levels between 2008 and 2012. The proposal also called for "advanced" developing countries to agree to voluntary emissions cuts. Japan supported differentiation; some countries would have targets higher or lower than 5% based on a formula that considered economic and population factors. Its own proposed target was 2.5%.

As the host country for the climate talks, Japan was anxious to avoid a spectacular flop in Kyoto and it hoped this intermediate approach would find favor among the warring factions. It didn't. While the U.S. was cautiously supportive, the Europeans were not impressed. "Not nearly ambitious enough" was the verdict from the European Union. Environmental groups were scathing in their criticism, calling the proposal "scandalous."Australia, on the other hand, rejected the targets it would face under the proposal, even though, at 2 to 3%, they would be lower than the average.

In fact, Australia announced that it did not propose to reduce its emissions at all, but rather to increase them by 18% by 2010, arguing it was a "special case" among developed countries, almost uniquely dependent on energy-intensive exports such as coal and aluminum. Its exports are, in fact, about twice as "carbon-intensive" as its imports. "We are not prepared to see Australian jobs sacrificed and efficient Australian industries, particularly those in the resources sector, robbed of their hard-earned competitive advantage," said Howard.

By the end of October, after an aggressive public education blitz aimed at getting American public opinion on his side, Clinton announced that the U.S. government would propose a target of stabilizing greenhouse gas emissions at 1990 levels by sometime between 2008 and 2012, with further unspecified reductions below 1990 levels by 2017. "If we do not change our course now, the consequences, sooner or later, will be destructive for America and for the world," Clinton said. Although the president said the U.S. was prepared to "take the lead" in the debate, this proposal in fact amounted to little more than an extension of the failed commitments made at Rio—although it was the first time the U.S. accepted the concept of binding limits.

Despite the failure of voluntary cuts, Clinton rejected immediate imposition of mandatory measures at home. Instead, he said there would be US$5 billion in incentives to encourage U.S. companies to develop clean technologies and improve energy efficiency. "It is a strategy that if properly implemented, will create a wealth of new opportunities for entrepreneurs at home, uphold our leadership abroad, and harness the power of free markets to free our planet from an unacceptable risk," he said in a speech. "If we do it right, protecting the climate will yield not costs, but profits, not burdens but benefits, not sacrifice, but a higher standard of living."

Clinton proposed that mandatory limits would not go into effect until 2008, but industries that reduced their emissions voluntarily before that would receive credits as an incentive to start early. After 2008, industries and countries that failed to meet their limits could buy "emissions rights" from others who had received credits for reducing emissions more than required. Cognizant of the political minefields at home, Clinton reiterated that the U.S. would not accept binding limits without "meaningful" participation in the treaty from developing countries, although he did not specify what that participation should be.

Predictably, Clinton's plan pleased no one; the *Washington Post* described the response as "almost universally negative." While some environmentalists were glad the president had depicted global warming as an urgent problem, most found his proposal weak. The Natural Resources Defense Council said it was a "minimum starting point" that did not go far enough "to get us out of harm's way." One environmentalist predicted the Kyoto talks could collapse unless the U.S. did better.

On the other hand, groups opposed to the climate treaty thought Clinton's proposal was courting economic disaster. The U.S. Chamber of Commerce described it as "a one-way ticket to ship America's industrial capacity overseas." The Global Climate Coalition said the target would "cost huge sums of money, loss of jobs, and disrupt many industries and all energy users." However, the Council for Sustainable Energy said Clinton's proposal was reasonable and would be welcomed by many businesses.

Just one day before Clinton's announcement, a group of 77 developing countries called for developed countries to make cuts much deeper than anyone had previously broached—35% below 1990 levels by 2020—and to allow exemptions for developing nations to continue. Clearly, the prospects for smooth sailing at Kyoto were growing dimmer by the hour.

With Clinton's announcement, Canada was the only major industrial country that had not stated its negotiating position on emissions targets. However, reacting to the U.S. proposal, Chrétien told the House of Commons: "We want our position to be better than the one by the Americans." Achieving that goal was proving difficult in the face of fierce opposition from the oil-producing province of Alberta (which produces about 80% of Canada's oil, gas, and coal and about 30% of its greenhouse gas emissions), the energy industry, much of the business community, and political foes. Worse, a divisive battle was raging within Cabinet ranks; in fact, one news report described ministers as being "at war" with each other. In response to Chrétien's "beat the Americans" marching orders, a special committee of the Cabinet met to discuss the matter, but failed again to reach consensus.

Chrétien's approach provoked derisive jibes from political opponents, the media, and environmental groups. It was described as "a little ego war" and a "juvenile game of one-upmanship." *Globe and Mail* political columnist Jeffrey Simpson wrote scathingly that the government was in "palpable disarray" on global warming and engaged in "frenzied inaction," reduced to a policy that amounted to little more than "Beat Clinton."

To make matters worse, preliminary estimates of Canada's 1996 greenhouse gas emissions had recently been released, revealing that they'd grown faster than expected and were roughly 13% higher than 1990 levels. Environmental groups saw the emissions figures as more proof that voluntary measures were failing, but government representatives argued that these measures hadn't had time to produce much effect. Clearly, however, the Rio commitment was fading further into the mists—a source of embarrassment in a country that prided itself on being an environmental good guy.

UN Conference, Bonn, October–November 1997

This was the context in which 150 nations gathered in Germany for what was billed as a "dress rehearsal" for Kyoto. They were trying to hammer out the text of the Kyoto Protocol—a difficult process when the major players were either poles apart or sitting on the fence. Everyone was in a disputatious mood and all the contentious issues that had erupted at the UN in June were reprised. The closer Kyoto approached, the further consensus retreated.

The newly minted U.S. proposal was attacked by the Europeans and developing countries. The European Commission said it was "not an adequate response" and "a significant step in the wrong direction." "Disappointing and insufficient," said German environment minister

Angela Merkel. Only Australia offered lackluster support. It welcomed the proposal but, still intent on increasing its emissions, said it wouldn't adopt the U.S. measures because of the economic harm they would cause.

For its part, the U.S. argued that the Japanese and European proposals were unrealistic. It also pushed hard for joint implementation and emissions trading, as well as some kind of commitment from developing countries. However, its efforts to get developing countries to agree by 2005 to start setting a timetable for future cutbacks went nowhere. Developing countries also fought the idea of emissions trading and assailed the developed countries for their intransigence. "The short-term economic interests seen to be driving some proposals are clearly unacceptable," said Tuiloma Neroni Slade of Western Samoa, head of the Alliance of Small Island States.

It was, as the saying goes, déjà vu all over again. The conference ended in deadlock. The U.S. and Europeans remained wide apart, Australia continued to reject uniform, binding targets, and developing countries successfully resisted U.S. attempts to include targets for Annex II countries in the draft protocol that would go to Kyoto. The journal *Nature* commented that the draft contained mostly "blocks of square brackets— denoting text that has yet to be agreed."

In the theater, a bad dress rehearsal is often considered a harbinger of success on opening night. Not so with Bonn. For many negotiators, the fact that consensus was still so elusive at this late date portended disaster in Kyoto.

With the clock ticking, the Canadian government was still trying to forge enough of a consensus at home to come up with a proposal to take to Kyoto. Federal and provincial energy and environment ministers met in mid-November and agreed that Canada should stabilize its emissions at 1990 levels by 2010. There were no recommendations on further cuts after 2010. Quebec was the only dissenter; having argued for stabilization by 2000 and a 7% cut by 2010, it refused to sign the agreement. This plan hardly qualified as progress; in fact, it amounted to a 10-year rollback of the Rio commitment. And since it was essentially identical to the U.S. position, it failed to meet even the "beat the Americans" criterion. However, Environment Minister Christine Stewart said the agreement gave the federal government some flexibility to enhance the targets if needed to get a deal in Kyoto.

This did not, however, put a stop to Cabinet discord, which persisted literally down to the wire. As a result, the government didn't announce its official negotiating position until December 1, the first day of the Kyoto conference. Surprising nearly everyone, the policy moved beyond the agreement reached by the federal and provincial ministers in November, calling for a reduction of 3% below 1990 levels by 2010 and a further 5% drop by 2015. (Since Canada's emissions were already 13% above 1990 levels, these proposals represented effective cuts of 16 to 21%.) While more aggressive than expected, they still fell short of the promised cuts to 20% below 1990 levels made during the 1993 federal election.

The Canadian position paper also called for "maximum flexibility" in achieving reductions, including joint implementation, emissions trading, and "recognition for exports of energy with low carbon content" (a reference to Canada's desire to obtain credit for exporting nuclear energy technology). Finally, the announcement said that Canada would promote the "constructive and timely engagement of developing countries" in the climate treaty. While acknowledging that they needed time, it said, "their ultimate participation will be necessary for any global agreement to be successful."

There were few specifics on how reductions would be achieved, other than a stated intention to build on measures already included in the National Action Plan. Carbon taxes were explicitly disavowed. Typically, the proposal also attempted to appease regional sensibilities; the proposed 3% emissions cut was almost exactly halfway between the stabilization target advocated by most of the provinces and the 7% reduction pushed by Quebec. At the same time, in a clear reference to the energy sector and the western provinces, the government emphasized the need to "address our international commitments in such a way that no region or sector is asked to bear an unreasonable share of the burden." In short, the policy was a tiny microcosm of the Canadian political zeitgeist.

Canada also tried to position itself between the U.S. and Japan in the hopes of brokering a compromise at Kyoto. Predictably, however, the proposal drew salvos from both sides of the political spectrum. Environmentalists said it was a weak proposal sandwiched between two other weak proposals. Those who opposed moving beyond a stabilization target reacted with outrage. Alberta's Energy Minister Steve West, described in news reports as "visibly angry," accused the federal government of having betrayed the consultation process. "You can take it to the bank that this is going to get noisy," he said. The story of global warming, unlike that of the Three Bears, never seems to involve an option that is "just right."

COP-3, Kyoto, Japan, December 1997

It was show time. All the usual suspects gathered in Kyoto; six years of haggling came to a head and produced...more haggling. This time it was déjà vu to the tenth power, as delegates from 150 countries hammered each other yet again over issues that had divided them since Rio. This time, the stakes were higher because in the end, they would be left either with legally binding commitments or with complete failure.

Going in, there were six major unresolved issues:

- *Binding targets and timetables for emissions reductions by developed countries.* The proposals on the table were still far apart. The European Union and many developing countries were asking for reductions of 15% or more below 1990 levels by 2010, while the U.S. was sticking to stabilization at 1990 levels. Australia and the oil-producing Arab countries were fighting the concept of binding targets, while Canada and Japan were wedged in between with their proposed 3 and 5% targets, respectively.

- *Differentiation.* Some countries wanted to set different emissions targets for different countries, based on their economic circumstances. Others argued for a one-size-fits-all target. The European Union opposed differentiation, yet wanted to apply a similar concept, known as a "bubble," to average out emissions among its own member states—a position that did not sit well with others.

- *The number of greenhouse gases to be included.* Some countries wanted only the three major gases included: CO_2, methane, and nitrous oxide. Others wanted to add three halocarbons, arguing that while these chemicals are less plentiful than the others, they're much more powerful greenhouse gases.

- *"Flexibility" mechanisms.* The U.S. and Canada favored mechanisms like emissions trading and joint implementation, which would give them maneuvering room in fulfilling their commitments and enable them to achieve reductions in the most cost-effective way. The Europeans and developing countries were suspicious that these measures would be used to avoid cutting emissions at home. The U.S. wanted to meet 50% of its target by buying emissions rights elsewhere; the EU said trading should be restricted to no more than one-third of a country's target. The U.S. also wanted to count "sinks" (e.g., forests) that absorb greenhouse gases in calculating their emissions. Others rejected this, in part because it's difficult to calculate accurately how much CO_2 is actually absorbed by sinks. Environmental groups argued against giving developed countries credit for tree-planting programs, saying they wouldn't compensate for deforestation occurring around the world.

- *Compliance.* Since the targets were to be legally binding, a mechanism was needed to monitor compliance and sanction those who failed to meet their commitments.
- *Participation by developing countries.* This was perhaps the thorniest issue of all, which was ironic because it wasn't even supposed to be on the agenda. The whole point of Kyoto was for developed countries to make a good-faith showing that they were prepared to get serious about reducing emissions after their failure to keep the promises made at Rio. The parties had specifically agreed in Berlin that the question of when and how developing countries would reduce their emissions would be addressed *after* Kyoto.

What changed the agenda was the concerted campaign against the climate treaty by U.S. lobby groups. They succeeded in making the exemption of developing countries a major bone of contention, thus putting the question of equity on a collision course with U.S. domestic politics. Aware that the U.S. Senate was unlikely to ratify a treaty that exempted developing countries, the U.S. delegation came to Kyoto demanding "meaningful participation," including a willingness to accept joint implementation and emissions trading measures. In particular, they wanted countries like China, India, Brazil, and Mexico, which were growing rapidly, to commit to taking action to reduce emissions in the future. Vice President Al Gore warned that the U.S. was "perfectly prepared to walk away from an agreement that we don't think will work."

Since the international community believed the treaty would fail without U.S. participation, delegates were forced to confront this matter even though it was not within the mandate of the Kyoto conference. Keeping the U.S. in the game meant increasing pressure for concessions from developing countries.

They weren't having it. Most members of the G77 group of developing countries and China resisted U.S. attempts to force them to commit to taking future action to reduce emissions and warned that pursuit of this issue could break up the conference. "We have said categorically no," said Mark Mwandosya of Tanzania, chairman of the G77 group. "We need to see leadership from [developed nations] who have been historically responsible for the majority of emissions." He said the failure of developed countries to meet their Rio commitments "does not give a moral platform to ask anyone in developing countries to take up new commitments." Richard Mott of the World Wildlife Fund described it as "the nation of sports-utility vehicles...lecturing a nation of bicycles."

By the second day of the conference, delegates were already talking about how to word a statement if the conference failed to reach agreement. Halfway through the meeting, the only thing that had been agreed on was where to hold the next meeting. "It's time for some adult supervision," one frustrated U.S. official told the *Washington Post*.

Whether they were the "adults" or not, it took the arrival of big political guns toward the end of the conference to break the logjam. With the conference at a near standstill, U.S. Vice President Al Gore flew in and out within 24 hours and announced that he'd instructed the U.S. delegation "to show increased negotiating flexibility if a comprehensive plan can be put in place, with realistic targets and timetables, market mechanisms and the meaningful participation of key developing countries." Gore acknowledged that developing countries needed to grow economically, but urged them to accept emissions trading and other mechanisms that developed countries wanted to help them to reduce their own emissions.

However, Gore was upstaged by an impassioned speech by Kinza Clodumar, the leader of Nauru, a small island state. In a voice described as "soft and solemn," he described how 80% of his country has already been destroyed by mining started by colonial powers. "My people have been confined to the narrow coastal fringe that separates this wasteland from our mother the sea. The coastal fringe where my people live is but two metres above the sea surface. We are trapped, a wasteland at our back, and to our front, a terrifying, rising flood of biblical proportions. We submit respectfully that the willful destruction of entire countries and cultures, with foreknowledge, would represent an unspeakable crime against humanity. No nation has the right to place its own misconstrued national interest before the physical and cultural survival of whole countries."

In the end, as is typical of these conferences, the wrangling went down to the wire and then some. The U.S. moved to accept targets below 1990 levels, but arguments over emissions trading and acceptance of voluntary limits by developing countries raged on. Delegates worked right through the last night; media reports described them asleep in their chairs after two days of nonstop haggling. U.S. delegate David Doniger, who was described at one frenzied juncture as "leaping over sleeping conferees," commented to the *Washington Post* afterward that it was "not a day for any of us to operate heavy machinery." There were many times when delegates thought the cause was lost, but finally, on December 11, they announced to the world that they had an agreement. It included the following terms:

- Thirty-eight developed countries would collectively reduce their emissions by an average of 5.2% by 2008 to 2012.
- Six greenhouse gases are included: CO_2, methane, nitrous oxide, and three halocarbons. The first three must be cut relative to 1990; the last three can be cut relative to either 1990 or 1995.
- Different countries have different targets. Those who agreed to reduce their emissions below 1990 levels included the European Union, Switzerland, and many central and eastern European states (8%); the U.S. (7%); and Canada, Japan, Hungary, and Poland (6%). New Zealand, Russia, and the Ukraine agreed to stabilize their emissions at 1990 levels. Countries allowed to increase their emissions were Norway (up to 1%), Australia (up to 8%), and Iceland (up to 10%).
- A clean development mechanism is established that allows industrialized countries to obtain credit against their own targets by funding emissions reduction projects in developing countries.
- Emissions trading is allowed among industrialized countries. However, the Protocol calls for developed countries to achieve most of their reductions through domestic measures, not via emissions trading and other flexibility options.
- Carbon emissions from deforestation and carbon sequestration from reforestation will be factored into a country's emissions equation.

A UN statement notes that targets in the Protocol mean that greenhouse gas levels will actually be 30% lower in 2010 than they would otherwise be because most countries have increased their emissions since 1990. "The Protocol should therefore send a powerful signal to business that it needs to accelerate the delivery of climate-friendly products and services."

The Protocol does not require developing countries to make any commitment to reducing emissions.

At Kyoto, Greenpeace sculpted a "Carbonosaurus," a dinosaur made of old auto parts, fuel tanks, and gasoline pumps. Their message clearly was that the days of the carbon industry are numbered. If what happened after Kyoto is any indication, however, reports of the death of Carbonosaurus are definitely premature. It's clear the Protocol faces a long haul before it will be ratified and implemented in a meaningful way.

The protocol means nothing without ratification by the national governments of at least 55 countries, including countries that produced at least 55% of total greenhouse gas emissions from Annex 1 countries in 1990. The biggest stumbling block could be the U.S. The ink wasn't even dry on the agreement when Congressional Republicans said it would be "dead on arrival" if presented for ratification. The Clinton administration didn't even try, saying it would continue negotiations with developing countries and would not submit the treaty for ratification until they agreed to meaningful participation.

The skeptics were, predictably, hopping mad. "We gave the store away," William O'Keefe, then-chairman of the Global Climate Coalition told ABC news. "We conceded everything. We got nothing." He added that the treaty was impractical and unachievable—there was no way Americans were going to reduce their energy consumption that much in so little time. A week after the Kyoto conference, Thomas Stallkamp, president of Chrysler Corp., said the company intended to turn out as many trucks, sport utilities, and minivans as consumers wanted to buy. (These larger vehicles generally emit more CO_2 than lighter cars; light trucks accounted for about 70% of Chrysler's output.) Stallkamp questioned whether American consumers would change their buying habits because of global warming. "They're closer to the global warming of their electric blanket than they are to the emissions of the air," he told Reuters. "They haven't made the connection yet, if that connection's even there."

By the end of 1998, the treaty had virtually disappeared from political radar screens, as gone as El Niño was from weather satellite pictures, blasted into oblivion by turmoil on global markets and the fulminating sex scandal that threatened to bring down the Clinton presidency.

Nor was the treaty much more visible in Canada. In theory, the Canadian government could ratify the treaty with less trouble than the U.S. government; having a majority in the House of Commons it could, in essence, just do it. In practice, however, the federal government faces political problems of nearly equal magnitude. Canada's provincial governments can be just as cantankerous as the U.S. Senate, especially when it comes to anything they perceive as high-handedness on the part of the feds. Miffed by the federal government's last-minute decision to take a proposal for a 3% reduction to Kyoto—instead of the stabilization target the provincial governments had agreed to in Regina—they were even more annoyed when Canada came back from Kyoto committed to cuts of 6%.

The question of how the targets would be achieved caused a great deal of dissension. Prior to Kyoto, the federal government had said that implementation measures would be negotiated in Canada afterward, but Alberta Premier Ralph Klein argued it should have been done before that.

Oil industry representatives also expressed apprehension about implementation. "Nobody in Canada knows how we can do this," said Eric Newell, CEO of Syncrude Canada. "I don't know why we would ratify a protocol we can't achieve."

Just two days after the Kyoto meeting ended, at a first ministers' conference in Ottawa, provincial leaders forced the climate treaty onto the agenda as an emergency item and emerged with a communiqué that stated: "First ministers agreed to establish a process, in advance of Canada's ratification of the Kyoto Protocol, that will examine the consequences. They also agreed that it is important to achieve a thorough understanding of the impact, the cost and the benefits of its implementation and of the various implementation options open to Canada." Klein said the communiqué indicates that implementing the treaty is not a foregone conclusion. In fact, even the prime minister left that impression. The *Edmonton Journal* noted that he used the word "if" when talking about implementation and suggested Canada would not be among the first countries to ratify the treaty, which he described as "an agreement in principle. We presume it will be ratified. But we have seen some treaties in the past that have not been ratified." He added that ratification was going to take a long time "in all jurisdictions."

Less than a month after the Kyoto conference, according to documents obtained by the *Toronto Star* under the Freedom of Information Act, the government revised projections of Canada's emissions by 2010, saying they could be nearly 40% higher than the estimates used in the Kyoto negotiations. The implication was that reducing emissions to 6% below 1990 levels would require an effective cut as high as 32% by 2010. The *Star* quoted an internal memo in which a government official expressed concern that publicly releasing the revised figures would complicate consultation with the provinces on measures to implement the Kyoto Protocol.

There's been much speculation that Canada will not ratify the treaty in any event if the U.S. fails to do so. Jeffrey Simpson, for one, has argued it's doubtful Canada would "plunge ahead with emission reductions while the United States did little." Natural Resources Minister Ralph Goodale has suggested Canada might consider backing off if this were to happen because the U.S. is Canada's major trading partner, accounting for about 80% of exports. "The provinces are particularly concerned that they not be put at a competitive disadvantage should the United States not go along with a Kyoto agreement."

By April 1998, it was apparent the Kyoto Protocol was not on the fast track in Canada. At the first meeting of federal and provincial environment and energy ministers since the Kyoto conference, the ministers decided to put off doing anything for two years to conduct further studies on the economic and environmental impact of compliance. "We will not do anything that would jeopardize our economy," said federal Environment Minister Christine Stewart at a news conference.

This decision was, not surprisingly, denounced by environmental groups as another delaying tactic and an act of "political cowardice." But they weren't surprised. Canada's reputation as an environmental leader had been on a long, slow slide for years. The British scientific journal *Nature* commented that while Canada had assumed "strong and influential leadership" at Rio, environmentalists at Kyoto were no longer counting on Canada and had turned to the European Union. James Bruce, a prominent Canadian participant in many international climate organizations, agrees that Canada is no longer perceived as a world leader on the environment. This is a perception that's taking hold within the country as well. In May 1998 Canada's environmental commissioner Brian Emmett issued a report saying the federal government is all talk and no action. "It's difficult to give the government a good grade....While Canada has demonstrated vision, it is failing in implementing it."

In November 1998, the next round of negotiations took place in Buenos Aires, Argentina. The disputes that had plagued the previous conferences were reprised. The U.S. continued to demand meaningful participation from developing countries and the two largest, China and India, remained intransigent. Some developing countries, however, expressed interest in participating in economic mechanisms such as emissions trading and the clean development mechanism. Believing it could benefit from such programs, Argentina became the first developing country to agree to set targets and timetables to reduce its emissions. Many of the thorniest issues weren't resolved, however, including deciding on the extent to which developed countries can use emissions trading to meet their domestic targets, determining whether countries will get credit for carbon sinks and setting emissions targets and timetables for developing countries. What finally emerged was a "work plan" that set a deadline of late 2000 for nations to agree on the final details of the mechanisms needed to achieve the goals set out in the Protocol and to establish an enforcement mechanism. A Reuters report described the talks as "arduous," and indicated these divisive issues "will not be so easily papered over" at the next conference in late 1999. In a reference to the melting of glaciers around the world, Christopher Flavin of the Worldwatch Institute told the *Washington Post*: "For the first time, the glaciers are moving literally faster than the negotiations."

During the conference, the U.S. signed the Kyoto Protocol. The gesture was largely symbolic, since ratification by the U.S. Senate is still required before the U.S. would be bound by the treaty—and that seemed as unlikely as ever. By the close of the Buenos Aires conference, 60 countries had signed the treaty but only two—the small island states of Fiji, and Antigua and Barbuda—had ratified it.

Too Little Too Late

All the debate about whether the Kyoto Protocol will be ratified obscures the more important question of how much good it will do even if it is. Robert Watson, chairman of the IPCC, said the agreement is important because, without it, emissions would rise even higher. But he said it would be a grave mistake for people to assume that nothing more is needed. "We have a long way to go in finding long-term solutions."

Most supporters of the Kyoto Protocol know it's not enough to get the job done, but they often argue that at least it has turned the ship in the right direction. It's hard to take too much solace in that—the *Titanic* was turning when it hit the iceberg. And all the wrangling over joint implementation, emissions trading, and ratification brings to mind visions of rearranging deck chairs. A disturbing pattern has emerged at international climate conferences—thousands of people repeatedly gather in one city or another and spend a week or two squabbling until it all comes down to the wire. They grudgingly hammer out an agreement (often promising little more than to squabble again in the future) to avoid the political embarrassment of spending all that time and money to come up with nothing. Between the conferences, there appears to be little progress. One wonders if they wouldn't accomplish more for the climate simply by not burning the jet fuel required to transport thousands of delegates to meetings all over the world.

As William Hooke of NOAA's Weather Research Program once commented: "Global change is a rapid-onset phenomenon compared with the time required for five billion people to reach consensus on how to respond." The fact is, we have yet to decide whether we really want to solve the global warming problem or simply get an A for effort. Self-congratulatory hype about forging a climate treaty isn't going to change what's going on in nature. CO_2 levels will continue to rise inexorably, with all that implies for the climate. As Christopher Flavin points out: "We would do well to remember that we are ultimately negotiating with the oceans and the atmosphere, far less flexible dealmakers, to whom paper promises mean nothing."

If we continue doing what we have been doing—making half-hearted promises to take half-hearted measures that won't get the job done even if implemented—it might be argued that we'd be better off not to waste our time and money on avoidance and to concentrate on adapting to climate change instead. This would be a mistake, because if we don't at least attempt to slow the progress of global warming, adaptation will be that much more difficult and perhaps ultimately impossible. But nor should we assume that we're going to be able to turn this ship around fast enough to avoid catastrophe. It might be a good time to start checking what shape the lifeboats are in.

RIDING OUT THE STORM:
Adapting to Climate Change

The concept of adapting to the climate is both very old and very new. Humanity has always been forced to adapt as a matter of survival, and much of our technology is a response to that imperative. Technologies that provide us with shelter, food and water, transportation—all of these and more are to some extent designed to enable us to transcend barriers imposed by climate and weather. The battle is not always successful, however and, for those times, we've also developed techniques for responding to and recovering from natural disasters. This, too, is an adaptive response.

In the context of climate change, adaptation is defined as any adjustment that enhances the ability of our socioeconomic systems to survive changes in the climate and weather; that reduces our vulnerability to climate variability, extreme weather events, and natural disasters; and that enables us to take advantage of opportunities provided by changes in climatic conditions. Examples of adaptation include improving weather forecasting, developing emergency preparedness programs, building dikes to protect against rising sea levels, planting crops and trees that grow better in a changed climate, and discouraging development in regions vulnerable to extreme weather events.

Since adaptation has been so much a part of our history and everyday life, it may seem surprising that until recently, it attracted little attention from scientists and policymakers concerned with climate change. "There has been little systematic attention to adaptation...because the focus has been primarily on ways of mitigating climate change by the reduction of greenhouse gas emissions," according to Ian Burton, the first director of Environment Canada's Environmental Adaptation Research Group (EARG). But it's now clear that avoidance measures are unlikely to be strong enough or implemented soon enough to prevent further global warming. There's also concern that global warming may alter climate variability and increase the incidence of extreme weather events, which will increase our vulnerability. As one study noted, while "past climate has generally been considered a reliable guide for planning the future ...the instrumental climate record may no longer be a suitable indicator of future climate." These concerns have increased the urgency for research on adapting to climate change. In a paper outlining the rationale for an adaptation research program, Burton noted: "No matter how successful mitigation attempts might be in the future (and nobody suggests that it can be done easily or quickly) some impacts of climate change are going to occur and probably are occurring now. Some adaptation response is therefore required."

Giving greater prominence to adaptation in policymaking is controversial, however. Burton says the idea has been branded "politically incorrect." Those who favor strong avoidance measures are wary of increasing the emphasis on adaptation because, to them, it smacks of giving up on cutting greenhouse gas emissions; they say it plays into the hands of opponents of emissions controls and warn that overconfidence about our ability to adapt may lull us into a dangerous complacency about the consequences of climate change until it's too late. In his book *Earth in Balance: Ecology and the Human Spirit*, U.S. Vice President Al Gore argued that "believing that we can adapt to just about anything is ultimately a kind of laziness, an arrogant faith in our ability to react in time to save our skins."

Thomas Homer-Dixon, director of the Peace and Conflict Studies Program at the University of Toronto, has argued that it's risky to assume that human scientific and technical ingenuity can always find a way to overcome resource scarcities in the future. "The optimism of those who have great faith in the potential of human ingenuity when spurred by necessity is...imprudent. We are taking a huge gamble if we follow the path they suggest, which is to wait until scarcities are critical and watch human ingenuity burst forth in response. Should it turn out that this strategy was wrong, we will not be able to return to a world resembling the one we have today. We will have burned our bridges: the soils, waters, and forests will be irreversibly damaged, and our poorest societies will be so riven with discord that even heroic efforts at social renovation will fail." In short, it may simply be impossible to undo the damage once it's been done. Therefore, it's better to act early to prevent scarcity than to adapt to it after it happens.

It's true that some proponents of adaptation contend that human ingenuity can cope with just about anything the climate may throw at us and that economic growth, if not derailed by curbs on greenhouse gas emissions, will provide the resources needed to do so. (The *Economist*, for example, argued that *if recent growth continues*, [emphasis added], the world economy will be 300% larger in a century and thus "much better able to bear the costs of coping with climate change.") Many skeptics belong to this camp, but it should be remembered that their confidence in our adaptive abilities is often colored by a conviction that global warming is unlikely to throw very much trouble our way.

Other proponents of adaptation—including many scientists involved in impacts research—take a different view, seeing adaptation not as a substitute for avoidance but as a necessary complement to it. Those who belong to this group accept that global warming is a serious problem and that we must press ahead with avoidance measures as quickly as possible. However, current political realities dictate that we cannot avoid some warming over the next century and therefore, as a practical matter, we must consider how to protect ourselves from the consequences. This holds true even if the provisions of the Kyoto Protocol are implemented. Roger Street, head of EARG, said the Kyoto agreement may "change the rate and magnitude of climate change and that will have an impact on how much we will have to adapt. But even if we were to cut emissions to zero, we'd still have to adapt."

David Runnalls of the International Institute for Sustainable Development in Ottawa is among those who are dubious about what he refers to as the "adaptation craze." He thinks that going "full bore" on adaptation in the near term spells trouble for a long-term commitment to reducing emissions. "It's perfectly obvious that climate change is on its way, that it's a bad thing, and we have to do something about it. Adaptation offers people what sounds like an easy way out." Adaptation will always seem cheaper than avoidance measures like carbon taxes, he says, and politicians will find it more palatable to do things like building dams and reservoirs than to impose new taxes on gasoline. "All the political brownie points are on the wrong side of the line. We might have to retreat to adaptation if we find ourselves short of political vision—at some stage we may have to give up and batten down the hatches—but it's too early to do that."

As a leading proponent of adaptation, Burton denies that adaptation research is a disguised attempt to sabotage efforts to reduce greenhouse gas emissions; instead, he says, it's aimed at trying to reduce the overall losses and impacts from climate change. When he first started talking about the importance of adaptation, many people thought he was proposing adaptation as an alternative to mitigation. "I tried to explain that we ought to push as hard as we could on both those doors but it's a difficult message to get across." The fact that many environmentalists see this as "playing into the hands of the polluters," has unfortunately helped to create an "unhappy, excessive concentration on one side of the question" that has made it difficult to get adaptation on the agenda.

Burton agrees that controlling greenhouse gas emissions is important, because "the long term consequences may be quite horrific, the risks are quite substantial," but he argues that "adaptation is a device for buying time. We're so far away from getting a handle on this problem that we need to do both." While concentrating only on avoidance is "nice in theory, in practice, you do what you can do. The situation is such that we really need to pay a lot more attention to adaptation, on how to do it better, and we ought not to be put off by the concern that it will be used as an excuse."

In fact, Willam Thorsell, editor of the *Globe and Mail*, has argued that adaptive efforts may in fact encourage greater support for avoidance measures. "The more visibly we prepare for the possibility of global warming, the greater the political constituency may be for possibly mitigating it through actual CO_2 reductions." Surprisingly (given the *Globe*'s usually dismissive editorial attitude toward global warming), Thorsell also argued for another "IPCC-like study focused solely on adaptation, which could well be the pressing issue for humanity in the next 50 years."

Roger Pielke, Jr. of the Environmental and Societal Impacts Group of the U.S. National Center for Atmospheric Research (NCAR) notes that adaptation traditionally has been associated with a passive acceptance or fatalism about the human impact on environment. Writing in the journal *Global Environmental Change*, he says: "There is little wonder that adaptation has been out of favor: who wants to be viewed, at best, as working prematurely on adaptation studies and, at worst, obstructionist, lazy, arrogant and anti-environmental?" These objections are misplaced, he says. "To be honest, I think we have our priorities backward. If I had a dollar to spend on climate change policy with an aim to reduce climate impacts on society and the environment, I'd spend 90% on the adaptive responses and 10% on emissions reduction."

Pielke argues that the outlook for reducing greenhouse gas emissions (much less their atmospheric concentrations) is dismal. But even if we stabilize concentrations and even if this means that fewer changes in the climate occur, we'll still be faced with increasing adverse impacts from the climate for two reasons: first, we may experience climate surprises like those seen in the past and, second, we keep putting ourselves and our property in harm's way. Population growth, increasing urbanization, migration to coastal areas, and the increasing value of property all contribute to the growing cost of weather-related damages. In fact, Pielke found it is *primarily* socioeconomic factors, and not an increased frequency of extreme events, that accounts for the mounting economic losses to hurricanes in the U.S. during the latter half of 20th century. "No matter what climate policy we adopt regarding emissions reductions, the societal impacts of extreme events are going to rise anyway. Social factors are more of a problem than climate change."

Policies that focus on reducing emissions are putting all the eggs in one basket, Pielke argues. Adaptation is "backup strategy that would provide complementary benefits even if mitigation efforts do succeed." And if they fail, it doesn't really matter that much because reducing greenhouse gas emissions will address only a very small part of what's causing the socioeconomic impacts of climate and weather. For most of these impacts, "the most effective steps are adaptive."

Pielke argues that giving adaptation a more prominent role in climate policy would completely sidestep the unproductive debate over whether increasing greenhouse gas emissions from human activities will cause a dangerous interference with the climate in the future. Society is, in many ways, poorly adapted to the present climate and we need to do something about our vulnerability to climate per se, not just to future climate *changes*. "Unless we give adaptation more explicit attention, the way we've responded to extreme events in [the] last 100 years doesn't give me confidence that we'll do much better in the future," he said. And he feels that developing countries will almost certainly experience growing vulnerability and social impacts as well. "The human toll of disasters have to go up as these countries strive to achieve development. It will be difficult for them not to make the same mistakes that we made."

In his paper, Pielke concludes that "there are *no* situations...existing or predicted, in which some type of adaptive measures do not make sense. Further, to the extent to which societies around the world are maladapted to climatic variability, these adaptive measures will almost certainly provide benefits under the entire spectrum of climate change scenarios offered by the IPCC."

There's little doubt it would make sense to improve our ability to cope with the present climate and any future changes, whether caused by nature variability or human activities. The real issue is whether giving a higher priority to adaptation would seriously undermine efforts to slow the progress of human-induced climate change by reducing greenhouse gas emissions and whether, as a result, we face changes in the future that may be beyond our ability to adapt or cope. Joel Scheraga, director of the climate and policy division of the U.S. Environmental Protection Agency, warned of this possibility in a speech: "I'm very concerned when people say, 'No problem, we can adapt.' Even in resilient economies like Canada and the U.S., we're going to be experiencing changes we haven't ever encountered before."

The policy debate over the relative weight that should be given to adaptation versus avoidance will go on, but there's no doubt that adaptation has gained a higher profile in recent years. While the Framework Convention on Climate Change treats adaptation as a poor cousin to avoidance, the IPCC report gives it more prominence, explicitly including it in the precautionary principle and recommending a no-regrets approach to adaptation as much as to avoidance. "Policymakers will have to decide to what degree they want to take precautionary measures by mitigating greenhouse gas emissions and enhancing the resilience of vulnerable systems by means of adaptation. Uncertainty does not mean that a nation or the world community cannot position itself better to cope with the broad range of possible climate changes or protect against potentially costly future outcomes. Delaying such measures may leave a nation or the world poorly prepared to deal with adverse changes and may increase the possibility of irreversible or very costly consequences. Options for adapting to change or mitigating change that can be justified for other reasons...and make society more flexible or resilient to anticipated adverse effects of climate change appear particularly desirable."

Of course, there will be wide disparities in adaptive capacity in different parts of the world. Many developing countries lack the scientific, technical, and socioeconomic resources to make adjustments that developed countries take for granted. Small island states, for example, simply cannot afford what it would take to protect themselves against sea level rise. Burton says most developing countries have more urgent near-term priorities like basic health care, education, clean water, and food supplies; they find the idea of spending money to adapt to future climate changes almost laughable. Ironically, he says, most of these countries are in low-latitude tropical regions where the climate is expected to change the least, yet they're likely to suffer the worst damages because of their lack of adaptive capacity.

While wealthier industrialized countries have a far greater adaptive capability, this won't insulate them from the impacts of climate change that occur elsewhere. "I've never seen adaptation scenarios for Canada that take account of the fact that the rest of the world may be in a fairly disastrous situation," says Runnalls. "I don't think a rich country that gets 40% of its revenue from foreign trade can pretend the rest of the world doesn't matter. If the rest of the world is in chaos or under pressure as a result of climate change there will be huge pressure on us."

Burton agrees that Canada could be more affected by the impacts of climate change that occur in other parts of the world than by those that take place within its own borders. He argues that industrialized nations should help developing countries improve their adaptive capabilities if for no other reason than self-interest. This view is echoed by Michael Toman, director of climate economics and policy for Resources for the Future: "Improving the capacity to adapt where it is weak, as in many developing countries, may be one of the most effective ways to respond to some climate change risks."

So far, according to Burton, this has not been happening. He says that the World Bank's Global Environmental Facility, the agency that transfers funds from industrialized countries to developing countries for climate projects, has focused mainly on projects aimed at reducing greenhouse gas emissions. Developing countries need to incorporate adaptation into their economic development plans or they risk "setting themselves up to increase their vulnerability rather than reduce it." For example, Caribbean countries anxious to expand their tourist trade have been building hotels and other facilities too close to the ocean. This destabilizes beaches, increases erosion, and also increases exposure to hurricane damage. "It would be lovely if they could develop their tourist industry to do less damage," said Burton. "Unfortunately, tourists want to be able to take their foot out of bed and put it directly into the water."

This kind of "maladaptation" is not confined to the developing world. It's common in industrialized countries, too—which may spell even more trouble if the incidence of extreme weather events increases as anticipated.

Maladaptation

Maladaptation refers to any activity that increases the vulnerability of our socioeconomic systems to climatic changes or weather extremes. In developed countries, we pride ourselves on being well adapted to climate and there's some justification for this; we've succeeded to a great extent in using technological and economic resources to carry on in spite of bad weather and to pick up the pieces when nature gets the best of us. But this comes at a price. Although there are no really firm data on annual adaptation costs, in Canada, they've been roughly estimated at about $10 to $14 billion (which is probably an underestimate), and the U.S. probably spends at least several times that.

However, scientists are questioning assumptions about how well we're adapted to climate today, notwithstanding the billions spent on the effort. A careful analysis of our vulnerability and the risks we're currently taking suggests we're not as well adapted as we'd like to think. The real question is whether the measures adopted so far have truly made us more resilient in coping with climate variability or, rather, have merely masked our vulnerability temporarily, allowing us to plow on with risky behavior, sheltered from the consequences until we're confronted by an event that so greatly exceeds our coping techniques that we face catastrophic losses.

This situation is described in a report on natural disasters and human activity by David Etkin and his colleagues. Extreme events have a "return period" that indicates how often they're expected to occur. Extreme flood levels, for example, are often described as having an expected 50- or 100-year return period. The report points out that society is more vulnerable to rare events of great magnitude than to smaller events that occur more frequently. To understand why this is so, consider what would happen to two groups of people who live on flood plains. The first group does nothing to prevent flooding and is exposed to it at all levels of severity. In this case, the impacts of flooding on their communities increase in direct proportion to the magnitude of the flood. Now take the second group. Say they've built dams and levees that protect them from flooding up to the level of a 100-year event. As long as flooding remains below that level, their vulnerability remains low, but once that 100-year level is breached, their losses can be enormous—possibly greater than those faced by the communities that did not build dams and levees. Their losses may be greater because the protection provided by dams and levees against low levels of flooding often encourages unwise development in flood-prone regions, putting more property at risk when "the big one" hits. "When people perceive that their risks have been reduced as a result of technology, they will tend to act in riskier ways if voluntary actions are available to them," the report notes. "Since people tend to develop a false sense of security, they will tend to develop high risk areas."

This is a maladaptive response even with respect to the current climate. It becomes even riskier if climate change causes an increase in the frequency and severity of extreme events—for example, reducing the return period of what is now a 100-year flood to 50 years. This would result in an overall increase in vulnerability and could transform an adapted society into a maladapted one. We would then have three choices: invest even more in protective measures, pay the price of being wiped out more often, or reduce our exposure by retreating from the zone of vulnerability.

Kevin Trenberth of NCAR notes that if climate change alters the return period of extreme events, the criteria used to design all kinds of structures will require serious rethinking. Dams, levees, bridges, buildings, electrical distribution systems—all of these and more are based on certain assumptions about the weather. "People are beginning to realize that they have to question a lot of those assumptions. The risk of a heavy build-up of ice may have increased, the risk of flooding exceeding the capacity that a levee was designed to deal with may have increased."

The alarming growth in losses to climatic disasters in recent decades suggests that our adaptive capabilities are already eroding. The fact that much of this increased loss is due to socioeconomic factors and *not* to climate change is, ironically, one of the more worrying things about it. It demonstrates that we are deliberately, even blindly, exposing ourselves to greater risk by engaging in maladaptive practices that increase our vulnerability to the *present* climate. If we continue to do this, we will be just that much more vulnerable as the climate changes in the future.

One of the best ways to improve adaptation, therefore, is to discourage maladaptive behaviors. Two in particular have come under scrutiny: land use practices and disaster compensation.

As we've seen, one of the major causes of increased losses to natural disasters is the increased concentration of people and property in areas vulnerable to weather-related disasters, particularly flood plains and coastal regions. During the 1970s and 1980s, nearly half of all construction in the U.S. took place in coastal regions. Growth in the value of exposed property is also a contributing factor; it has been estimated, for example, that more than US$1 trillion in property along the U.S. east coast is potentially vulnerable to damage from hurricanes and storm surges. This trend is expected to increase as more people (especially, in North America, retiring baby boomers) move to coastal regions. It's been estimated that by 2010, the coastal population in the U.S. will have grown to more than 127 million people, a 60% increase over 50 years.

In many countries, dikes have been built to hold back the ocean and have allowed development on what used to be salt marshes. According to John Shaw of the Geological Survey of Canada, these marshes used to provide a natural buffer zone, rising as the sea level rose, but "now because they're behind the dikes, they get lower and lower relative to sea level in the region." Many marshes have been converted to agriculture or development; for example, in Nova Scotia's Bay of Fundy there's an "extensive system of dikes and behind that we've got railroads, agricultural lands, and some development of industry and residences in low-lying areas." He says that if there's a coincidence of high tide with a major storm, the dikes could be breached. "The dikes are maintained about one foot above the highest predicted tides, but that doesn't include the possibility of having a storm surge of one to two metres."

Shaw said that in some European countries, where much of the coastline is already behind large engineering structures, dikes are being thrown open and salt marshes are being returned to their natural state to provide natural buffering against sea level rise and storm surges. "We should try to allow the coast to function as much as possible as it does in nature," he said. "We should not be building there and we should be prepared to move property back when the beach moves back." However, measures such as those undertaken by European countries are not even being studied in Atlantic Canada, he said. "We already have a problem and it will be very much worse if the predictions of global warming come true." He added that, unlike Europe, Canada still has large areas of pristine coastline. "The way we adapt to future sea level rise is to recognize the natural coast and plan development activities on that basis, allowing the coastal system to function close to its natural way."

In the U.S., there has been ongoing controversy over development on barrier islands along the Atlantic and Gulf coasts. These islands are thin strips of sandy beaches that, in the past, shielded the coastlines from storm surges and erosion; in recent decades they've become extremely valuable real estate. In an effort to discourage development in these flood-prone regions, the U.S. government in 1982 passed the Coastal Barrier Resources Act, which prohibits the sale of federally sponsored flood insurance to designated high-risk areas. Property owners can still build there, but they assume the entire financial risk of doing so. Florida has also banned the use of state funds to encourage growth on coastal barrier islands or to provide assistance after storms.

However, there are many flood-prone areas that are eligible for coverage under the National Flood Insurance Plan sponsored by the U.S. Federal Emergency Management Agency (FEMA). (Private homeowners' insurance doesn't cover flooding.) The premiums are very reasonable—on average, about US$300 a year. Property owners in designated "special flood hazard areas" are required to have this insurance if they want to get a federally backed mortgage loan. Moreover, since 1994, uninsured people who have received federal disaster assistance after a flood aren't eligible for future assistance unless they buy flood insurance.

The flood insurance program has been criticized for encouraging development and redevelopment in risky areas, but FEMA argues this isn't so. Communities in high-risk regions can't qualify for the program unless they have a flood plain management program and impose building standards intended to reduce flood damage. FEMA says it also endeavors to discourage redevelopment in risky areas after flooding. Speaking on CNN, FEMA director James Lee Witt said that, since 1993, the agency has bought out some 20,000 properties and prohibited further building on those lands. It will continue doing so, he said. "It not only helps the environment, but it gets people out of harm's way."

Witt later announced that FEMA would take stronger measures to reduce its payments to communities and homeowners who make repeated claims to repair damage from flooding. When the flood insurance program started, it was projected that by 1990, only 10% of homes in risky areas would be subsidized—by 1998, 40% of these homes were still subsidized. These structures accounted for 96% of the estimated US$200 million spent on repetitive losses. Witt said that communities and homeowners who stay in risky areas must "take a more appropriate share of the responsibility for their natural hazard risks. Neither flood policyholders nor the taxpayer should subsidize repetitive losses....There is no reason to stay in harm's way when we can break the damage-repair cycle."

Still, it's difficult to get people to abandon development in high-risk areas that are often highly prized for their location and beauty. Californians love clifftop houses that overlook the ocean, but the price they pay is watching multimillion dollar structures collapse into the water or be carried away by mud slides when heavy rains destabilize the steep hillsides. Jacinthe Lacroix, a natural hazard specialist with Quebec's Ministry of Public Security, said it's difficult to dissuade people from building near rivers and lakes because "people love to be around the water and will do anything to be there. Usually when people build, it's summertime and the river is very nice." She said the province does at times relocate people whose homes have been destroyed by floods, particularly in high-risk areas such as the Saguenay region. Often this is less expensive than rebuilding the houses, especially if they're vulnerable to being damaged again. "We try to explain that they have to do it. We're paying for it, so when a house is at risk, we have to move it because we don't want to pay again in two or five years." But she admits it's often hard to persuade people to move. "People tend to stay where they were born and raised."

A town called Folly Beach in South Carolina was nearly wiped out by Hurricane Hugo in 1989, but the devastation caused a rebuilding boom that resulted in even more development on the beach, financed largely by federal insurance and assistance funds. Orrin Pilkey, of Duke University's Program for the Study of Developed Shorelines, described Hugo as "an urban renewal project" and said that local residents, rather than learning from the storm, thumbed their nose at it.

"In many coastal resorts, beaches are disappearing because houses are being built where the dunes used to be," writes James Titus of the U.S. Environmental Protection Agency. "If the houses weren't there, the dunes could move inland. Beaches are also getting narrower because bulldozers are excavating the coastline, often to build dunes....Along estuarine shores like Chesapeake Bay, the natural shoreline is being replaced by walls of concrete, rock and wood....In many cases, we are not yet even dealing with the fact that sea level is rising at all. For example, wetlands protection laws and other U.S. policies are based on the assumption that the sea isn't rising and shores aren't eroding, even though we know that they are."

Sometimes, however, the damage is so severe that it forces political leaders to act to prevent a reoccurrence of the disaster. After Hurricane Pauline devastated parts of Acapulco, Mexican President Ernesto Zedillo acknowledged that local authorities should not have permitted shanty towns to be built on steep hillsides around the beaches. It was these buildings, not the lavish tourist hotels, that took the brunt of the flooding and mud slides that killed hundreds of people. Zedillo said he would not allow homes to be reconstructed in these highly vulnerable locations.

Given the large populations and the amount of infrastructure now firmly entrenched in coastal zones and along major rivers, it may not be feasible to retreat from all vulnerable areas. The IPCC report notes that this could result in loss of property, costly resettlement of populations, and in some cases refugee problems. It may be that measures like building codes, new materials, design changes, and engineered structures will, of necessity, become the first line of defense in many of these regions. Unfortunately, societies that cannot afford such costly measures may face a rising toll from weather disasters unless they receive financial and technical assistance from outside.

The issue of government assistance and compensation for losses to natural disasters has come under scrutiny because of concerns that these, too, may have a maladaptive impact. Both the Canadian and U.S. governments provide disaster assistance, but in different ways. In Canada, federal payments kick in when the dollar amount of a disaster exceeds the provincial population; federal payments equal one-half of the costs between $1 and $3 per capita, three-quarters of costs between $3 and $5 per capita; and 90% of the rest. In the U.S., federal disaster relief, which usually consists of low-interest loans and/or temporary housing plus partial compensation for damaged public facilities, is given only when the U.S. president officially declares a region a disaster area. (It should be remembered that most of these disaster payments, which have climbed into the tens of billions in recent years, are added to the public debt.)

Disaster compensation programs are a political necessity, but whatever else they accomplish, they enable people to continue engaging in behavior that puts lives and property at risk. Burton comments that many people who work in sectors that are most affected by climate and weather—such as water managers and farmers—don't take climate change very seriously. He says they tell him that there's nothing they've heard about so far they can't cope with, that adaptation is a way of life for them, and that "climate change is just one more wrinkle on top of it." But he says that "part of the calculation is that if they suffer severely enough from extreme weather, they will get some help—and experience shows they do." It's politically difficult for governments to tell these people they're on their own. "Once a disaster strikes, the public pressure for giving relief is very strong and it takes a strong politician to refuse it."

University of Guelph geography professor Barry Smit says that government subsidies and disaster assistance reduce the incentive for farmers to adapt to climate variability and extremes. If farmers know they have a safety net, there's little motivation to diversify their operations; instead, they engage in risky behaviors and push their farms to the limits, knowing they won't have to assume the financial consequences. For example, farmers will grow crops they know are covered by compensation rather than planting other crops that might be more suitable to a changing climate. Thus, Smit says, government support programs tend to "perpetuate a system of agriculture which guarantees future crop losses and hardship to farm families and communities and guarantees a continued drain on public funds. The package is currently structured to shift climate-related risks from the farmers to the general public." He argues there should be no compensation for climate-related risks that are known to recur frequently, such as drought and hail, and for which adaptive responses are available. These risks should essentially be considered part of the "business environment" that farmers should be factoring into their planning and government compensation should be limited to very rare events that are unpredictable and for which adaptation is impractical.

This is the approach taken by the New Zealand government, which withdrew support in the late 1980s because of a debt crisis, arguing that "farmers choose to farm where they farm, and should accept the risks that go with it, pay a land price accordingly, and farm to the conditions." As a result, the percentage of farm income derived from subsidies dropped from 35% in 1983 to 2% in 1990. Smit notes that New Zealand's farm economy did not collapse. Instead, farmers made their operations less vulnerable to climate variability and extremes; they diversified their operations, stopped planting vulnerable crops in risky areas, and no longer pushed their land to its limits. There was increased demand for private insurance, climate data, and adaptive technologies such as drought-resistant grasses. New Zealand's experience shows that removing or modifying government support programs "could result both in an agriculture less susceptible to climatic variations and an agriculture which does not continue to drain the public purse," Smit notes in a scientific paper.

Adapting to Climate Change

Concern about our current levels of adaptation and our ability to adapt to future climate changes is one of the major reasons for the increase in impacts studies in recent years. We need to gain a better understanding of how our social systems and different sectors of the economy are likely to be affected by climate change to determine our adaptation priorities, to help us decide how best to deploy our adaptive resources, both financial and technological, and to get a handle on what they will cost. This includes positioning ourselves to take advantage of any opportunities that climate change might send our way, as well as protecting ourselves from adverse effects.

The Canada Country Study (CCS) grew out of a recognition that there were serious deficiencies in adaptation research, in part because of the stronger focus on avoidance measures, but also because we have so little information about the environmental and socioeconomic impacts of climate change. "Impacts knowledge is clearly in an unsatisfactory state, resulting from the uncertainty about the rate and extent of climate change and its regional variation," notes a report from the workshop that initiated the study. Until we do such studies, optimistic assertions that we'll cope somehow (because we always do) are nothing more than statements of blind faith.

Adaptation involves costs, just as avoidance does, but we still don't have a very good understanding of what those costs will be. According to the IPCC report, "systematic estimates of the costs of adaptation to cope with impacts on agriculture, human health, water supplies and other changes are not available." Burton says there are no reliable data, nor even any reliable estimates, of the current costs of adaptation and what their trends might be in the future. "It is a continual source of frustration that we are prepared to spend so much effort trying to guesstimate the costs of an uncertain future climate on an unknown future economy without any sort of reliable baseline data."

According to the CCS, adaptation costs fall into three categories:

• *Technical adaptation*, which involves engineering measures such as building dikes, protecting pipelines from permafrost melting, and building structures to withstand weather extremes.

• *Social adaptation*, which involves changes in lifestyle, employment, and the workings of the social system. These costs include dislocations caused by the impact of climate change on economically important natural resources such as fisheries, forestry, and agriculture.

• *Environmental adaptation*, which involves changes to and redistribution of natural ecosystems in response to climate change. The costs associated with this kind of adaptation "are largely unknown," the CCS says.

The range of possible adaptive measures within these three categories is enormous. Because different regions face different climatic challenges, their choice of adaptations will also vary greatly. In many places, however—particularly in developing countries—choices are limited because the money or technology to implement adaptive measures is not available. Even developed countries may have difficulty implementing all the adaptive strategies that might be called for; if climate change occurs rapidly or greatly increases the frequency and severity of extreme weather events, it could strain the resources of even the richest developed countries.

The IPCC report outlines six basic strategies for adapting to climate change: preventing the loss, tolerating the loss, spreading or sharing the loss, changing the use or activity, changing the location, and, finally, reconstruction. In their paper on climate change and insurance, Rodney White and David Etkin note that, in North America, the emphasis has been on preventing loss with technology, on sharing the loss (i.e., insurance), and on rebuilding. This approach assumes that the current rate of losses is sustainable and that losses resulting from extreme events that overwhelm our protective measures will not be prohibitively large. However, given what's happened in recent years, they say these assumptions are debatable.

In another paper on human activity and natural hazards, Etkin and his colleagues summarize three basic approaches to mitigating or reducing natural hazards: modifying the hazard (e.g., hail suppression programs), sharing the risk (e.g., insurance, disaster assistance), and reducing vulnerability (e.g., changing the built environment, disaster preparedness). Some representative examples of these and other measures are described below; though this is far from an exhaustive list, the examples illustrate some of the potential benefits and pitfalls associated with trying to "ride out the storm" of climate change.

As previously noted, researchers concerned with adapting to climate and weather often refer to this as "mitigation." Unfortunately, this word is also used by climate scientists to refer to *avoidance* measures (i.e., reducing greenhouse gases). In short, the same word is used to describe almost exactly opposite approaches to dealing with climate change—something that could be confusing to the public, policymakers, and other nonscientists trying to understand expert opinion about how to cope with climate change. Moreover, as Pielke notes, the conflicting meanings of the word may inhibit climate scientists and natural hazard experts—two groups that have traditionally functioned separately—from working more closely together. This is unfortunate, he says, because "without a doubt the knowledge gained by the hazards community has an important role to play in the climate policies of the future." The participation of natural hazard experts is especially important considering the fact that climate change may be accompanied by an increase in extreme weather events.)

Adaptation Measures

Infrastructure: There are two ways in which adaptation affects infrastructure: first, engineered structures (e.g., dams, dikes) provide a means of protecting people and property from natural hazards and second, structures themselves (e.g., homes, businesses, roads, bridges) need protection from natural hazards.

As White and Etkin have noted, using technology to protect us from natural hazards (as opposed to changing our activities or location) is one of the favored strategies in North America. For example, we've built dams, reservoirs, canals, dikes, and irrigation systems in an attempt to smooth out the vagaries of nature when it comes to water resources. People responsible for such programs say they're doing a good job of adaptation. At a 1997 conference, Eugene Stakhiv of the U.S. Army Corps of Engineers' Institute for Water Resources said: "We are constantly adapting in the water management field. Do not tell me that adaptation is not working. We are spending a lot of money, $125-billion per year, to restore the environment, adapt it, and make it a livable place. We are not simply sitting waiting for things to happen....We are following a 'no regrets' adaptation policy in the water management field."

Even so, engineered structures can be defeated by extreme weather events. Jacinthe Lacroix gives one example: a series of dams on a river is typically built with the biggest dam at the top and smaller dams downstream. In a situation like that which caused the Saguenay flooding, when the top dam was opened, the smaller dams downstream could not take the huge influx of water. To better adapt new dam systems we should "consider that we have to be able to open the doors in a row and make sure all the dams can accept water from the top."

It may become increasingly difficult to use engineered structures for adaptation as the climate changes in the future, however. They're already very costly and are likely to be even more expensive if they're expected to provide protection against larger and more frequent extreme events. One study cited in the IPCC report estimated that it would cost Japan about US$92 billion to protect its coastal infrastructure from flooding due to rising sea levels; about two-thirds of that would be needed to raise port facilities like quays, wharves, and sea walls.

Canada spends more than $1 billion a year to adapt its water resources to current climatic conditions, according to the CCS. "These measures include the construction of dams, sewers, drainage ditches, floodways and other structures. Adapting to future climate change is expected to increase these costs even more." Moreover, projected reductions in water levels and increasing drought in the prairies may make expansion of the irrigation system an impractical solution. "Irrigation is expensive and uses vast quantities of water. With a projected reduction in water supply in this region, irrigation may not be a viable way to adapt to the projected climate change."

A report on water resources by Kenneth Frederick of Resources for the Future says that in recent years, increased costs and limited opportunities for expansion have shifted the emphasis from engineered structures, such as dams and levees, that manage the water supply toward social and economic measures that encourage conservation and reduce the demand for water. While new infrastructure still may be appropriate in some future circumstances, he says it's difficult to plan for and justify such projects when there are so many uncertainties about what will happen to water resources as the climate changes. "The prospect that global warming will alter in unknown ways local and regional supplies and demands reinforces the need for institutions that can facilitate adaptation to whatever the future brings and promote more efficient water management and use," says Frederick, who advocates measures such as elimination of irrigation subsidies, pricing water at its true social cost, and developing transferable water property rights. Unlike engineering measures, he says, these approaches do not require accurate information about future climate changes, large amounts of cash, or long lead times.

A report on water resources by the Worldwatch Institute also says it will be far less costly to implement conservation and efficiency measures (e.g., repairing leaks, water recycling) than to build new dams and reservoirs. It points out that a conservation program in southern California is providing water at about 10 cents per cubic meter, far less than the cost of water from new supplies. "A host of such measures with the potential to save vast quantities of water remain untapped because of inadequate incentives to encourage their use," the report says.

Protecting infrastructure from the impact of climate change is the other side of the coin. The design of buildings, roads, bridges, airports, pipelines, coastal facilities, and, indeed, nearly everything else we build should be reevaluated in light of their increasing vulnerability to climate and weather and, in many cases, new approaches or retrofitting may be needed.

The debate that followed the 1998 ice storm provides a good example of the reassessment that occurs when extreme events unmask our vulnerability. Quebec's Public Security Minister Pierre Belanger said the province must be better prepared next time: "There are climatic realities that I think we can no longer deny. We have to put our efforts into being ready for these events." *Toronto Star* political columnist Rosemary Spiers wrote that when Canada's federal Cabinet ministers met during the height of the storm, some conceded in private that it might be more than just a "freak of nature." They reportedly ribbed Environment Minister Christine Stewart about going overboard to gain acceptance of the Kyoto treaty, saying, "Enough already, you've made your point."

The storm prompted a great deal of discussion about burying electrical transmission lines and about the need to decentralize and diversify the energy system. Quebec is unusually dependent on electricity for heating, and most of it comes long distances from hydroelectric installations in James Bay—factors that made the crisis worse that it might otherwise have been. Hydro-Québec officials weren't enthusiastic about calls to bury transmission lines, however. A spokesman, Guy Versailles, told the CBC that the expense would be enormous; the company has tens of thousands of kilometers of wires, and the cost of its $7 billion distribution system would triple if they all were buried. But he acknowledged that "events such as this one will bring everyone in Quebec to rethink this."

However, even before the crisis was over, Hydro-Québec announced a $650 million plan to build new power lines to Montreal and strengthen the existing ones. In fact, it decided to push ahead with a new aboveground transmission line to the east end of Montreal Island that had previously been vetoed by the province's environmental bureau. Hydro-Québec argued that putting this line underground would cost up to $189 million and take three to five years, while the aboveground plan would cost about $28 million and take one year.

This plan was widely criticized as being more of the same and just as vulnerable to failure. "New power lines will crumble under ice storms just as much as these ones," said Steven Guilbeault of Greenpeace. Philippe Dunsky, director of the Helios Centre for Sustainable Energy Strategies, said it was time to end the province's reliance on what is "essentially a very fragile system" by diversifying into other energy sources such as solar power or fuel cells.

The Montreal newspaper *The Gazette* also advocated greater diversification to reduce risks from extreme weather events. "If the storm is a sign of changing weather patterns, as Premier Bouchard contends, then why use the same technology that has failed Quebecers so miserably?" the *Gazette* editorialized. Not only did it recommend that new power lines be buried, but it also suggested that the province look at measures like using new fuel cell technology to power hospitals or building a gas-fired generating plant on Montreal Island to support essential services in an emergency. "Mr. Bouchard should spend what's required to ensure that Montreal will not be at the mercy of an ice storm in the future."

Northern regions of Canada may face major adaptive challenges caused by the melting of permafrost, which poses potentially serious problems for many structures such as pipelines, roads, bridges, air strips, reservoirs, and mining facilities. Permafrost loses strength as it warms. Ice-rich soils settle and heave, creating uneven surfaces and also become increasingly permeable, allowing fluids to pass through. This could result in "severe maintenance and repair problems," according to a report by the Environmental Adaptation Research Group, which says the operating life of many structures could be significantly shortened. "Some roads and structures might have to be rebuilt or abandoned." It recommends particular scrutiny of permafrost structures used to store hazardous wastes and to contain toxic mine tailings. One way of adapting to the problem of permafrost melting is to bury sensors in the soil to monitor its frozen state and to use refrigeration technology to extract heat from the soil.

Coastal facilities will also require reassessment as sea level rises, particularly if the incidence of storms also increases as the climate warms. Damage may increase in regions where ice cover traditionally has provided some protection from waves and storm surges. The insurance industry is pushing for improvements in building practices and codes in coastal communities vulnerable to extreme weather. The Insurance Bureau of Canada's Institute for Catastrophic Loss Reduction plans to collect information on safer building techniques and on structural designs and materials that can best withstand natural hazards. "Often incorrect construction is to blame for damage and, indeed, one of the great future issues will be how to retrofit substandard buildings," said Jim Harries, IBC's manager of policy development.

In the U.S., there's been an effort to improve the ability of mobile homes to withstand wind damage. Occupants of mobile homes, which provide the only affordable housing for many poor and elderly people, account for nearly half the annual deaths caused by tornadoes, according to Joseph Golden of NOAA's Weather Research Program. Mobile homes built to the improved standards survived Hurricane Andrew with less damage than those that were not. Of course, it's important that governments not only establish but strictly enforce these building codes. Some of the strongest codes for wind damage were in force in Dade County, Florida, when Hurricane Andrew blew in, but it was later discovered that these codes had been poorly enforced and many of the damaged buildings did not meet the codes. "If you have a good building code in law but it's not enforced, it won't do much for you," Pielke notes. Some buildings in eastern Canada have also been damaged by tornadoes because they did not meet code standards.

Angus Ross of SOREMA Insurance said that U.S. insurers are now rating municipalities according to their level of code compliance and adjusting their premiums accordingly. He suggests this trend is likely to grow. "Insurers may start grading towns on their municipal infrastructure." One area of particular concern is sewer backup, which occurs during heavy rainfall. Episodes in Ottawa and Winnipeg caused $200 million in damage to commercial and residential buildings. A suburb of Montreal experienced serious sewer backups three times in 18 months and insurers refused further coverage, believing that sewer backup was virtually inevitable because of the inadequacy of the local infrastructure, Ross said. "It cannot cope with growth in the area and any rainfall is going to back them up." In many municipalities, he said, green areas that once provided natural absorption and drainage are now covered with asphalt and buildings, and sewer systems just aren't up to the job.

Refusal of coverage is "a consequence the insurance industry is going to visit on people in areas where things can be done," he said. "It will highlight for municipalities the problems they can face in future; local politicians will have to answer to their citizens why they can't get insurance because of the failure of infrastructure." He noted that the National

Round Table on Water and Wastewater Management has estimated that about $41 billion in new capital spending will be required for municipal water and wastewater infrastructure in Canada over the next two decades, and another $38 to $49 billion will be needed to maintain and refurbish the existing stock.

By increasing premiums, setting higher deductibles, and, in some cases, denying coverage, insurance companies have forced communities and individuals to confront the need to improve adaptation to climate change. The industry's Institute for Catastrophic Loss Reduction states that "those who knowingly choose to assume greater risk must accept an increased degree of responsibility for their choice." Harries noted in a speech that Hurricane Andrew caused a major "tightening up" of Florida's property insurance market. "The majority of insurers withstood the hit but not without radical adjustments to future exposures they were prepared to maintain, premiums charged, deductibles allowed and coverages offered." This process left many property owners without insurance and led to the creation of a Joint Underwriting Association. However, Harries said there are doubts about the association's ability to handle another major hurricane. He added that while higher deductibles aren't popular with property owners, "they can discourage complacency" and provide an incentive to undertake adaptive measures.

The Canadian insurance industry is also involved in one of the few adaptation programs aimed at modifying weather hazards. For several years, they have funded a cloud-seeding program in Alberta to reduce the size of hailstones. (In 1991, a hailstorm in Calgary cost nearly $400 million in damage to houses and cars, which set a record for insured losses at the time.) Ross said the annual cost of seeding ranges from $1.4 to $2.5 million and the industry feels it's getting its money's worth. In 1996, they seeded 84 storms; two developed damaging hail because they were seeded late and caused about $100 to $120 million in damage. But Ross says one storm was estimated to have been big enough to cause what would then have been Canada's first $1 billion loss. In 1997, 111 storms were seeded and there were no major losses to hail.

Hail suppression programs have been tried in nearly a dozen countries and while some reported successes, others (including a US$25 million NCAR project) did not have encouraging results. According to the report on natural disasters and human activity, support for cloud seeding in the U.S. has waned for several reasons, including the fact that 50 years of research has provided no rigorous evidence that it's effective. It notes that the Alberta program does not involve scientific research and "its perceived success is based on unproven hypotheses." However, it says that analysis of a cloud-seeding program in North Dakota by the state's Atmospheric Resource Board suggests it has been successful in reducing hail damage to crops.

Warnings and forecasts: The development of technologies like radar, satellites, and computer models during the 20th century has increased our ability to forecast extreme weather events and warn people in their path. For example, progress has been made in predicting the number of hurricanes that will occur during the June to November Atlantic hurricane season by analyzing diverse indicators such as rainfall in Africa, sea surface temperatures, atmospheric high pressure zones, winds in the stratosphere, and El Niño.

Data from satellites, aircraft, and improved computer models have also improved the ability of scientists to track hurricanes and estimate where they'll hit land. During the very active 1995 season, with 19 storms, some right on top of each other, these tools improved forecasting of storm tracks by 10% despite the unusually heavy workload, said Bob Burpee, director of the U.S. National Hurricane Center (NHC). Accurate track forecasting helps to minimize unnecessary, costly, and disruptive protective measures and evacuations. However, researchers say it's still hard getting the public to take forecasts seriously; Jerry Jarrell, deputy director of the NHC, said that many people—even those living in areas already devastated by hurricanes—do little to prepare and rush to stores at the last minute to buy protective equipment and supplies.

In the U.S.—the tornado capital of the world, with an average of 100 a year—tornado tracking has also improved with the deployment of Doppler radars in many high-risk regions. Joseph Golden of NOAA'S Weather Research Program notes that until the early 1980s, the lead times for tornado warnings were negative—in other words, the twister had already touched down before the warning was issued. In the 1990s, however, the lead times have been positive, averaging about 18 minutes prior to touchdown.

Perhaps one of the most important tools for warning people about impending weather disasters is a special weather radio that turns on automatically when it receives an emergency signal. This alerts people who are unaware of warnings issued via conventional media. One of the reasons that a series of severe tornadoes caused so many injuries in the southeastern U.S. in early 1998 was because they hit at night when many people were asleep.

Emergency preparedness: The point of forecasting is to give people time to prepare for disasters and thereby reduce their social and economic impact. At an "El Niño summit" held in California in October 1997, James Lee Witt, director of FEMA, launched a national campaign, called Project Impact, to help communities reduce the impacts of disasters through education and planning. "Our goal...is to change the way America prevents and prepares for disasters," he said, adding that failure to do so would mean paying "over and over for our lack of preparation."

California communities did, in fact, take measures to reduce El Niño's impact; beaches were built up; storm drains, gutters, and runoff areas were cleared of debris; and roofing contractors enjoyed a booming business as property owners fortified their homes against wind and rain. "It's not possible to overprepare," said Los Angeles Mayor Richard Riodan. (This, it turned out, was prophetic. While the preparations undoubtedly did some good, the relentlessness of the rains defeated many of these adaptation efforts. It doesn't help much to clear gutters and reinforce roofing tiles when the whole house slides down the hillside in a sea of mud.)

However, preparedness can produce substantial benefits in reducing damages from extreme events. The report on natural disasters and human activity prepared by David Etkin and his colleagues cited an analysis of two similar floods that occurred in Michigan and Ontario. In Michigan the nonagricultural damages were US$200 million, while the Ontario flood caused $500,000 in damage—a difference attributed to the fact that Ontario had a stricter flood damage reduction program.

The report notes that disaster preparedness requires the development of a capability to manage an emergency in advance of a hazardous event. Responding to emergencies involves several major tasks: detecting the hazard and warning people; evacuating those threatened by the disaster; providing food, shelter, and medical care for victims; search and rescue; and protection of property. Typically, numerous agencies are involved, from the local to the national level, so coordination is always a major requirement and is something that must be worked out ahead of time, not in the midst of crisis.

Preparedness is not solely the responsibility of government agencies, however. The entire community should be involved in reducing vulnerability to extreme weather events. One example is the proactive approach to heat waves adopted in Memphis. People with air conditioning are urged to take in those who don't have it, particularly the elderly. Various organizations and service providers—letter carriers, Meals on Wheels delivery people, visiting nurses, the department of health, and the Office on Aging—participate in actively seeking out elderly people and others who are susceptible to heat stress and transporting them to air-conditioned shelters.

Writing in the *New England Journal of Medicine*, Arthur Kellermann and Knox Todd of Emory University note that the county in which these measures were implemented has experienced an average of only two heat-related deaths per year since 1980 and a maximum of 11 deaths in any one summer. "Communities must make a determined effort to seek out the elderly, the infirm, and the shut-in," they said. "Aggressive efforts to reach out to the most vulnerable can markedly reduce the toll of heat-related deaths." They recommended that local churches, synagogues, and mosques open their air-conditioned facilities during the hottest times of the day and that local governments in high-risk areas modify building codes to require hotels and apartment buildings to have air-conditioned lobbies.

This is only a small sampling of the adaptation measures that could be undertaken to reduce society's vulnerability to future climate change and weather extremes. A great deal more research must be done to enhance our prediction and warning skills and the technological and social mechanisms that will help us cope as the climate warms during the next century. Efforts to this end are already being made through the United Nations' program known as the International Decade for Natural Disaster Reduction (IDNDR). This program, which ran throughout the 1990s, aims "to reduce through concerted international action, especially in developing countries, the loss of life, property damage, and social and economic disruption caused by natural disasters." The fact that deaths from natural disasters have been rising at a slower rate than economic losses suggests there have been improvements in warning and preparedness systems. But deaths are still rising, especially in developing countries, which accounted for 88% of the world total between 1985 and 1992.

The U.S. government, through FEMA, has developed a National Mitigation Strategy with two goals to achieve by 2010: first, to greatly increase public awareness of the risks of natural hazards so the public will demand safer communities and, second, to significantly reduce the loss of life, injuries, and socioeconomic costs caused by natural hazards. This program is based on the principle that "current dollars spent on mitigation [adaptation] will save a significantly greater amount of future dollars by loss reduction." This program signals that FEMA no longer confines its activities solely to responding to disasters after the fact; it's taking a more proactive approach by encouraging advance preparation to reduce the damages wrought by natural disasters when they do strike. Since 1994, FEMA has spent about 15% of its disaster assistance funds on adaptive measures.

Canada has not established a similar comprehensive national program, although many federal and provincial departments and agencies have programs aimed at reducing the impacts of natural hazards. According to Etkin's natural hazards report, the reason for this is that "Canada has, until recently, been relatively disaster-free compared to other places."

This situation may have lulled us into complacency, but we should not count on its continuation, given the projections that climate change will affect higher latitudes more than lower latitudes. Certainly the huge and costly weather disasters of recent years—the Saguenay and Red River floods and the 1998 ice storm in particular—are warning signs that not only are we not immune to such disasters, but also we're not as well equipped to deal with them as we'd like to think. It also may be time to reassess our approach to adaptation, in particular our tendency to rely on technology and engineered structures to hold back nature. While these measures will probably always have an important place in any adaptation plan, we must consider other options, such as choosing to remove ourselves from harm's way and to stop encouraging risky maladaptive behavior through government subsidies and compensation programs.

Finally, while it would not be prudent to reject adaptation as a response to climate change, it would be equally imprudent to assume that our adaptive skills are so remarkable that we can dispense with the effort to reduce greenhouse gas emissions. Reducing emissions is critical if for no other reason than it will slow the rate of climate change and buy time to improve our adaptive capabilities. Both strategies are needed but both involve substantial costs and uncertainties. Finding the right balance between adaptation and avoidance is one of the key challenges we'll face in our efforts to come to grips with the impacts of a changing climate over the next century.

twelve

THE WRONG
QUESTION

When the history of global climate change is finally written, it will show that the debate was badly derailed early in the game by three simple words: proof, uncertainty, and risk. More specifically, it was sidetracked by a preoccupation with inappropriate legalistic definitions of proof, a refusal to deal realistically with scientific uncertainty, and a failure to address the question of risk.

This was not entirely accidental. The demand for proof that global warming was happening, that human activities were causing it, and that it would have negative consequences was a key strategy the skeptics deployed very effectively, especially in the U.S. It's a delaying tactic that has worked well in other scientific controversies, such as those surrounding CFCs and ozone depletion or smoking and cancer.

The inevitable and unavoidable existence of scientific uncertainty is the fuel that powers disputes over global warming. It's an effective mechanism for diverting the energies of the scientific community into unproductive wrangling over what is, ultimately, the wrong question. It also diverts the attention of the media, the public, and policymakers from the most important question, which is *not* whether global warming projections will come true but whether we—or, more likely, our children and grandchildren—will be prepared to deal with the consequences if they do.

The real question, therefore, has little to do with proof. It has to do with the precautionary principle—in its simplest terms, "better safe than sorry"—or, more precisely, our willingness and ability to embrace it. This is the question we've barely begun to answer because, thanks largely to fruitless debates over the issue of proof, we've barely begun to ask it.

The Question of Proof

It may strike many people as strange to suggest that proving the validity of the global warming theory should not be our primary objective as we struggle to come to grips with climate change and its impacts. It seems more than reasonable, even prudent, to demand such proof before taking steps to combat the problem, especially since they may involve substantial up-front costs. Certainly no one is advocating that we embark on expensive and draconian efforts to stave off global warming without any *evidence* that the phenomenon is happening or that it will have adverse impacts. But scientific research to date has provided compelling evidence of the reality of global warming and more than adequate grounds for serious concern about its potential for causing negative social, economic, and environmental consequences.

Many skeptics, of course, deny that this evidence is strong enough to qualify as proof. In fact, some of the more extreme skeptics dismiss virtually every scientific finding that tends to support the global warming theory as "junk science," "voodoo science," or "paparazzi science," while citing approvingly, though often selectively, other research they believe casts doubt on the theory. One example concerns the paper on the 1995 Chicago heat wave by Thomas Karl and Richard Knight (discussed in Chapter Seven). An article in a skeptical newsletter called *Calmer Weather* was critical of environmentalists and President Clinton for citing the heat wave as evidence of global warming, pointing out that Karl and Knight had found no increasing temperature trend in the Chicago area between 1948 and 1995 and that the urban heat island had played a role in causing the heat wave.

This was true as far as it went. However, the article failed to acknowledge other findings that did not fit the skeptics' agenda quite so well. For example, Karl and Knight concluded that the Chicago heat wave was unprecedented mainly because the nighttime (minimum) temperatures remained so high, which prolonged the heat wave by not giving people any respite at night. This directly undermines a key skeptical claim that global warming will not increase the incidence of heat waves because temperatures are rising faster at night than during the day. Moreover, while it's true that Karl and Knight did not attribute the Chicago event to global warming, they did say that future climatic changes could increase the incidence of extreme events of this kind. Neither of these conclusions was mentioned in the *Calmer Weather* article.

What's most ironic about this game of scientific citation is that many research papers that are mentioned approvingly by the skeptics have been conducted using the same scientific methods and published in the same peer-reviewed scientific journals as those that are dismissed as "junk." What criteria, then, do they use to distinguish junk research from good research? It appears that for many skeptics there is only one criterion: findings that support global warming are junk and those that don't are not. In short, among skeptics given to cries of "junk science," it appears there is simply no evidence supporting the theory of global warming that would be accepted as proof.

Good examples of this are provided by the newsletter *World Climate Report*; the Global Climate Coalition, a lobby group opposed to reducing greenhouse gas emissions; and the Science and Environmental Policy Project (SEPP) (described as a non-profit educational and policy research group founded to "foster greater reliance on sound science in decisions affecting health and the environment"), which have been highly vocal opponents of the global warming theory. These organizations clearly have substantial resources to follow, comment on, and react to practically everything published in the scientific literature and the media about global warming and to everything going on in the political arena as well. They immediately jump on any new information and spin it according to their unique perspective. For example, the same day that NOAA scientists presented data that 1997 was the warmest year on record, SEPP put out a news release contending that 1997 was one of the coolest years in two decades, based on satellite measurements, which it depicted as the "most reliable and only global temperature data."

As for *World Climate Report*, it argues with absolutely *everything* that lends even the smallest support to the concept of global warming. And although its articles contain a lot of scientific jargon, its tone is so aggressively glib and sneering that it could never be confused with a scientific journal. One example: "By now you know the drill: Warmer air evaporates more water, a moister atmosphere produces more clouds and precipitation, more precipitation yields more blizzards...yadda, yadda, yadda...pay more taxes." Hectoring sarcasm is directed not only at climate science, but also at any individual or group that expresses concern about the impacts of global warming. For example, it attacked the Physicians for Social Responsibility for urging reductions in greenhouse gas emissions, saying "these docs" didn't have degrees in climatology. "Our advice about future death and disease caused by global warming? Take two aspirin and call us in 2050."

WCR also accused the insurance industry of hyping global warming to justify raising premiums. "That's the cha-ching of cash registers you hear ringing up the increased premiums insurance companies may charge because of the weather....Go ahead insurance companies, make our day. Try to raise our rates based on global warming." Of course, insurers make absolutely no bones about the fact that they're concerned about the impact of extreme weather on their bottom line. They're in the business of assessing risk—and, yes, profiting from it—but they do not need to manufacture spurious weather trends. If demographic and socioeconomic trends were the whole explanation for increasing losses, they would simply adjust their premiums based on those factors. Are we to assume from *WCR*'s derision about their profit motive that the fossil fuel industry's opposition to greenhouse gas cuts has absolutely nothing to do with concerns about *their* bottom line?

It's important to remember that the debate over scientific proof of climate change involves two important underlying issues: the *burden* of proof and the *standard* of proof.

The burden of proof: One of the more interesting aspects of the climate debate is that opponents of greenhouse gas cuts have so successfully placed the burden of proof on the scientific community, environmental activists, and politicians who advocate immediate action to curb global warming. This arbitrary assignment of responsibility by one of the combatants in the dispute has gone largely unchallenged. Yet it would be equally reasonable to expect advocates of business-as-usual emissions to prove that such emissions will do no harm. This alternative is not unprecedented; after all, many businesses are routinely required to demonstrate in advance, at least to some extent, the safety of products they bring to market and to recall products if there's evidence they endanger public safety. Consumers expect this with everything from cars and aircraft to drugs and baby seats, and they're angered when companies are perceived to be skimping on safety and using the public as guinea pigs to protect their profits. Is it such a stretch to expect as much for products that we have good reason to believe can seriously harm the earth's climate?

Greg Paoli, an Ottawa consultant whose firm, Decisionanalysis, specializes in risk assessment, argues that there's little justification for the skeptics' views about who should bear the burden of proof. The burden should go with "the beneficiary of inactivity," he said—in short, those who benefit from delaying action to reduce greenhouse gas emissions. "The bearers of the risk should not also be the bearers of the burden of proof." In this context, the public and the natural environment are the bearers of the risk associated with warming.

In counterpoint, the skeptics argue that the fossil fuel industry and, indeed, the public bear the economic risks associated with limiting greenhouse gas emissions. Clearly, they do not intend to accept the burden of proof; one reason they so fiercely oppose the Kyoto Protocol is precisely because it represents a move in that direction. In this sense, the Protocol is a positive step, even though the emissions reductions it mandates will not, in themselves, do much to reverse global warming. But it will take ratification of the treaty and its full implementation in developed countries to signal that a shift in the burden of proof has truly taken place.

The standard of proof: In the climate debate, as in a legal proceeding, it's important to define what is meant by "proof." Many skeptics appear to demand a standard of proof analogous to that associated with a criminal trial (i.e., "beyond a reasonable doubt"), and they have successfully thrust upon the scientific community the burden of proving the global warming "case" to this high legal standard. The "accused" greenhouse gases are to be considered innocent until proven guilty.

In a society that eats up saturation coverage of courtroom spectacles, it's not surprising that environmental debates get tangled up with notions derived from the criminal justice system, but their relevance in the environmental arena is questionable. Applying legal concepts to complex environmental problems is not only inappropriate but actually dangerously misguided because it virtually guarantees a lengthy period of political and social paralysis before we finally accept the irritating but unavoidable reality that with environmental problems, *policy decisions must almost always be made in the absence of scientific proof beyond a reasonable doubt.* This is not, however, the same as saying decisions must be made in the absence of scientific *evidence.* The fact that we don't know everything does not mean we don't know anything. We may not have absolute proof, but there comes a time when the evidence tips the balance of the scales. What we should be concerned with is whether the projected impacts of global warming are *more likely to occur than not.* If we insist on being legalistic about it, the criterion that should apply is that of the civil case: *preponderance of evidence.* This is the criterion reflected in the IPCC's statement that the "balance of evidence" suggests a discernible human influence on the climate. It provides the foundation for the precautionary principle, since it's generally considered sensible to take precautions against adverse consequences that are more likely to happen than not.

The concept of *scientific uncertainty* is a key element in the debate over proof of global warming. The phrase is bandied about a great deal—by scientists and the media as often as by skeptics—and yet it may be one of the most misunderstood elements of the controversy. What many nonscientists do not realize is that when scientists talk about "uncertainty," they're typically referring to findings in which they have a level of confidence *below a range of about 90 to 99%*. In short, they classify things as "uncertain" at a much higher level than most of us in do in our daily lives; to the average person, saying that something is uncertain generally means it has less than a fifty-fifty chance of happening.

Disputes among scientists are usually over uncertainties at this extreme end of the scale (there's no need to argue about things they mostly agree on), but the public often concludes from these debates that scientists can't agree on anything. Worse, they generally assume that if scientists can't agree on the problem, the problem must not be that bad after all. What they often fail to realize is that uncertainty goes in both directions—that it's just as likely that the situation will be *worse* than scientists project as that it will be better. Uncertainty wasn't properly taken into account during the 1997 Red River flood, when people in Grand Forks, North Dakota, fixated on the prediction that the river would crest at 15 meters and were caught unprepared when it crested at about 16.5 meters.

These misunderstandings are compounded by the frequent failure of the scientific community to explain more explicitly that they define uncertainty far more stringently than most people do. In fact, scientists who try to convey the state of scientific knowledge in terms that nonscientists can understand are often criticized by other scientists "who feel the science has been oversimplified and uncertainties trivialized," says Gordon McBean, assistant deputy minister of Canada's Atmospheric Environment Service. This stems from their rejection of findings that fail to reach the 95 to 99% level of confidence—"a preoccupation that has its place within the world of scientific research, but is seldom found in other domains of human endeavor."

As we've seen, there are elements of the global warming theory that are more uncertain than others, but this does not negate the high levels of confidence that most climate researchers have concerning the fundamental elements of the global warming problem. They're already giving us better than even odds on many projections and as much as 90 to 99 out of 100 odds on some.

We make decisions every day with levels of uncertainty greater than this. If we insisted on a 90 to 99% probability for everything we do, most economic and political activity would grind to a halt. Does an oil company refuse to search for new deposits unless they're certain of finding oil? Does the business community guarantee positive returns when they exhort us to buy mutual funds? The simple fact is that most of life involves making choices with incomplete knowledge about what the outcome will be. It's absurd to refuse to deal with global warming because there are no ironclad guarantees about how it's going to turn out. If that were the criterion for living our lives, no one would go to school, get a job, get married, get in a car or on a plane, or, for that matter, even get out of bed in the morning.

Why do we expect so much more of science? Perhaps, in part, because many people don't understand the fundamental nature of the scientific endeavor—that it's not so much a case of finding the "right" answers as it is a laborious give-and-take process of reducing uncertainty. Science works by proving things wrong, not by proving them right. Conflict is a fundamental and essential part of this process, says Tom Yulsman, a professor at the University of Colorado's Center for Environmental Journalism. "Somebody proposes a hypothesis, finds some evidence. The whole purpose of his peers is to find all the ways that this new finding is wrong. I don't think the public understand that's how science works. They see conflict in the scientific process and say okay, we don't have to think about it until they all agree."

A 1996 statement issued by delegates at the climate talks in Geneva stated that "research is needed to reduce uncertainty, but cannot eliminate it entirely....It is not possible to quantify anticipated economic and social impacts of climate change fully before taking action." Yet many laypeople still expect that we can and must wait for science to tell us exactly what's happening before we can figure out what to do about it. This is essentially what the skeptics are advocating when they exhort us to "wait for proof." What we should be asking is this: What kind of proof are we waiting for? Do we really expect 100% odds before acting to prevent global warming? Are we holding out for a smoking gun? This would be foolishly unrealistic. "We must not expect a single, dramatic discovery to confirm 'global warming' once and for all," warns a statement from the United Nations Environment Programme. "If we wait for that discovery, we will wait for a long time—until well after it is too late to do much about it. There is no climatic counterpart to the Antarctic ozone hole."

The message is clear: Uncertainty exists. It won't go away. Deal with it.

In the end, it doesn't even matter whether the skeptics win the debate over waiting for proof. Simply *creating* debate is sufficient to delay measures to reduce greenhouse gas emissions.

It's unfortunate that the scientific community fell into the trap, but it's difficult to see how they could have avoided it. Many scientists do not like the rough and tumble of political and media wars—and this one is rougher than most—but at the same time, they were urged to take on the skeptics with their taunts of "junk science." When he was U.S. Interior Secretary, Bruce Babbitt spoke at a conference of the Ecological Society of America, arguing that scientists had a civic duty to make the case for global warming. "There are times that cry out for scientists' involvement, because the public doesn't have anywhere else to turn to," he said. "We can't get the message through by speeches from people like me. It is all of you that have that obligation." Jerry Melillo of the U.S. Office of Science and Technology Policy said that scientists have a responsibility to "straighten out" misleading statements about global warming. Kevin Trenberth, a scientist who has responded to the skeptics, has warned that "disinformation campaigns" have effectively confused both policymakers and the public.

Unfortunately, these encounters tend to reinforce the public's view that no one really knows anything. Competing petitions from different camps, which take on the air of "I'll see your 2000 climatologists and raise you 4000 biologists," simply add to the confusion. Media coverage generally exacerbates the problem; some critics argue that, in trying to appear balanced and fair, the media often create the misleading impression that the weight of scientific opinion is more evenly divided than it really is. McBean, for example, says the media "often portray a polarization of the debate about climate change science that is simply not evident amongst the experts actually working on these issues." Runnalls says bluntly that it's "a dereliction of duty by mainstream media" to simply pit Scientist A against Scientist B without examining their credentials and without providing the context that "every major climate scientist in the world has been part of the IPCC process."

Yulsman also argues that journalists covering complex environmental issues "should not give equal weight to both sides of an issue if, in fact, the balance of scientific opinion weights on one side over another." Journalists have a responsibility to "assess where the balance lies and report what they learn to their readers," he commented in a letter published in Columbia University's *21stC* magazine. "The skeptics should get time, but not equal time."

Yulsman says that the media—television in particular—are "very uncomfortable in dealing with uncertainty. They just don't seem to want to do it, maybe because it's difficult to get at nuance and subtlety." Another problem is that many editors in charge of science coverage "don't know anything about science. They handle science the way they handle politics—people on two sides of an issue screaming at each other." And they have difficulty dealing with the way science really works. "Science is an incremental process of uncovering new clues about how nature works. The media tend to focus on the great discoveries." They like to cover what Yulsman calls "frontier science," research that breaks new ground. This kind of science is "new and sexy, but it's also where things are most likely to change and the findings are least reliable. Editors want the new and sexy and they don't want any of the caveats."

Less appealing to the media is the next stage, consensus science, which occurs when "a body of opinion begins to form that a certain way of looking at things is probably true," Yulsman says. This doesn't mean the matter has been proved, but consensus science reflects what appears to be the truth based on the best currently available evidence. "The global warming debate has lots of elements of frontier science but there is an emerging element of consensus." This doesn't necessarily mean the skeptics are wrong, but the more salient issue is this: "Based on what we know and given the uncertainties, what is the best assessment that can be made about global warming now? I don't see journalists focussing on that; they're still focussing on proof. " Yulsman says the media should be giving the public the full picture: what scientists agree on, what they disagree on, and the level of uncertainty associated with findings and hypotheses, but unfortunately much of the media may be incapable of such sophistication.

The Risk of Climate Change

The time has come to shift the climate debate from the question of *proof* to the question of *risk*. Risk involves two components: the *probability* that something will happen and the *consequences* if it does. We've been so preoccupied with the first component that we've badly neglected the second.

Environmental groups and some climate scientists have recently begun to emphasize the need to reframe the questions we're asking about global warming. For example, in the November/December 1997 issue of *Worldwatch*, the magazine of the Worldwatch Institute, both the lead article and the editorial focus on the question of risk, arguing persuasively that it's sheer foolhardiness to refuse to protect ourselves against the risks we face on the grounds that no one has yet proven climate change will lead to "certain calamity."

The concept of protecting against risk is not difficult for the average person to grasp. We already have an excellent example in our daily lives— insurance. The interesting thing about insurance is that we buy it *despite the considerable uncertainties that the disasters they protect us against will ever happen.* We do not expect *proof* that our house is going to burn down, that we'll be in a car crash, or that we'll die early of a heart attack before we buy insurance. The reason is simple and it's found in the second part of the risk equation—*consequences.* We buy insurance—and spend quite a lot of money doing it—not because we're sure something terrible will happen but because we don't want to face a catastrophic loss if it does. It's precisely because the consequences are potentially devastating that we try to protect ourselves against such events *even if we believe their probability of happening is very low.* In fact, in cases where the probability of severe loss is relatively high, precautionary measures may be deemed mandatory, as in the case of vehicle insurance.

We take anticipatory protective measures in other ways as well. Wearing seat belts is one example. Diversifying an investment portfolio is another. Moreover, we expect governments to take such measures on our behalf—for example, by requiring testing and inspection of prescription drugs or vehicles used for public transportation. Governments don't refuse to do tests on the grounds that no one has proved in advance that these products will harm the public. The potential consequences if they do are large enough to justify a preemptive approach.

One way of assessing the consequences of different approaches to dealing with climate change is examined in a report by the Environmental Adaptation Research Group for the Ontario Round Table on Environment and Economy. This matrix "portrays contrasting views of the impacts of climate change as either 'optimistic' or 'pessimistic'…[and] suggests that if we adopt an optimistic (business-as-usual) policy and the pessimists are correct, the 'payoff' could well be a global disaster. Alternatively, if we adopt a 'pessimistic' policy (taking strong actions to limit climate change, and to mitigate and adapt to its impacts), the payoff will be reduced to only 'moderate' if the optimists turn out to be correct, but will be rendered 'tolerable' if the pessimists' position is realized. This suggests, in accordance with the 'precautionary principle,' that if we want to avoid risking disaster…we should provisionally adopt a 'prudently pessimistic' policy," the report concludes. It notes that while this analysis can be applied to many environmental policy problems that involve uncertainties, with climate change, "the stakes are much greater, indeed global and permanent, in scope."

Henry Hengeveld emphasizes that value judgments will play a key role in any policy decisions we make about global warming. "Ideally, the decision maker wants to avoid error by taking the right action to match the problem," he notes. There are two types of errors that can be made in this situation. A "Type I" error involves taking action to prevent or adapt to a problem that either does not occur or is not as bad as anticipated. A "Type II" error involves failing to take action to prevent or prepare for a problem that turns out to be as bad or worse than anticipated.

Hengeveld notes that "information on the probability of the event occurring, the consequences if it does occur and the costs of taking various action options are all important to the decision making process. Ultimately, however, the choice between options for responsive action is generally based on a value judgement with respect to the consequences of making a Type I versus a Type II decision error." As a result, different decision makers, presented with the same data, may choose different options to respond to the problem. Hengeveld gives a simple example: the weather forecast indicates a 40% probability of rain. A woman dressed up for a wedding will probably carry an umbrella because a Type I error (unnecessarily carrying an umbrella) is less severe than a Type II error (getting good clothing wet). A man heading for the beach would make a different choice; he'd feel silly carrying an umbrella while wearing a swimsuit and since he plans to get wet anyway, rain is of little consequence.

In the context of global warming, a Type I error would mean taking measures to slow down and/or adapt to climate change that turn out to be unnecessary because climate change does not occur or is much less serious than projected. A Type II error describes what would happen if we adopt a "waiting for proof " approach and the projections of climate models turn out to be true. Since both approaches involve costs, we'll be forced to decide, on the basis of uncertain information, which type of error presents us with the worst consequences.

It's worth remembering that for most people in most circumstances, uncertainty generally triggers a sense of caution, a retreat to safer ground, or at least a slowing down to think things over and survey the territory before moving ahead. If you're driving along an unfamiliar, narrow mountain road in the dark and you've been warned the area is subject to falling rocks, the prudent and reasoned response is to slow down, not speed up. It is a perverse logic that uses uncertainty and ignorance about what lies ahead as the rationale for tearing on full tilt and damn the consequences. There *are* people who live their lives that way, and we call them daredevils or reckless fools. Yet it is precisely this daredevil logic we buy into when we adopt the "waiting for proof" approach to climate change.

There *is no proof* that global warming will cause adverse, even catastrophic, damages around the world. There probably *will be no proof* unless and until it happens. But the probability that it will do so is very high—certainly at least as high as many of the risks we routinely protect ourselves against every day. The real question we should be asking ourselves is this: If we're willing to invest in precautionary measures against catastrophic loss for the sake of our health, our families, our property, our homes, and our businesses, why is it so difficult to do the same for our planet, the life support system that sustains everything else?

Ultimately, the decision we make about what to do about climate change will come down to value judgments about how much risk we want to take. Though science can provide us with some of the tools we need to make a decision, it cannot give us the answer. The decisions are fundamentally social and political in nature. This is why we need to take stock of how the scientific debate over global warming is "playing" to public opinion and how that's likely to impact the political decision-making process.

Public Opinion on Global Warming

Public opinion surveys done in Canada, the United States, and around the world in recent years suggest that the public is concerned about climate change but not very knowledgeable about the science underpinning the global warming debate. And even though polls suggest that people want their governments to take more aggressive action against global warming, there are concerns about the depth of their awareness of the social and economic impacts that such policies would have. This situation presents a considerable dilemma for political leaders, who sense danger on all fronts and are proceeding with great caution as a result.

In Canada, several surveys done in 1996, 1997, and 1998 show a high level of public concern about environmental issues in general and about what the public perceives as the failure of governments to tackle environmental problems more aggressively. A November 1996 poll conducted for Environment Canada found that half the respondents believed the earth is warming and another quarter believed the climate is fluctuating dramatically. Only 15% did not think the climate was changing, while 6% believed it was cooling. The belief in climate warming was highest (above 50%) in Quebec, B.C., and Atlantic Canada and lowest (37%) in the prairies.

The fact that so few people said they believed the climate is cooling indicates that one of the major claims of the skeptics has not made much of an impression on Canadians. Nor has the argument that global warming would be a good thing: the vast majority of Canadians who believed the climate is changing thought the changes would have negative impacts over the next 50 years. Those who believed the climate is warming or that it's fluctuating were three to four times more likely to think the effects would be negative rather than positive.

The majority of Canadians also believe that these negative impacts will cause significant economic harm, according to a 1997 poll by

Environics International Ltd. of Toronto. Respondents were asked to rate the believability of several statements about the effect of climate change on the Canadian economy. Nearly half rated as "very believable" a statement that climate change will cause significant economic damage because of flooding and impacts on agriculture, forestry, and fisheries. Another 38% rated this statement as somewhat believable. Canadians were less convinced of the argument (often put forth by the skeptics) that efforts to reduce emissions would cause severe economic harm because of Canada's dependence on fossil fuel production. Only 15% found this statement very believable, while 45% found it somewhat believable.

Finally, this poll found that most Canadians believed the argument (often put forth by environmental groups) that Canada can reduce emissions without significant economic harm because new technologies and energy conservation programs would create investment and jobs. Thirty percent found this very believable and 51% found it somewhat believable.

A 1998 survey found that opinion was divided on whether strict environmental standards would make Canadian companies more competitive (40%) or less competitive (44%). This represented a more even balance than a poll in 1992, in which only 32% said companies would be more competitive and 47% felt they'd be less competitive.

By 1997, Canadians were also more inclined than previously to attribute extreme weather events to human activities. When the Environics poll asked an open-ended question about the cause of severe weather in Canada, 47% of respondents attributed these events to various human or environmental factors, according to Doug Miller, president of Environics International. This figure was up five points from a 1995 survey. Among this group were people who attributed the severe weather to climate change and CO_2 levels (18%), human causes in general (16%), environmental pollution (10%), air pollution (5%), and ozone depletion (5%). The remainder attributed the weather events to a variety of other causes including forestry, spaceships, and nuclear testing. (These numbers exceed 47% because respondents were allowed to give more than one answer.)

On the other hand, 43% of the respondents felt there was no particular reason for the weather changes or that they were random acts of nature and within normal climatic fluctuations. A small percentage of the people in this group said the weather had not been any different than in other years (2%) or attributed it to El Niño (1%). It should be noted that this survey was conducted before the extreme El Niño of 1997–98 had done its worst. By the spring of 1998, however, its public recognition factor had shot through the roof. When Environics asked Canadians about causes of extreme weather in the spring of 1998, 41% fingered El Niño, while 10% blamed climate change and another 5% cited global warming or the greenhouse effect. There was also a slight drop—from 5% in 1997 to 2% in 1998—in the number who blamed ozone depletion for extreme weather.

Interestingly, most people did not see any link between El Niño and climate change. Nearly seven in ten of those polled said El Niño resulted solely from natural causes, while 13% felt it was caused by humans and another 13% felt that both causes were involved.

According to Miller, the polls show that changes to the earth's atmosphere are at the "very top" of Canadians' list of environmental concerns. "They really believe that the global atmosphere is becoming unraveled." However, they're not very knowledgeable about the underlying scientific details; for example, there's a lack of understanding of the relationship between climate change and CO_2 emissions. "They're very confused about causal relationships," says Miller. "Many still think climate change is caused by ozone depletion." This view stems in part from the fact that the concept of the "ozone hole" is as powerful in their minds as the image of the earth from space.

The lack of connection in the public mind between climate change and CO_2 emissions perhaps accounts for one of the more interesting findings of the 1996 Environment Canada poll. Few of those questioned had any suggestions about what they could personally do about climate change. In fact, nearly half said there was nothing they could do. (The pollsters noted, however, that this answer may have been influenced by the public's relative unfamiliarity with the term "climate change" as opposed to "global warming" or "greenhouse effect." The report notes that "very few Canadians are comfortable yet with the term 'climate change.'") Nevertheless, only 7% or fewer of the respondents offered any suggestions about actions they could undertake to combat climate change. The suggested remedies included cutting emissions of CO_2, CFCs, and pollutants; reducing vehicle use; and planting trees, as well as more general suggestions such as "recycling/clean the earth/be environmentally friendly." However, ozone popped up again; a number of people suggested that reducing ozone depletion would help slow down climate change.

Findings such as these probably account for the ad campaign later launched by Environment Canada that featured the headline "Make the energy link to climate change." Not only did it push the message that changing the way we use energy "on the road, at work and at home" is what's needed to deal with global warming, it suggested a few simple (and not-very-painful) actions people could personally undertake to do their bit: "Restarting your engine uses less fuel than 10 seconds of idling" and "Washing your laundry with cold instead of hot water uses 93% less energy." But they failed to even hint at any politically risky measures such as carbon taxes or restrictions on vehicle use. While it would certainly help if more people adopted such minor voluntary conservation measures, there's a danger that ads like these may leave the misleading impression that nothing more is needed.

Miller attributes the lack of knowledge about climate change to the fact that the environment has not been a "top of the mind" concern for Canadian since the beginning of the 1990s and they were learning very little about it. "Canadians haven't learned anything about the environment for six years," he said in late 1997. "They're frozen in time with the rapid learning and perceptions that were achieved around turn of decade."

One (seemingly paradoxical) reason for this is that "top of the mind" problems are those that are considered urgent in the short term, and by the end of the 1980s, many people began to realize that there weren't going to be quick or simple solutions for environmental problems. "People realized that there was a lot more to it than buying green products, that a lot of change was needed and it was going to take time. The environment went from an urgent short-term issue to urgent long-term issue."

But another major reason why the environment went off the "top of the mind" chart was that most Canadians thought "their institutions were on the case," Miller said. "They were convinced people were going to do something, that it was in good hands and they had day-care to worry about." Certainly, at the end of 1996, the environment did not make the top tier of items that worried Canadians; in a *Maclean's*/CBC poll that asked respondents to identify the most important issues facing Canada, the environment was ranked tenth on a list of 16 issues, well behind unemployment, national unity, the state of the economy, and health care.

What put the environment on the public's map in the late 1980s was "a breach of trust issue," Miller said. "They woke up to the realization that the people they thought were on the case were not. They thought their institutions were failing them. Our research indicates this is about to happen again." The data indicate that Canadians are starting to learn about environmental issues again and "we're predicting the next 'green wave' of top-of-mind concern in Canada within the next four years."

Evidence that global warming was already becoming an important global environmental issue in the minds of Canadians was apparent in a 1998 poll conducted by the Angus Reid Group. This survey found a threefold increase since 1996 in the number of people who ranked global warming as one of the two most serious environmental issues facing the world (22% in 1998 compared with just 7% in 1996). Global warming was tied with water pollution and was slightly below air pollution (24%). Ozone depletion (20%), general pollution (15%), and deforestation (14%) followed. (Since respondents were able to name more than one item, the total exceeds 100%.)

The increase in ranking that occurred in the national survey between 1996 and early 1998 no doubt reflected to some extent the extensive media coverage and public debate over climate issues, notably El Niño–related extreme weather events and the buildup to the Kyoto climate conference. It would be difficult to find another period during which climate issues and weather events made headlines more frequently or competed so strongly with top political and economic news.

According to Miller, the polling results show that Canadians are "hugely concerned about the atmosphere and they are totally committed to Canada showing leadership on the environment in the world." However, it appears that a great deal of work remains to be done to educate them about the scientific specifics of climate change and how it may affect their lives. It's a cause for concern that many Canadians appear to be ill-informed or, in fact, badly misinformed about the true nature of the global warming problem. It also appears that many people do not fully comprehend the choices they may be called on to make, or the changes in lifestyle, technology, and energy use that might be necessary. The *Globe and Mail*'s political columnist Jeffrey Simpson has argued that both Canadians and Americans "are almost completely unaware of the lifestyle changes and new buying patterns, tax incentives and technologies that will be required for targets agreed to at Kyoto to be met." He says the costs of taking action against global warming will involve "wrenching changes" and argues that the Canadian government has "at no time…spoken truthfully to Canadians about what lies ahead.…Only when people appreciate how precisely their lives will change, what those changes will cost and when the changes must be implemented, can a reasonable measure be taken of public opinion. For now, people have only the haziest idea, if they have any at all, about how their lives must change." While one can (and many do) dispute just how "wrenching" the changes may be, Simpson is right that this lack of understanding raises questions about how enthusiastically Canadians would embrace an all-out fight against global warming if it should require serious personal economic sacrifice.

In a 1997 survey, most Canadians and the majority of people in several other countries said they want their political leaders to take more aggressive action on climate change even at the cost of economic growth. The international survey, directed by Environics International, asked 27,000 people in 24 countries (representing about 60% of the global population) about their preferences for action on climate change, considering the scientific uncertainties involved and the potential economic costs. Doug Miller, who was in charge of the project, said the question was framed in such a way as to emphasize the economic impacts of taking action immediately. At one extreme were those who said we should "assume the worst" and take major action now. At the other extreme were those who felt no major action should be taken until we know more about the problem. In between were people whose answers were classified as "other/depends/don't know" or who didn't answer. In 62% of the countries—15 out of the 24—more than half the population supported the idea of immediate action. In none of the countries did a majority of the population support the idea of waiting for more information before taking action; in every country except one—the U.S.—only one-third or less of the population chose the wait-and-see option.

For political leaders, "the day of reckoning and accountability is fast approaching," Environics said when they released the poll results in June 1997. Miller commented at the time that despite the preoccupation of world leaders with economic matters, "citizens in most countries are even more likely today than five years ago to say that economic growth should be sacrificed if necessary to ensure the environment is protected. It is as if the survival instinct of our species has been activated. Government and industry leaders ignore these deepening environmental and health concerns at their peril."

The degree of commitment to the precautionary principle varied among the 24 countries. People in western European countries were most in favor of immediate action, followed by Japan, Korea, Australia, and New Zealand, with two-thirds or more of respondents favoring action. In eastern European nations, less than half the population opted for this strategy. Significantly, China and Russia were at the bottom of the list, with roughly one-third to one-quarter of their populations supporting immediate action. However, few opted for the wait-and-see approach either; more than half of the Chinese and Russian respondents fell into the middle category, indicating a high degree of uncertainty about what to do.

Canada was about in the middle of the pack, with 61% supporting immediate action, 32% against it, and one of the smallest percentages (7%) in the undecided middle category. The U.S. results were unique and interesting. It was among the lowest-ranked countries in terms of supporting immediate action (46%) and it was the highest-ranked country, by a considerable margin, in the number of people advocating a wait-and-see approach—also 46%. Only 8% of the population was in the uncertain category. "They're very pragmatic people when the dollar comes into things," said Miller. "They're the only superpower and they know they're going to be paying for most of it. Clearly the U.S. is going to be the battleground."

Other polls done in the U.S. show varying and, to some extent, contradictory results. They indicate that most Americans believe global warming is happening, but are divided on how serious a problem it is and what should be done about it. Several surveys taken in early 1996 indicated that Americans ranked the environment low on their list of national problems, according to an article in *Resources,* the newsletter of Resources for the Future, a Washington-based research organization. It cited a study in which the environment was ranked 14th out of 15 top problems, and another in which environmental issues didn't even make the list.

Somewhat paradoxically, the authors of the article, Everett Carll Ladd of the University of Connecticut and Karlyn Bowman of the American Enterprise Institute, conclude that these findings reflect the fact that most Americans "are deeply committed to a safe and healthful environment." They regard the need for protecting the environment as a "settled" issue and expect the government to take care of achieving that goal. "The transformation of the environment from an issue of limited concern to one of universal concern is complete, and, today, survey after survey shows that most Americans have turned their attention to other things." However, they warned that environmental issues could go back on the agenda if "political leaders are not mindful of their concerns."

It appears this started happening in 1997, when several polls suggested an increase in public concern about the environment and specifically about global warming, which may have resulted from extensive media coverage of El Niño's depredations (which hit the United States particularly hard) and intensifying political conflict over the Kyoto conference.

A survey done in August 1997 for the World Wildlife Fund found that two-thirds of those polled viewed global warming as a "serious" or "very serious" threat and 61% felt it would continue to get worse. Three in four of the respondents agreed that the only scientists who don't believe global warming is happening are those paid by fossil fuel companies "to find results that will protect business interests." Sixty percent of those polled also rejected the view that reducing greenhouse gas emissions would cost jobs and hurt the economy, compared with only 18% who said it would.

On the other hand, many people didn't want to pay for such measures at the gas pump; 58% were opposed to a 50-cent increase, in gasoline taxes even with a US$250 income tax rebate. Eight in 10 of the respondents said electric utility companies should be required to offer wind and solar power.

However, another poll conducted in November 1997 (just before the Kyoto conference) by the Pew Research Center for the People and the Press found that only a quarter of respondents said they worry a lot about global warming; more respondents (ranging from 46 to 61%) ranked other environmental issues such as water pollution, loss of habitat, and toxic wastes as more urgent. Nevertheless, nearly three-quarters said they'd be willing to pay five cents a gallon more for gas to combat global warming, and 60% said they'd go as high as 25 cents more—but only if other countries shared equally in the fight against global warming.

The poll found that one tenet of the skeptics' agenda—the demand that developing countries participate fully in the battle against global warming—was firmly entrenched as a requirement in the minds of most Americans. Seventy percent said both developed and developing countries should make equal efforts, notwithstanding the fact that most greenhouse gas emissions come from industrialized countries. The idea that developing countries should be permitted to do less was supported by only 19% of the respondents.

A Harris poll conducted immediately after the Kyoto conference found that three-quarters of Americans who were familiar with the climate treaty supported it, while 18% of those familiar with it said it went too far. However, only 55% of those questioned said they were familiar with the treaty so, in fact, only about four in 10 Americans support the Kyoto Protocol, according to this poll. However, two-thirds of the respondents said they believe greenhouse gas emissions cause warming and half said they felt that global warming is a very serious problem. This latter number is considerably higher than that found in the Pew poll, which was done prior to the Kyoto conference. If this increase reflects greater public awareness of the global warming issue as a result of the fierce debate and massive coverage associated with the Kyoto conference, it suggests that the aggressive media campaign launched by the skeptics against the Protocol may have had at least one effect they did not intend.

Given that the political debate over reducing greenhouse gas emissions has been fiercest in the U.S., which is home to the most vociferous and well-organized opposition to the international climate protocol, it's not surprising that public opinion there is so evenly divided. The intense media campaign mounted by skeptics since the release of the IPCC report in late 1995 has clearly met with considerable success in promoting public acceptance of their point of view. Many of their arguments are echoed in comments made in media forums such as letters to the editor and Internet discussion groups. While these are not scientifically selected samples, their comments do reflect the polarization of the U.S. population on the global warming issue.

A correspondent to the *Wall Street Journal*, using phrases common in skeptical rhetoric, accused Clinton and Gore of invoking "scientific gibberish and junk science" to encourage "fear-mongering of this so-called problem." A participant in a forum on CNN's Web page argued that global warming science "ranks right up there with Astrology or Alien Abduction." Another correspondent argued that any action humans take to reduce global warming—what he described as "pitiful little changes"— would be so small compared with natural processes as to "go completely unnoticed"—a statement that demonstrates a lack of understanding that it's only the human-induced greenhouse gas emissions that count here, since natural processes are nearly balanced.

One unique aspect of the debate in the U.S. is the extent to which people there perceive emissions cuts as an ill-disguised plot by other countries and the United Nations to gain political and economic advantage over the U.S., a point of view vigorously promoted by conservative politicians and media. The *New York Post* described the Kyoto treaty as "a massive U.N.-orchestrated transfer of wealth from the First World to the Third." Republican Congressman Benjamin A. Gilman, chairman of the House International Relations Committee, was quoted in the *Washington Post* saying that proposals to cut emissions would lead to "carbon taxes, restrictions on the operation of business and homes and payments to Third World countries for the right to build new factories or drive cars here in America."

On CNN's *Crossfire*, conservative commentator Pat Buchanan described climate scientists as "ideologues...interested in power [and] running things" who were trying to "get control of the whole planet....These guys are far more dangerous than any lobbyist." (The accusation of power-seeking, which inspires considerable hilarity among scientists, seems rather a curious taunt from a man who once aspired to become president of the United States.)

The idea that global warming is a plot to control or damage the U.S. economy also seems to have taken root in the public mind. One participant in CNN's Internet forum argued that support for the global warming theory comes from "a bunch of guilt ridden liberals who wish to pull us down to...third world conditions so they can feel less guilty about our success as a nation." He suggested they move to the Third World if they want to live in such conditions. Another writer argued that less powerful countries "would love the chance to handicap us a little more." A third described global warming as a socialist plot to stop the "rising standards of living afforded by the free enterprise system." Many commentators reiterated the view that the U.S. should reject any treaty that didn't require equal sacrifice from developing nations, but at least one warned against "knee-jerk nationalistic attitudes," saying that developed countries must scale down their emissions and their standard of living and developing countries must accept that they can't engage in the same wasteful energy consumption found in developed countries in the past.

Polls and the Politicians

What is perhaps most curious about the polls on global warming is that they've had so little real political impact, despite the findings that people in industrialized countries are concerned about the problem and appear not to have bought the argument that we should wait for more definitive proof before acting. These surveys indicate that in most countries, most people believe the preponderance of scientific evidence now provides sufficient justification for governments to take immediate and aggressive action, even if it means paying a price in terms of economic growth. Even in the U.S., where opinion is most evenly divided, poll results provide the government with considerable ammunition in its quest to ratify the Kyoto Protocol, according to Humphrey Taylor, CEO of the polling organization Louis Harris and Associates.

Moreover, the surveys clearly indicate that people expect their governments to show leadership and to "handle" the problem. With such a mandate, introducing policies to combat global warming should be easier than it appears to be. Yet, except for some European governments, most political leaders treat the issue like a live hand grenade. They're loathe to undertake the kind of aggressive action the public says it wants; indeed, their actions suggest they're far from persuaded the public is truly ready for an all-out assault on the global warming problem, no matter what they tell pollsters. The fact that both the Canadian and U.S. governments felt it necessary to launch public education programs before and after the Kyoto conference indicates an uneasiness about the strength of the public support needed to shepherd the Protocol through the political minefields.

During the summer and fall of 1997, President Clinton embarked on a public education campaign to convince Americans and the U.S. Congress that the climate change problem is "real and imminent." He sponsored several events designed "to lay the scientific facts before our people to understand that we must act, and to lay the economic facts there so that they will understand the benefits and the costs."

These events came at a time when the blitz of media ads sponsored by the Global Climate Coalition had been bombarding the American public for several weeks and the U.S. Congress was increasing pressure on Clinton not to sign a climate agreement in Kyoto that would harm the U.S. economy. Various business groups, such as the U.S. Chamber of Commerce, also weighed in against the Protocol. The journal *Nature* noted that "it remains to be seen how far [Clinton] will take the effort in the face of public apathy on the issue and widespread hostility to taxation or any other government action aimed at reducing energy consumption." *Time* magazine wondered why the president could not "speak out at least as forcefully as an oil baron"—a reference to the endorsement of greenhouse gas cuts made by John Browne of British Petroleum a month earlier.

In October 1997, the White House sponsored a climate briefing for TV weather forecasters around the country, an event that inspired considerable mockery from the political media. The event was described by White House communications director Ann Lewis as a way of reaching "a set of people...who professionally pay attention to issues like climate change and have great local credibility. They can give the public more information and help us communicate the issue." Nevertheless, it was viewed with suspicion, even by some of the attending forecasters, as a public relations tactic.

The media found it hard to take the event seriously. Canada's CTV News quipped that Clinton was seeking help from people who are "right only half the time." There were unoriginal barbs about all the "hot air" rising from the White House lawn (from which many of the forecasters made their weather broadcasts that day). The *Washington Post* commented derisively: "Armed with climatological factoids, the well-coiffed legions...deployed to bombard their home stations with the latest data on greenhouse gases." Allowing that "TV weathermen can be kind of goofy," the *Associated Press* observed that this had been the "oddest day" on the White House lawn since the Easter egg roll.

There were also suggestions that the forecasters had been unduly flattered by the unaccustomed attention from the White House and had been invited precisely because they were unlikely to be critical of Clinton's policies. *Reuters* reported that "while Clinton has had trouble getting news coverage for the [global warming] issue from the normal stable of White House reporters, he clearly saw an eager audience in the weather forecasters."

292 ■ STORM WARNING

In response, then-press secretary Mike McCurry argued that the White House was providing a service in educating the forecasters, who appreciated being treated as "something other than airheads. We want these folks to get the best education we can make available on the subject. Maybe they can make the subject of global warming a little more lively for their audiences." Speaking to the regular White House press corps, he added: "We try to do that for people in this room sometimes and you're not usually as receptive, so we go find some other audiences to talk to."

A couple of weeks later Clinton opened a White House–sponsored climate conference by saying that "the potential for serious climate disruption is real." While he acknowledged that not everyone agreed on how serious the risks might be, he said he was "completely persuaded" that dealing with global warming was necessary to "avoid leaving our children and grandchildren with a catastrophe....Although we do not know everything, what we do know is more than enough to warrant action....It would clearly be a grave mistake to bury our head in the sand and pretend the issue will go away."

Clinton nevertheless shied away from the idea of carbon taxes, claiming that increasing the price of coal and oil "won't pass muster with the American people." Instead, he said, the U.S. must look to new energy-efficient technologies for the solution. These remarks suggest that the U.S. government is not putting a lot of faith in surveys that indicate Americans would willingly pay more for fossil fuel energy to protect the climate.

The Canadian government also launched an education campaign prior to the Kyoto conference, although there were no forecasters littering the lawn in front of Parliament Hill. Several educational initiatives had already been started as part of the federal action plan, but in the fall of 1997, activity stepped up with a series of public meetings across the country to release several reports on the impacts of climate change on different regions of Canada. Environment Minister Christine Stewart said it was necessary to prepare Canadians for the "dramatic measures" required to deal with global warming, just as they'd had to deal with the war on the deficit. Her objective, she said, was to ensure the public understood the climate change issue and that "they have the information and knowledge they need to be supportive of the measures that have to be taken."

This effort was somewhat undermined, however, by the escalating political drama leading up to the Kyoto conference. At the time, the federal government was under intense fire from virtually all quarters for its reluctance to reveal its negotiating position for Kyoto. Reports of fierce infighting among federal Cabinet ministers and between the federal and provincial governments received much more media attention than climate science.

In the middle of the political wrangling it was easy to lose sight of a simple question: Why did the government feel that a public education effort was even necessary? After all, there were plenty of polls telling them that the public already supported strong measures to combat global warming, that, in fact, they were disappointed with and even ashamed of the foot-dragging that had caused Canada's failure to meet its Rio commitments.

Two possible reasons for this approach come to mind (aside from the fact that calling for increased public education gives the appearance that governments are leading rather than following the public on environmental issues). First, in both Canada and the United States, it's difficult for governments to implement politically volatile environmental policies without solid public support because they're often engaged in hand-to-hand combat with powerful interests opposing such measures, not the least of which come from within the ranks of government itself. Unfortunately, however, while public support—or at least the appearance of it—may be a necessary condition, it's not sufficient to win battles engaged in the back rooms of power. Public opinion is an amorphous thing and, in the case of global warming, often confused. It's a rather weak bargaining chip when pitted against the well-oiled opposition of groups such as the fossil fuel industry, the big business lobby, and their political supporters, all of whom have a well-defined agenda and the means to imprint their message where it counts. Governments are far more attuned to listening to these voices and, unless public opinion becomes much more insistent, focused, and well-informed than it has been to date, the public will be reduced to extras in the global warming drama, a background chorus of lamentation and disapprobation, comprising of bit players without speaking parts.

The second, and perhaps even more salient, reason why political leaders felt the need for a public education campaign is that, from all appearances, they don't really trust the polls that say the public is willing to sacrifice economic growth for climate protection. Even though people say they want their governments to take strong action, there are legitimate questions as to whether they fully comprehend how those actions will affect their lifestyles and pocketbooks.

"It isn't that the public doesn't mean what they're saying," said Alex Manson, executive director of Environment Canada's Climate Change Bureau. "But there are some concerns that the public doesn't understand it completely. We need this education and outreach to make sure they understand the nature of the actions that need to be taken by the government, the private sector and citizens and that they support those kinds of actions. We've got to make some fairly significant changes in the way we produce and use energy, the way we transport people and goods around and we've got to get it right."

The extreme wariness with which politicians view public opinion on the global warming issue is probably well advised. There's every indication that most people are stuck in the early stages of the transition from public opinion to what U.S. pollster Daniel Yankelovich calls "public judgment"—the point at which they understand the complexities of an issue and are committed to doing something about it. Yankelovich defines several stages of this journey; during the first three stages, people become aware of an issue, develop a sense of urgency about it, and consider a wide range of possible actions, but "without really coming to grips with any of them....They don't understand all the costs and the tradeoffs or the conflicts of values."

The fourth stage is the most problematic; this is where people engage in wishful thinking and avoidance stratagems to put off dealing with the complexities of an issue and making hard decisions. "Most important issues entail cost and tradeoffs, hard choices, maybe even sacrifices. And people put up a real battle to avoid confronting such unpleasant realities," Yankelovich says. A favored technique is to attack the institutions mandated to deal with the problem as ineffectual, wasteful, and corrupt.

The public can remain stalled at stage four for years, even decades. Only when they move beyond it—by evaluating the problem objectively and realistically and figuring out where they really stand—can they exercise public *judgment*, as opposed to public *opinion*. Yankelovich says that with any controversial issue, gridlock will result unless political leaders can move people beyond stage four, because it can be political suicide to attempt to introduce bold measures that the public is not ready to accept.

Furthermore, he notes, it's important to understand that public opinion polls that focus on issues are not the same as political election polls. With the latter, there is a clearly defined end to the decision-making process—an election—and people are forced to make up their minds one way or the other. With complex social or environmental issues, there generally is no clearly defined end and polls do not show what stage respondents have reached in their journey toward understanding and realistic decision making. Many people questioned in such polls may be bogged down in the fantasy/denial phase of stage four and remain so for long periods of time.

A case can be made that, on the global warming issue, most of the public has not moved beyond stage four. Indeed, there's evidence that many are barely out of the first three stages. There are widespread mis-conceptions about the basic nature of the problem and little real com-prehension of the scientific and economic complexities it entails. There is also a failure to comprehend the true nature and extent of the measures that may be necessary to solve it and how those measures will affect indi-viduals' daily lives. Among other things, people stalled in this state of confusion and indecision are extremely susceptible to being pulled back and forth in the tug of war between dueling experts.

To some extent, this state of affairs is not the public's fault. No one has articulated the consequences and costs of the measures to combat global warming very well, in part because no one is really sure yet exactly what they will be. With all the fireworks over scientific uncertainty and all the rhetoric about trade-offs between economic and environmental disasters, it's perhaps not surprising to find people implicitly—and often explic-itly—expressing the attitude: "So, what do you expect from me?"

There are other indicators that many people are still behind the learn-ing curve on global warming. An expressed willingness to pay a few more cents at the gas pump—even if genuine—isn't going to get the job done, and it's disturbing if large numbers of people think minor efforts of this sort will suffice. When Richard Berk, a sociologist with the University of California, Los Angeles, asked people if they'd be willing to pay between $25 and $500 extra a year to prevent certain climatic changes from hap-pening, it was not surprising that he found the willingness to pay dropped as the price got higher. What's more interesting is that the amounts involved were not particularly large, suggesting that the public tolerance for price hikes in aid of climate protection has a fairly rapid attrition rate.

Berk also found that the way people respond to surveys on climate change is influenced by the climatic conditions they're experiencing at the time. Interestingly, the results showed that, while most people find major departures from the climate they're used to to be unacceptable, it nevertheless takes "big changes to budge them" to take action. These findings suggest why extreme weather events play such a potentially sig-nificant role in galvanizing action on global warming; to many people, they are tangible manifestations of "big changes" in the climate. The losses caused by El Niño in 1997–98—whether or not they had anything to do with climate change—served a useful educational purpose by pro-viding a kind of "dress rehearsal" for what we might expect in a warmer, wetter world. Millions of people experienced the extreme weather events personally or vicariously through extensive media coverage, a collective experience that shook comfortable assumptions and technological opti-mism about our relative invulnerability to nature.

It will be interesting to see if such experiences reduce the disparities between what people say in surveys and polls and what they do in their daily lives. While many people say they're concerned about greenhouse gas emissions, it's not hard to find ample evidence of contradictory behavior. Our attitude toward vehicles is a particularly instructive case in point.

In early 1998, Environics found that more than half of Canadians listed vehicles as the major source of air pollution and nearly a third identified transportation as the sector that has the furthest to go to become environmentally sustainable. These findings might lead one to conclude that Canadians would be trying to find ways to reduce vehicle use, but such is not the case. According to Statistics Canada's 1996 census report, nearly three-quarters of employed Canadians drive their vehicles to work, compared with only 10% who take public transportation. Significantly, the survey found that the use of public transportation, which had been declining steadily since 1988, had dropped almost to the record low set in 1970. The drop was attributed to several economic and demographic factors—not the least of which was the influence of sprawling automobile-dependent suburbs and bedroom communities—but personal preferences also played a big role. A *Globe and Mail* article quoted a Toronto subway rider complaining about the smell, the crowding, and being elbowed by "obnoxious people." Another woman was quoted as saying she'd paid her dues when she was younger and now felt entitled to the convenience and privacy of her car. This view is implicit in a commercial for an upscale car targeted at the well-heeled, middle-aged baby-boomer market, which states that "it's not meant for carpooling, but then neither are you."

These habits and attitudes will be hard to break, particularly with the erosion in the quality of public transit systems that has been occurring in many cities as a result of budgetary cutbacks, price hikes, and declining numbers of riders. Richard Soberman of the University of Toronto, co-author of *Full Cost Pricing and Sustainable Urban Transportation*, says "our research shows people are prepared to pay a lot more for things related to automobile use like gasoline, tolls and parking before they will agree to change their behavior."

Even more instructive are purchasing trends. Vehicles are highly symbolic lifestyle icons and thus serve as an excellent bellwether of the extent to which people in industrialized countries are willing to make a personal investment in tackling the global warming problem. Although they're not the only source of greenhouse gases, vehicles are major contributors and, more importantly, their use is very much influenced by individual choice and behavior. If actions speak louder than words, the data are not particularly encouraging, at least in North America, where sales of larger, more expensive gas-guzzlers like vans and sport utility vehicles are increasing rapidly. In mid-1997, CNN reported that four of the top 10-selling cars in the U.S. were SUVs. Owners interviewed in the piece admitted they'd given barely a thought to the cost of gas or the mileage the cars obtain; what was most important, as one put it, was that the vehicles are "the coolest car you can possibly buy." An industry analyst commented that when consumers look for "the vehicle that is 'emotionally me', they're buying sport utilities."

It's clear that government regulations are the only thing preventing the environmental impact of these vehicle-purchasing decisions from being worse than it might otherwise be. In the U.S., for example, automakers are required to meet a certain average fuel efficiency for all cars they have on the road, so if they sell a lot of profitable gas-guzzlers, they're forced to offer good deals to get other consumers to buy smaller, more fuel-efficient cars. One U.S. consumer survey found that while more than 70% of respondents said they were interested in reducing air pollution from vehicles and 62% said governments should force car makers to sell electric or other low-emission vehicles, 39% nevertheless reported their intention to buy less energy-efficient vehicles like trucks, sport utilities, vans, or minivans. The survey, conducted by the Dohring Co. of California, a provider of market research for the automotive industry, also found that less than half of the respondents would consider buying an electric car *even if they knew it would cost less to operate than a gasoline-powered car*, without the cost advantage, less than one-third said they would consider buying an electric vehicle.

A Dohring spokesperson, Sara West, said the survey shows that while consumers are concerned about the environmental impact of their vehicles and most express a willingness to do something about it, "in order to effect a change in the habits of America's highly mobile and car-loving public, citizens must be educated on the choices available to them that can help curb the energy gorge currently underway. Carbon dioxide restrictions will likely be a tough sell to the American public, who have long enjoyed some of the world's cheapest fuel prices."

Asked about the growing popularity of less fuel-efficient vehicles, Clinton's science adviser Jack Gibbons commented that people have to "learn to better understand the long-term implications of our short-term decisions." It's an admonition that applies to more than just decisions about purchasing vehicles. The issue of "short-term pain for long-term gain" is one of the central philosophical dilemmas of the global warming debate. One of the biggest impediments to dealing with this issue is the fact that costs tend to be front-end loaded while benefits tend to be back-end loaded. And because the costs associated with fighting global warming are near-term, they're much easier for people to see than the costs of *not* fighting it, which are long-term and more uncertain. Moreover, the short-term pain will be borne for the most part by people who'll be long gone by the time the "long-term gain" becomes apparent. This inevitably makes it harder for many people to bite the bullet, notwithstanding their often-expressed views about the importance of leaving a clean environment for future generations. As noted in a paper published by Resources for the Future: "Decisions are easiest when threats and benefits are immediately visible to the naked eye....But much of the modern environmental protection movement has been a response to menaces that are invisible, indirect, and detectable only through advanced technology."

What makes the choice more daunting is the difficulty of weighing accurately the more easily quantified near-term costs against the rather fuzzier future benefits. There are many uncertainties about how serious the up-front costs will be, but even assuming they aren't as onerous as the skeptics claim, there will be an initial period of adjustment that probably will test our resolve. It doesn't help that the more successful we are in warding off the worst that global warming can do, the less "proof" we'll have that the battle was even necessary in the first place, since we'll have no way of knowing how bad things might have been had we not intervened.

It remains very much open to question whether scientific evidence, international treaties, or public education campaigns have sufficiently primed humanity to make potentially unpalatable near-term choices to protect the climate for the benefit of future generations. We must, however, stop deluding ourselves about the nature of the decision that confronts us at this time. Any near-term choices we make about reducing greenhouse gases will have to be made in the absence of absolute proof that the currently projected socioeconomic and environmental impacts of climate change will occur—or, for that matter, proof that they *won't* occur. Failure to make a choice is a choice in itself—a choice to continue the experiment we're conducting on the earth's climate.

We may think we still have time to make a decision but the longer we wait, the more limited and expensive the options left to us and the harder they'll be to accept. It's like a cancer that continues to grow as you try to make up your mind what to do about it. The longer you wait, the harder the disease will be to defeat and the more aggressively you'll have to go after it. And if you're holding out for absolute proof that the treatment will cure you, the cancer will probably win the waiting game.

The argument that we're just "waiting for proof" of global warming is the comforting fiction we use to avoid facing up to the hard choices that confront us. It is a cleverly misleading way of presenting an option that does not involve "waiting" at all. It is, in reality, an affirmative decision to allow our experiment on the earth's climate to proceed virtually unchecked. The consequences—whatever they may be, however poorly we may understand them—are not "waiting" while we make up our minds what to do.

Understanding the true nature of this choice is an important step in reframing the global warming question, shifting our perceptions away from the issue of proof and toward the issue of risk. We must recognize the strategy of "waiting for proof" for what it is: a form of gambling. We're betting that projections about the negative impacts of global warming won't come true. In the end, we're forced to make a judgment—based on the best scientific evidence available, imperfect though it may be—whether to hedge our bets and play it safe or to ignore the odds and play on. But, then, disregarding probability and ignoring the odds is what keeps the casinos full and the lottery tickets selling, so perhaps it's not so surprising that we're willing to throw the global climate system into the pot.

If we're intent on playing this game of chance, however, it's important that we ask ourselves a few hard questions: What are the stakes? What are we risking by continuing to play? What do we stand to lose if the worst does come true? Finally, and most importantly, are we so wedded to the choices that have forced us into this game that we cannot bring ourselves to even *think* about the stakes or to question the risks? Are we like addicted gamblers who can't walk away no matter what it costs to stay in the game?

Nobel laureate Sherwood Rowland has said: "After all, what's the use of having developed a science well enough to make predictions, if in the end, all we're willing to do is stand around and wait for them to come true?" It would appear that, in the end, this is all we *are* willing to do. Diverted by unrealistic, impossible expectations of proof that will come only if and when the worst does, in fact, happen, we are failing to protect ourselves against that eventuality by utterly refusing to consider what we stand to lose if it occurs.

ACRONYMS

AGCI	Aspen Global Change Institute (www.gcrio.org/agci-home.html)
API	American Petroleum Institute (www.api.org/)
CCS	Canada Country Study (www.ec.gc.ca/climate/ccs/index.htm)
CGCP	Canadian Global Change Program (www.cgcp.rsc.ca/)
DOE	U.S. Department of Energy (www.doe.gov/)
EARG	Environmental Adaptation Research Group (www1.tor.ec.gc.ca/earg/)
EDF	Environmental Defense Fund (www.edf.org/)
EC	Environment Canada (www.doe.ca/)
EPA	U.S. Environmental Protection Agency (www.epa.gov/)
EPC	Emergency Preparedness Canada (hoshi.cic.sfu.ca/epc/en_home.html)
FAPCC	Federal Action Plan on Climate Change (www2.ec.gc.ca/climate/resource/fapcc/index.html)
FCCC	See UNFCCC
FEMA	U.S. Federal Emergency Management Agency (www.fema.gov/)
GCC	Global Climate Coalition (www.globalclimate.org/)
GCCIP	Global Climate Change Information Programme, Atmospheric Research & Information Centre, Manchester Metropolitan University (www.doc.mmu.ac.uk/aric/gcciphm.html)
GEF	Global Environment Facility (www.worldbank.org/)
GLSLB	Great Lakes–St. Lawrence Basin Project (www.cciw.ca/glimr/metadata/stlawrence-basin-project/intro.html)
IBC	Insurance Bureau of Canada (www.ibc.ca/)
ICLEI	International Council for Local Environmental Initiatives (www.iclei.org/)
ICLR	Institute for Catastrophic Loss Reduction (www.ibc.ca/)
IPCC	Intergovernmental Panel on Climate Change (www.ipcc.ch/)

JAMA	*Journal of the American Medical Association* (www.ama-assn.org/public/journals/jama/jamahome.htm)
MBIS	Mackenzie Basin Impact Study (www1.tor.ec.gc.ca/comm/basin.htm)
NAS	National Academy of Science (www.nas.edu/)
NAPCC	National Action Plan on Climate Change (www2.ec.gc.ca/climate/resource/cnapcc/)
NASA	U.S. National Aeronautics and Space Administration (www.nasa.gov/)
NCAR	U.S. National Center for Atmospheric Research (www.ncar.ucar.edu/)
NCDC	U.S. National Climatic Data Center (www.ncdc.noaa.gov/)
NEJM	*New England Journal of Medicine* (www.nejm.org/content/index.asp)
NIEHS	National Institute of Environmental Health Sciences (www.niehs.nih.gov/)
NOAA	U.S. National Oceanic and Atmospheric Administration (www.noaa.gov/)
NRDC	Natural Resources Defense Council (www.nrdc.org/)
NSF	National Science Foundation (www.nsf.gov/)
OISM	Oregon Institute of Science & Medicine (www.oism.org/)
RFF	Resources for the Future (www.rff.org/)
SEPP	Science & Environmental Policy Project (www.sepp.org/)
UCAR	University Corporation for Atmospheric Research (www.ucar.edu/)
UN-IDNDR	United Nations International Decade for Natural Disaster Reduction (hoshi.cic.sfu.ca/idndr/)
UNEP	United Nations Environment Programme (www.unep.ch/)
UNFCCC	United Nations Framework Convention on Climate Change (www.unfccc.de/)
USGCRP	U.S. Global Change Research Program (www.usgcrp.gov/)
WCR	*World Climate Report* (www.nhes.com/home.html)
WEC	World Energy Council (www.wec.co.uk/)
WHO	World Health Organization (www.who.org/)
WMO	World Meteorological Organization (www.wmo.ch/)
WRI	World Resources Institute (www.wri.org/)
WWF	World Wide Fund for Nature (www.panda.org/)
WI	Worldwatch Institute (www.worldwatch.org/)

NOTES

A note on chapter references:
References are from online sources unless indicated[*]. Information concerning public events (e.g., weather disasters, press conferences, scientific/political meetings etc.) were derived from multiple news sources. Information attributed to individuals in the text and not otherwise referenced is from personal communication. Multiple uses of a particular news source are listed only once per chapter, and show the various dates when stories appeared.

Preface:
– Env. Can. "Worst Ice Storm."
– Scanlon, J. Carleton Univ., Ottawa pers. comm.
– Phillips, D., Env. Can. pers. comm.
– NCDC, Jan. 20/98. "Eastern U.S. Flooding and Ice Storm."
– CBC *National,* Jan. 12, 14, 19, 28/98.
– CPAC, Ottawa–Carleton EPC press conf., Jan. 10/98.
– *CBC Newsworld, Jan. 9, 14/98.
– IBC, Sept. 11/98. Can. News Wire.
– *Gazette,* Jan. 8, 9/98.
– CNN, Mar. 13, Apr. 9/98.
– NCDC, Jul. 14/98, "Climate of 1998."
– NCDC, Oct. 14/98, "Climate of 1998 through September."
– Bove, M., Fla. State Univ.

Chapter One
[no references]

Chapter Two
– Env. Can. "Worst Ice Storm."
– *Global Change,* Nov./95, Sep./95.
– NCDC, Aug./95, "January & March 1995: A California Cloudburst."
– NCDC, June 17/97, "Billion Dollar U.S. Weather Disasters."

– Phillips, D., "Summer '95: One For The Record." Env. Can.

– Hengeveld, H.G., Aug./96, "Climate Change & the Frequency & Intensity of Extreme Weather Events in Canada," Env. Can.

– GCCIP newsletter, Jan./96, Nov./97.

– Env. Can., "The Science of Climate Change."

– *"Climate Variability, Atmospheric Change & Human Health," Nov./96, Env. Can.

– NCDC, Mar./96, "1995 Atlantic Tropical Storms."

– Nat. Hurricane Center, Dec. 31/96, NOAA.

– iSKI News, 1997, "Weather Cancels World Cups Around the World."

– WMO/WCP, 1997, "Statement on the Status of the Global Climate in 1996."

– NCDC, May/96, "Winter of 95–96: A Season of Extremes."

– *Lacroix, J., "Climate Variability, Atmospheric Change & Human Health," Nov./96., Env. Can.

– Munich Re., Dec. 23/96, "1996 Another Year of Natural Catastrophes."

– Gray, W.M, et. al, Jun. 6/97, "Forecast of Atlantic Seasonal Hurricane Activity for 1997," Colorado State Univ.

– UCAR, May 19/97, "New Findings Blame Jump in Hurricane Toll on Coastal Growth, Not Climate Change."

– White, R. and D.A. Etkin, 1997. *J. Nat. Haz.* 16:135–163.

– NCDC, May/97, "1996 Atlantic Tropical Storms."

– CNN, Apr. 6/98.

– CCS: Summary for Policymakers. 1997. Env. Can.

– NCDC, Jan./97, "Winter of 96–97: West Coast Flooding."

– *Wash. Post,* Aug. 16/97, Jan. 5/98.

– CBC *National,* Jan. 12/98.

– *The Economist,* Oct. 14/97.

– NCDC, 1997, "El Niño-related floods hit South America."

– WMO, 1998, "El Niño contributed to record global warmth in 1997."

– Col. State Univ., Jun. 6/97, "Above-Average Hurricane Season Forecast Stands."

– Col. State Univ., Nov. 26/97, "Strongest El Niño in History Dampens 97 Hurricane Season."

– Bove, M.C., Florida State Univ., pers. comm.

– NOAA, Nov. 11/97, "The 1997 Hurricane Season."

– CANOE/CP, Apr. 1/98.

– NCDC, Mar. 9/98, "January & February 1998 Warmest & Wettest on Record."

– NCDC, Jul. 14/98, "Climate of 1998: Recent Climate Extremes."

– NCDC, Oct. 14/98, "Climate of 1998 through September."

– *Commission for Environmental Cooperation. 1999. "On Track? Sustainability and the State of the North American Environment. Part B." Montreal: CEC.

– NCDC, "The Great Flood of 1993: Post-flood Report."

– NCDC, May/93, "The Big One! A Review of the March 12–14, 1993 'Storm of the Century'."

– UCAR, "Summary of Roger Pielke Jr.'s Flood Report."

– *Hengeveld, H.G., "Climate Change & Extreme Events," Env. Can.

– *Insight Can. Res., Nov./96, "A National Public Opinion Survey of Current Environmental Issues," Env. Can.

– *CBC *National*, Nov./98.

– Karl, T., "Global Warming and Changes in Flood Potential," NCDC.

Chapter Three

– *Harvey, D.L., 1996. *Clim. Chg.* 34:1–71.

– GCCIP, Fact Series. Manchester Met. Univ.

– IPCC, 1996. 2nd Assessment Synthesis of Scientific-Technical Information relevant to interpreting Article 2 of the UNFCCC. UNEP

– *Climate Change 1995. 1996. IPCC 2nd Asmnt. Rept. Vol. 1 & 2, Cambr. Univ. Pr.

– *Hengeveld, H.G., 1997, "Recent policy related questions about the science of climate change," Env. Can.

– UNFCCC, 1992.

– USGCRP National Assessment Missions & Goals, U.S. Gov.

– FAPCC, 1995, Gov. Can.

– Primer on Climate Change. Env. Can.

– *San Fran. Chron.*, Jul. 11/96: A1.

– *San. Fran. Ex.*, Mar. 2/95: A18.

– Crowley, T.J., *Consequences*, 2:1.

– Meehl, G.A., Jul./97, Colloquium on El Niño–Southern Oscillation, NCAR.

– UCAR, Oct. 31/97, "Plant Growth Surges 1–3 Years After Global Temperature Spikes."

– Braswell, B.H., et. al., 1997, *Science* 278:870–872.

– Toman, M.A., et al., "Climate Change & Its Consequences," *Resources* 124.

– McBean, G., Jun./95, "Status of the Global Warming Hypothesis," WMO 12th Cong.

– CNN, Jul. 6, Sept. 9/97, Jan. 8/98.

– AP, Nov. 27, Dec. 3/97.

– AGCI, "Coral Bleaching."

– Epstein, P., "Saving Scarce Public Health Resources and Saving Lives: Health Sector Applications of Climate Forecasting," *ENSO Signal,* NOAA.

– Trenberth, K.E., & T.J. Hoar, *Geophy. Res. Lett.*, 23:1, 57–60.

– CCS, 1997. Highlights. Env. Can.

– WMO/WCP, 1997, "Statement on the Status of the Global Climate in 1996."

– Mann, M.E., et. al. 1998. *Nature* 392:779.

– NCDC, Aug. 10/98, "Climate of 1998."

– NCDC, Oct. 14/98, "Climate of 1998 through September."

– Trenberth, K.E., 1997. *naturalSCIENCE,* 1:9.

– *Jacoby, G.C., et al, 1996. *Science* 273:771–773.

– Quayle, R.G. & T.R. Karl, 1996, "The State of the Climate 1996," NOAA.

– Broecker, W.S., "The coming warmup," *21st C,* 1:3.

– NCAR, Mar./97, "When Models and Satellites Mislead."

– SEPP, Jan./98, "1997 Registers on the Cool Side, According to Satellite Global Temperature Data."

– Wentz, F J & M. Schabel, 1998. *Nature* 394:661.

– Hecht, J., Aug. 15/98, "The heat is on," *New Scientist.*

– *Hansen, J.E., et al, 1998. *Science* 281:930–932.

– Spencer, Roy, "Even with Needed Corrections, Data Still Don't Show the Expected Signature of Global Warming," NASA, Aug. 14/98.

– Spencer, Roy. *Washington Times,* Sept. 3/98.

– Easterling, D.R., et. al., 1997. *Science* 277: 364–367.

– Karl, T.R., "Global Warming and Weather Extremes," NCDC.

– *McBean, G., Nov. 3/97, *Globe & Mail.*

– MBIS Backgrounder, 1997. Env. Can.

Chapter Four

– Primer on Climate Change. Env. Can.

– Colloquium on El Niño-Southern Oscillation, 1997. NCAR.

– *Trenberth, K.E., & T.J. Hoar, *Geophy. Res. Lett.* 23:1, 57–60.

– CNN, Jan. 8, Apr. 6, Jun. 11/98.

– Trenberth, K.E., 1997. Colloquium on El Niño-Southern Oscillation. NCAR.

– GCCIP, Jan./95, Sep./97.

– Bove, M.C., "Impacts Of ENSO On United States Tornadic Activity," Fla. State Univ.

– Lambert, S., Env. Can., pers. comm.

– Univ. of Mich., Oct. 24/97, "U-M Researchers Find Links Between El Niño Cycle And Weather In Great Lakes Region."

– CBC, Jan. 14/98.

– NOAA, Jun. 17/97, "El Niño plays prominent role in weather patterns for coming seasons."

– NASA/JPL, Sep. 23/98, "Satellite shows Pacific running hot and cold."

– *Climate Change, Cross-Canada Briefings Before Kyoto, 1997. CGCP.

– *Workshop on Extreme Weather & Climate Change, 1997. Env. Can.

– Trenberth, K.E., NCAR, pers. comm.

– *Sun, D–Z, 1997. *Geophy. Res. Lett.* 24:16, 2031–2034.

– Hengeveld, H.G., Aug./96, "Climate Change & the Frequency & Intensity of Extreme Weather Events in Canada," Env. Can.

– *Tor. Star*, May 8/98.

– Hazards Research & Applications Workshop, 1996. Nat. Haz. Cent.

– Suchocki, C., 1998. FEMA., pers. comm.

– NCDC, June 17/97, "Billion Dollar U.S. Weather Disasters."

– Golden, J., 1997, "Tornadoes," NOAA.

– *Commission for Environmental Cooperation. 1999. "On Track? Sustainability and the State of the North American Environment." Part B. Montreal: CEC.

– IBC, Sep. 11/98, "Ice Storm Insurance Claims Boost Economy," Can. News Wire.

– CCS, 1997. Highlights. Env. Can.

– Munich Re., 1996, "Natural catastrophe losses will continue to increase."

– Sigma, Apr./98. "Exceptionally few high losses from natural & man-made catastrophes in 1997," Swiss Re.

– *Nature*, July 30/98.

– Munich Re., 1997, "Urbanization worldwide creates extreme risks."

– Pielke, Jr., R.A. & C.W. Landsea, 1997, "Normalized Hurricane Damages in the United States: 1925–1995," NCAR.

– Bruce, J.P., 1994. Wkshp on Improving Responses to Atmospheric Extremes.

– *Bruce, J.P., 1996, "Disaster Frequency & Economic Impacts," Pan-Pacific Haz. Conf.

– Munich Re., Dec. 23/96, "1996 Another Year of Natural Catastrophes."

– Insurance Industry Initiative for the Environment, 1997. UNEP.

– IBC, Oct. 9/97, "IBC & American Association to Cooperate on Natural Disaster Loss Prevention."

– *Global Change*, May 15/95. "Extreme Weather Exacts Rising Costs from Insurers."

– Sigma, Nov./98. "Too little reinsurance of natural disasters in many markets." Swiss Re.

– *Bertz, G. 1997. Wkshp Indices & Indicators for Clim. Ext.

– *White, R. and D.A. Etkin, 1997. *J. Nat. Haz.* 16: 135–163.

– *Parker, D.E., et. al., 1997. Wkshp Indices & Indicators for Clim. Ext.

– *WCR*, "Is Our Climate Becoming More Variable?"

– Plummer, N. et. al., 1997. Wkshp Indices & Indicators for Clim. Ext.

– IPCC Report: Summary for Policymakers, 1996. UNEP.

– IPCC, 1996. 2nd Assessment Synthesis of Scientific-Technical Information relevant to interpreting Article 2 of the UNFCCC. UNEP.

– Karl, T.R., "Global Warming and Weather Extremes," NCDC.

– *Karl, T. et. al., 1996. *Bull. Am. Met. Soc.* 77–2: 279–291.

– *Climate Change 1995. 1996. IPCC 2nd Asmnt. Rept. Vol. 1. Cambr. Univ. Pr.

– *Climate Variability, Atmospheric Change & Human Health, Nov./96. Env. Can.

– *Rogers, J.C., 1997. Wkshp Indices & Indicators for Clim. Ext.

– *Easterling, D.R., et. al., 1997. *Science.* 277:364–367

– *Jones, P.D. et. al., 1997. Wkshp Indices & Indicators for Clim. Ext.

– Trenberth, K.E., 1997. *naturalSCIENCE*, 1:9.

– Quayle, R.G. & T.R. Karl, 1996, "The State of the Climate 1996," NOAA.

– *Mahlman, J.D., 1997. *Science* 278:1416–1417.

– *Groisman, P.Y. et. al., 1997. Wkshp Indices & Indicators for Clim. Ext.

– *Trenberth, K.E., 1997. Wkshp Indices & Indicators for Clim. Ext.

– *Karl, T.R. & R.W. Knight, 1997. Wkshp Indices & Indicators for Clim. Ext.

– NCDC, Apr. 15, Jul. 14/98, "Climate of 1998."

– *Ramasamy, S., et. al., 1997. Wkshp Indices & Indicators for Clim. Ext.

– *Whetton, P.H., et. al., 1993. *Clim. Chg.* 25:289–317.

– *Harvey, D.L., 1996. *Clim. Chg.* 34:1–71.

– Env. Can. 1996, "Water and Climate."

– Phillips, D., 1996, "Summer '95 — One For the Record," Env. Can.

– *Chagnon, D. 1997. Wkshp Indices & Indicators for Clim. Ext.

– *Angel, J.R., & S.A. Isard, 1998. *J. Clim.* 11:26–36

– *Harries, J. 1996. "An Insurance Industry Perspective on Climate Change & Extreme Events."

– Knutson, T.R., et. al., 1998. *Science* 279:1018–1020.

– Landsea, C.W., 1997. Wkshp Indices & Indicators for Clim. Ext.

– Landsea, C.W., 1996. *Geophy. Res. Lett.* 23:13, 1697–1700.

– NCDC, Mar/96, "1995 Atlantic Tropical Storms."

– NCDC, May/97, "1996 Atlantic Tropical Storms."

– NOAA, Nov. 11/97, "The 1997 Hurricane Season."

– NCDC, Oct. 14/98, "Climate of 1998 through September."

– NHC, Nov. 1/98, "Monthly Tropical Weather Summary."

– CBC Newsworld, Nov. 5/98.

– Gray, W.M, et. al., Jun. 6/97, "Forecast of Atlantic Seasonal Hurricane Activity for 1997," Colorado State Univ.

– *Hengeveld, H.G., 1997. "Recent policy related questions about the science of climate change." Env. Can.

– Env. Can., 1997. "Climate Change report for the Atlantic provinces released."

– CCS, Vol 1. *Ex. Summ.* 1997. Env. Can.

– Titus, J.G. 1996, "Probabilities in Sea Level Rise Projections," AGCI.

– *Scheraga, J.D., 1997. Adapting to Climatic Change & Variability in the Great Lakes–St. Lawrence Basin, Proceedings. Env. Can.

– Titus, J.G., "The probability of Sea Level Rise," EPA.

– Leatherman, S.P. in *Globe & Mail,* Sep. 19/95.

– CBC, Dec. 9/97.

– McBean, G., Jun./95, "Status of the Global Warming Hypothesis," WMO 12th Cong.

– Ecologists' Statement on the Consequences of Rapid Climatic Change, May 21/97.

– GCCIP Fact Series, Manchester. Met. Univ.

– Columbia Univ., May 21/96, "Water Vapor Seen as a Cause of Rapid Climate Change."

– Crowley, T.J., 1996, *Consequences* 2:1.

– Columbia Univ., Dec. 11/97, "Columbia Scientist Finds Abrupt Changes In African Climate."

– Broecker, W.S., "The coming warmup,"*21st C,* Vol. 1, No. 3.

– Columbia Univ., May 29/97, "Human Activity Could Cause Dramatic Climate Change."

– Univ. of Col., Nov. 12/97, "Climate Switch 118,000 Years Ago May Hold Clues to Earth's Future."

– *Wash. Post.,* Dec. 1/97.

– *NYT,* Sep. 9/97.

Chapter Five

– IPCC Report: Summary for Policymakers. 1996. UNEP.

– *Climate Change 1995. 1996. IPCC 2nd Assessment Rept. Vol. 1. Cambr. Univ. Pr.

– Trenberth, K.E., 1997, *naturalSCIENCE,* 1:9.

– Primer on Climate Change, Env. Can.

– USGCRP, Forum on Scientific Foundation for Climate Change and Impacts Studies, 1995. *Consequences* 1:3.

– UNFCCC, 1992.

– *Harvey, D.L., 1996. *Clim. Chg.* 34:1–71.

– Canada's National Environmental Indicators., 1996. Env. Can.

– *O'Meara, M., "The Risks of Disrupting Climate," *Worldwatch,* Nov.–Dec./97.

– GCCIP, Jul./97.

– Ramaswamy et. al. 1992. *Nature* 355:810–812.

– Molnar, K.O., et al. 1994. *J. Geophy. Res.* 99:25755–25760.

– Goddard Space Flight Center, Apr. 8/98, "Increasing Greenhouse

Gases May Be Worsening Arctic Ozone Depletion."
– Eaton, R.J., Jul. 17/97. *Wash. Post.*
– Browne, J., May 19/97, Stanford Univ.
– McBean, G., Jun./95, "Status of the Global Warming Hypothesis," WMO 12th Cong.
– *Changes,* "What is Climate Change?", CGCP.
– WI, 1997, "Vital Signs 1997."
– WI, 1996, "Vital Signs 1996."
– GCCIP, Fact Series. Manchester Met. Univ.
– EDF, 1995, "Global Warming: History of an International Scientific Consensus."
– *McBean, G., Nov. 3/97. *Globe & Mail.*
– Lean, J. & D. Rind, 1996, *Consequences* 2:1.
– *NYT,* Sep. 23/97.
– AP, Nov. 13, Dec. 5/97.
– *Hengeveld, H.G. "Climate Change & Extreme Events," Env. Can.
– Quayle, R.G. & T.R. Karl, 1996, "The State of the Climate 1996," NOAA.
– CNN, Dec. 1/97, Jan. 8/98.
– *Workshop on Extreme Weather & Climate Change, 1997. Env. Can.
– *Scheraga, J.D., 1997. Adapting to Climatic Change & Variability in the Great Lakes–St. Lawrence Basin, Proceedings. Env. Can.
– *Sarmiento, J.L., et al, 1998. *Nature* 393:245.
– Robinson, A.B. et al., 1998. *OISM.*
– *Braswell, B.H., et. al., 1997. *Science* 278:870–872.
– NCAR, Oct. 31/97, "Plant Growth Surges 1–3 Years After Global Temperature Spikes."
– Cao, M. & F.I. Woodward, 1998. *Nature* 393:249.
– *Wedin, D.A., & D. Tilman, 1996. *Science* 274:1720–1723.
– *Goulden, M.L., et al., 1998. *Science* 279:214–217.
– *WCR,* Vol. 2. No. 7.
– *WCR,* Sponsor Info.
– *Mahlman, J.D., 1997. *Science* 278:1416–1417.
– Hansen, J.E. & A.A. Lacis. 1990. *Nature* 346:713–719.
– *Wash. Post.*, Nov. 12/97.

Chapter Six

– UNFCCC, 1992.

– IPCC Report: Summary for Policymakers. 1996. UNEP.

– *Climate Change 1995. 1996. IPCC 2nd Assessment Rept., Vol. 2., Cambr. Univ. Pr.

– Burton, I., 1996. Wkshp on the Canada Country Study. Env. Can.

– *Hengeveld, H.G., "Scientific Tools to Help the Policy Debate on Greenhouse Gas Mitigation."

– Hengeveld, H.G., Aug./96, "Climate Change & the Frequency & Intensity of Extreme Weather Events in Canada," Env. Can.

– Ecologists' Statement on the Consequences of Rapid Climatic Change, May 21/97.

– CCS, 1997. Highlights. Env. Can.

– CCS, 1997. Vol. I – V. Env. Can.

– USGCRP, "National Assessment of the Potential Consequences of Climate Variability and Change."

– *Adapting to Climatic Change and Variability in the Great Lakes– St. Lawrence Basin, Proceedings. 1997. Env. Can.

– Mackenzie Basin Impact Study, 1997. Env. Can.

– *Climate Variability, Atmospheric Change & Human Health, Nov./96. Env. Can.

– Gleick, P.H. Nov./96, "State of the World's Water and the Implications for the Western United States," *Glob. Chg.*

– WI, Sept. 14/96, "Water Conflicts Looms as Supplies Tighten; Food Security Threatened; Ecosystems in Decline."

– WI, Jun. 3/97, "Governments Failing to Protect Societies from Spreading Water Scarcity."

– Env. Can. 1996, "Water and Climate."

– GLSLB Progress Report #1, 1996. Env. Can.

– RFF, Jun. 5/97, "Threats to Water Resources From Climate Change Addressed in RFF Issues Brief."

– EPA, Dec./93, "The Colorado River Basin and Climatic Change."

– *GCCIP*, Fact Series, Manchester Met. Univ.

– WHO, 1990, "Potential health effects of climate change."

– Homer-Dixon, T., Nov. 9/95, *Globe & Mail.*

– *Globe & Mail,* Apr. 27, May 11, 22/98.

– Etkin, D., Mar./98, "Climate Change Impacts on Permafrost Engineering Design," Env. Can.

– CNN, Oct. 8/97, Feb. 12, Apr. 6/98.

– McMurtrie, R., et al. 1994, CSIRO.

– "Sustaining Canada's Forests," Canada's National Environmental Indicators, 1996. Env. Can.

– Ontario Round Table on Environment & Economy, Nov/95, "Climate Change Impacts: An Ontario Perspective."

– EPA, 1996, "Economic Effects of Climate Change on U.S. Forests."

– Welch, D.W., et al., 1998. *Can. J. Fish. Aquat. Sci.* 55:937–948.

– Patz, J.A., "Climate Change & Health: Need for Expanded Scope of Occupational & Environmental Medicine." EPA.

– Rosenzweig, C. & D. Hillel., 1995, *Consequences* 1:2.

– *Street, R., et al. 1997. Wkshp on the Social & Economic Impacts of Weather. NCAR/UCAR/NSF.

– *Global Change*, Sep./95.

– *Keener, Jr., R.N., 1997. Wkshp on the Social & Economic Impacts of Weather. NCAR/UCAR/NSF.

– *Smith, M.R., 1997. Wkshp on the Social & Economic Impacts of Weather. NCAR/UCAR/NSF.

Chapter Seven

– Environics International Ltd., Jun. 23/97. "Global Poll Reveals Surging Health Concerns and Sharp Criticism of Inaction on Environmental Problems."

– *Climate Change 1995. 1996. IPCC 2nd Asmnt. Rept. Vol. 2. Cambr. Univ. Pr.

– WHO, 1990, "Potential health effects of climate change."

– Duncan, K., 1996. GLSLB Progress Report #1. Env. Can.

– *Global Change*, Nov./95, Jun./96.

– NIEHS, Mar. 9/98, "Global Warming Would Foster Spread Of Dengue Fever Into Some Temperate Regions."

– *Climate Variability, Atmospheric Change & Human Health, Nov./96. Env. Can.

– Hales, S. et al., 1996. *Lancet* 348:9042.

– *JAMA*, Jan. 17/96. "Global Warming Threatens World Health,"; "Deaths from Infectious Diseases on the Rise in U.S."

– Longstreth, J. 1991. *Env. Hlth. Persp.* 96:139–144.

– *Scheraga, J.D., 1997. Adapting to Climatic Change & Variability in the Great Lakes–St. Lawrence Basin, Proceedings. Env. Can.

– Epstein, P., "Saving Scarce Public Health Resources and Saving Lives: Health Sector Applications of Climate Forecasting," *ENSO Signal,* NOAA.

– Southam News, Jan. 11/97.

– Taubes, G., Nov. 7, 1997. *Science* 278:1004–1006.

– Colwell, R.R., et al., Feb. 13, 1998. Letts to *Science.* 279:963.

– *WCR,* 2:10, 2:20, 1:21.

– Hazards Research & Applications Workshop, 1996., Nat. Haz. Cent.

– Platt, A.E., 1995, "Infecting Ourselves: How Environmental and Social Disruptions Trigger Disease," Paper 129. WI.

– Haines, A. et al., 1993. *Lancet* 342:1464–69.

– NAS, 1995. Conference on Human Health & Global Climate Change, Proceedings.

– White, R. and D.A. Etkin, 1997. *J. Nat. Haz.* 16:135–163.

– Guest, C., et al. 1995, "The impact of greenhouse warming on extremes of temperature & human mortality in Australia," CSIRO.

– Tavares, D., 1996. GLSLB Progress Report #1. Env. Can.

– *Globe & Mail,* May 30/98.

– WRI, Nov. 7/97. "Greenhouse Gas Emissions Endanger the Public's Health Today."

– Semenza, J.C., 1996. *NEJM* 335:2, 84–90.

– Semenza, J.C., 1996. *NEJM* Letters 335:24.

– *Kalkstein, L.S., 1990. *Env. Impact Assessment Rev.* 10:383

– *Easterling, D.R., et. al., 1997. *Science.* 277:364–367.

– *Kalkstein, L.S, et al., 1986. in Vol. 3. Effects of changes in stratospheric ozone & global climate, ed. J. Titus, 273–93. UNEP/EPA.

– Karl, T.R., & R.W. Knight, 1996, "The 1995 Chicago Heat Wave," NCDC.

– Rippel, B., 1997, "Remarks on President Clinton's speech to the UN Environmental Conference, June 26, 1997," *Consumer Alert.*

– Kellerman, A.L. & K.H. Todd, 1996, "Killing Heat," *NEJM* 335:2.

– Jones, D.S., 1996. *NEJM* Letters 335:24

– Bruce, J.P., 1996, "Disaster Frequency & Economic Impacts," Pan-Pacific Haz. Conf.

– CBC, Mar. 20/98, Jul. 27/98.

– NCDC, "Climate Change & Human Health."

Chapter Eight

- *Climate Change 1995. 1996. IPCC 2nd Assessment Rept. Vol. 2. Cambr. Univ. Pr.
- Am. Red Cross, "Emotional Health Issues for Disaster Workers."
- *Tor. Star*, Jan. 14/98.
- CBC, Jan. 13/98.
- CNN, Jan. 16/98.
- Enarson, E., "Surviving Domestic Violence & Disasters," FREDA Centre for Research on Violence against Women & Children.
- Parrish, G., 1997. Wkshp on the Social & Economic Impacts of Weather. NCAR/UCAR/NSF.
- Hazards Research & Applications Workshop, 1996., Nat. Haz. Cent.
- CBC/CBS, Mar. 9/98.
- Enarson, E. & B. Morrow, "Hurricane Andrew Through Women's Eyes," Fla. Int. Univ.
- L.A. County Dept. Mental Health, "Steps you can take to cope in stressful situations."
- N. Dak. St. Univ., 1997. "Dealing with Stress after a Disaster," "Coping with Floods," "When Crisis Becomes Chronic."
- Hart, D.G., "Childhood Trauma: Tips for Caregivers," EPC.
- Krug, E.G. et al., 1998. *NEJM* 338:6.
- *Gazette*, Jan. 17/98
- Munich Re., 1997, "Urbanization worldwide creates extreme risks."
- WHO, 1990, "Potential health effects of climate change."
- White, R. and D.A. Etkin, 1997. *J. Nat. Haz.* 16:135–163.
- *Changes:* Iss. 5, 1997, "Environmental Change & Human Security," CGCP.
- Homer-Dixon, T., 1991. *Int. Sec.,* 16:2, 76–116.
- Primer on Climate Change, Env. Can.
- CCS: Summary for Policymakers. 1997. Env. Can.
- WI, Oct. 26/96, "New Threats to Human Security."

Chapter Nine

- Pielke, Jr., R.A, Aug. 15/97, "Two Faces of Mitigation," NCAR.
- *Climate Change 1995. 1996. IPCC 2nd Assessment Rept. Vol. 1 & 2. Cambr. Univ. Pr.
- IPCC Report: Summary for Policymakers. 1996. UNEP.
- FCCC Press Backgrounder, 1997. UNFCCC Secretariat.

– World Bank, "The World Bank & Climate Change: Issues & Opportunities."
– Canada's Second National Report on Climate Change. 1997. Gov. Can.
– EPA, Oct. 1/97, "Industry sees damage from limits."
– O'Keefe, W. F., "Spirit of Lewis & Clark Needed For Making Decisions on Climate Change." *Weathervane*, RFF.
– Moore, T.G., Apr. 25/97, "Global Chill," Barron's Online.
– AP, Sep. 7/97, Mar. 4/98.
– *WSJ*, Jul. 25/97.
– *Wash. Post*, Sep. 26, Dec. 1, 18/97.
– *Globe & Mail*, Oct. 10/95, Oct. 24, 31, Nov. 5, 28/97, Jul. 10/98.
– AFL-CIO, Feb. 20/97. Statement on U.N. Climate Change Negotiations.
– Reuters, Oct. 9/97.
– *Harvey, D.L., 1996. *Clim. Chg.* 34:1–71.
– *Globe & Mail/Economist*, Apr. 3/95.
– Stats. Can., May 12/98, "Total Income of Individuals."
– API, Oct. 22/97, "American Petroleum Institute on Global Climate Change."
– Adler, J.H., 1997, "Hurricane Hype," CEI.
– Repetto, R., & D. Austin, 1997, "The Costs of Climate Protection: A Guide for the Perplexed." WRI.
– DOE, Sep. 25/97, "Clean Technology Can Achieve Significant Greenhouse Gas Reductions."
– *Changes.* 1997, "Actions Yield Additional Benefits for Canada," Endnote 20, CGCP.
– GCCIP, Jul./97.
– Economists' Statement on Climate Change, 1997. *Redefining Progress.*
– NRDC, Jun. 17/98. "Energy Innovations: A Prosperous Path to a Clean Environment."
– Adams, J.H., 1997, "The True Cost of Curbing Global Warming," NRDC.
– *Flavin, C., 1997, "Kyoto: The Days of Reckoning," *Worldwatch.*
– *Nature*, Oct. 2/97, "No net cost in cutting carbon emissions."
– CGCP, "Global Change & Canadians."
– *Tor. Star*, May 13/98.
– NAPCC. 1995. Gov. of Canada
– *Ayers, E., Nov–Dec/97. "Flying Blind," *Worldwatch.*

– Eaton, R.J., Jul. 17/97. *Wash. Post.*
– Lashof, D. & Geller, H., Jul. 31/97. *Wash. Post.*
– *LA Times,* Aug. 13/97.
– *Worldwatch,* 1996, "A Billion Cars" WI.
– WI, 1997. "Vital Signs 1997."
– WI, 1998. "Vital Signs 1998."
– WI, Jan. 10/98. "Building a New Economy."
– CNN, Nov. 6/98.
– WI, Jul. 16/98. "Solar Power Market Booms."
– Browne, J., May 19/97, Stanford Univ.
– *Global Change,* Dec./95.
– *Climate Variability, Atmospheric Change & Human Health, Nov./96. Env. Can.
– Woods Hole Res. Cen., "Using Forests to Sequester Carbon."
– Parry, I.W., 1997, "Interactions with the Tax System Can Raise the Cost," RFF.
– *New Scientist,* Sept. 5/98, "There's no avoiding a carbon tax."
– Kyoto Protocol to the UNFCCC. 1997. Un. Nations.
– Primer on Climate Change, Env. Can.
– Stewart, Hon. C., 1998, "Opening remarks to Can-US greenhouse gas emissions trading forum."
– Nat. Res. Can., Mar. 5/98, "Goodale Applauds Emissions Trading Initiative by Suncor."
– Wiener, J.B. 1997. *Resources* 129. RFF.
– CBC *Newsworld,* Dec. 10/98.
– *Local Government Implementation of Climate Protection, 1997, ICLEI.
– Un. Nations. 1991. World urbanization prospects, 1990.
– Env. Can., 1997, "The State of Canada's Environment — 1996."

Chapter Ten

– EDF, 1995, "Global Warming: History of an International Scientific Consensus."
– IPCC Report: Summary for Policymakers. 1996. UNEP.
– FCCC Press Backgrounder, 1997. UNFCCC Secretariat.
– *Time,* Mar. 14/94, Jul. 7/97.
– UNFCCC, 1992.
– WEC *Journal,* Jul./97, "Carbon dioxide emissions 1990–1996."

– *Changes,* CGCP.
– WWI, 1997, "Vital Signs 1997."
– *U.N. Clim. Chg. Bulletin,* Issue 6, 1995.
– *Economist/Globe & Mail,* Apr. 3/95.
– GEF, Apr. 30/98, "Press Briefing by Chairman of Global Environment Facility."
– *NYT Serv.,* Apr. 12, Sep. 11/95, Nov. 15/98.
– Dyer, G., Apr 17/95 in *Vic. Times Col.*
– World Bank, "The World Bank & Climate Change: Issues & Opportunties."
– *Maclean's,* Apr. 24/95.
– *Nature,* 1997, "High Noon at Kyoto."
– WEC, Apr. 24/96, "World Energy Council reviews the 2nd Assessment Report of the IPCC & finds no real progress."
– *Nature,* Dec. 7 /95, "Climate panel confirms human role in warming, fights off oil states."
– Seitz, F., Jun. 12/96, *WSJ.*
– *San Fran. Exam.,* Dec. 1/95.
– Wamsted, D., May 22/96. *Energy Daily.* Letters, June 3, 20/96.
– Kerr, R.A., May 16/97. *Science.*
– *Insurance Industry Initiative for the Environment, 1997. UNEP.
– *Statement of Environmental Commitment by the Insurance Industry. 1996. UNEP.
– *UNEP Insurance Initiative: Position Paper on Climate Change, 1996. UNEP.
– *Wash. Post.,* Oct. 24, Dec. 1, 4, 7, 13/97, Nov. 2/98.
– Browne, J., May 19/97, *Stanford Univ.*
– WI, 1997, "Vital Signs 1997."
– WEC *Journal,* Jul./97, "Carbon dioxide emissions 1990–1996."
– Lashof, D. & Geller, H., Jul. 31/97. *Wash. Post.*
– GCCIP, Jul./97.
– *Changes,* "What is Climate Change?", CGCP.
– NAPCC. 1995. Gov. of Canada.
– *San Fran. Chron.* Jun. 20/96: A1.
– WEC, 1997, "Energy Complacency Threatens Sustainability."
– *Global Climate Information Proj. 1997, 1998. Advertising on CNN.
– Eaton, R.J., Jul. 17/97. *Wash. Post.*

– AFL–CIO, Feb. 20/97. Statement on U.N. Climate Change Negotiations.

– Scientists' Statement on Global Climate Disruption, Jun. 18/97. *Ozone Action.*

– *Aust. Fin. Rev.,* Sep. 17/97.

– O'Sullivan, P., Sep. 2/97, "The Case For Differentiation: Australia's Position on Climate Change," *RFF.*

– Reuters, Oct. 22, 23, 31, Dec. 15/97, Nov. 11, 14/98.

– AP, Oct. 22, 23, Dec. 9/97.

– GCC, Oct. 28/97, "Bonn Climate Negotiations Update."

– *Globe & Mail,* Oct. 11, 28, 31, Nov. 28/97.

– *Nature,* Nov. 6/97.

– *Ott. Cit.,* Nov. 13, 28/97.

– "Canada proposes targets for reductions in global greenhouse gas emissions," Dec. 1/97. Gov. Can.

– *Edm. Jrnl.,* Dec. 2, 12/97.

– WWF, 1996, "Looking for Loopholes."

– Un. Nat., Dec. 11/97, "Industrialized Countries to cut greenhouse gas emissions by 5.2%."

– CNN, Dec. 11, 14/97.

– *Tor. Star.,* Apr. 25, Sep. 11/98.

– Simpson, J., Dec. 18/97. *Globe & Mail.*

– United Nations, "Climate change meeting adopts Buenos Aires Plan of Action," Nov. 14/98.

– *Flavin, C., 1997. "Kyoto: The Days of Reckoning", *Worldwatch.*

Chapter Eleven

– *Mortsch, L. & B. Mills, Great Lakes–St. Lawrence Basin Project, Progress Report 1, 1996. Env. Can.

– Burton, I., 1996. Wkshp on the Canada Country Study. Env. Can.

– Pielke, Jr., R. A., 1997. Rethinking the Role of Adaptation in Climate Policy. *Gl. Env. Chg.,* NCAR.

– Homer-Dixon, T., 1991. *Int. Sec.,* 16:2, 76–116.

– Homer-Dixon, T., 1995. *Pop. & Dec. Rev.* 21:3, 587–612.

– *Economist/Globe & Mail,* Apr. 3/95.

– Thorsell, W., Aug. 29/98. *Globe & Mail.*

– *Scheraga, J.D., 1997. Adapting to Climatic Change & Variability in the Great Lakes–St. Lawrence Basin, Proceedings. Env. Can.

– IPCC: Summaries for policymakers. 1996. UNEP.

– *Commission for Environmental Cooperation. 1999. On Track? Sustainability and the State of the North American Environment. Part B. Montreal: CEC.

– CCS: Summary for Policymakers. 1997. Env. Can.

– Toman, M., "How to think about climate change," RFF.

– White, R. and D.A. Etkin, 1997. *J. Nat. Haz.* 16:135–163.

– NASA, "Congested Coastlines."

– FEMA, "How the National Flood Insurance Program Works."

– Witt, J. L., "Reducing Flood Losses." FEMA.

– *Wilm. Mrng. Star.*, Sep. 11/96.

– CNN, Apr. 6/98.

– *Tampa Trib.*, Sep. 25/97.

– Titus, J.G. 1996, "Probabilities in Sea Level Rise Projections," AGCI.

– *Climate Change 1995. 1996. IPCC 2nd Assessment Rept. Vol. 2, Cambr. Univ. Pr.

– Smit, B., 1994. Wkshp on Improving Responses to Atmospheric Extremes.

– Pielke, Jr., R.A, Aug. 15/97, "Two Faces of Mitigation," NCAR.

– CCS, 1997. Highlights. Env. Can.

– RFF, Jun. 5/97. "Threats to Water Resources from Climate Change Addressed in RFF Issues Brief."

– WI, Sep. 14/96. "Water Conflicts Loom as Supplies Tighten; Food Security Threatened; Ecosystems in Decline."

– Speirs, R., Jan. 10/98. *Tor. Star.*

– *Gazette*, Jan. 21, 23/98.

– *Tor. Star.*, Jan. 18/98.

– Ektin, D., Mar./98. Climate Change Impacts on Permafrost Engineering Design. Env. Can.

– ICLR, 1997. Business Plan.

– *Harries, J. 1996, "An Insurance Industry Perspective on Climate Change & Extreme Events."

– Golden, J., 1997, "Tornadoes," NOAA.

– Nat. Hurr. Cent., Dec. 31/95, "Final look at '95 hurricane season," NOAA.

– Kellerman, A.L. & K.H. Todd, 1996, "Killing Heat," *NEJM* 335:2.

– *Climate Change, Cross-Canada Briefings Before Kyoto, 1997. CGCP.

Chapter Twelve

– *Calmer Weather,* Aug. 28/97, "Anomalous Heat Wave," CEI.

– SEPP, "About the Project."

– SEPP, Jan/98, "1997 Registers on the Cool Side, According to Satellite Global Temperature Data."

– *WCR,* 2:17.

– *NYT,* Sep. 28/98.

– Wilson, J., & J.W. Anderson, 1997. *Resources* 128. RFF.

– *Statement of Environmental Commitment by the Insurance Industry. 1996. UNEP.

– UNEP, May 1/93, "Why three hot summers don't mean global warming."

– *Nature,* Aug. 21/97, "Ecologists urged to 'win climate debate'."

– Trenberth, K.E., 1997, *naturalSCIENCE,* 1:9.

– McBean, G., Jun./95, "Status of the Global Warming Hypothesis," WMO 12th Cong.

– Yulsman, T., *21st C., Letts,* Columbia Univ.

– *O'Meara, M., Nov-Dec 1997. "The Risks of Disrupting Climate" *Worldwatch.*

– *Ayers, E., Nov-Dec/97. "Flying Blind," *Worldwatch.*

– Ontario Round Table on Environment & Economy, Nov/95, "Climate Change Impacts: An Ontario Perspective."

– *Insight Can. Res., Nov/96, "A National Public Opinion Survey of Current Environmental Issues," Env. Can.

– Environics International Ltd., 1997. The Environmental Monitor survey of global public opinion on the environment.

– *Maclean's,* Dec. 30/96.

– Angus Reid Group, Feb. 16/98, "Public Split on Kyoto Environmental Accord."

– Simpson, J., Dec. 18/97, Feb. 10/98, *Globe & Mail.*

– Ladd, E.C. & K. Bowman, "Public Opinion & the Environment," *Resources* 124. RFF.

– EPA, Sep. 30/97, "WWF Poll Shows Voters Worried Now," *Greenwire.*

– AP, Oct. 1, Nov. 20/97.

– UPI, Dec. 17/97.

– *WSJ,* Aug. 12, Oct. 29/97.

– CNN, Oct. 6, 15/97.

– *NY Post,* Sep. 24/97.

– *Wash. Post,* Jul. 25, Sep. 18/97.

– *Time,* Jul. 7/97.

– Reuters, Oct. 1/97.

– *Ott. Cit.,* Sep. 2/97.

– Yankelovich, D., 1992. "Listenup," We the People, Congressional Institute Inc.

– Berk, R., 1994, "Public Perceptions of Global Warming," AGCI.

– Stats. Can., 1996, "Employed Labour Force by Sex, Showing Mode of Transportation to Work."

– *Globe & Mail,* Jun. 6/97.

– Univ. Tor., Feb. 23/98, "Drivers Should Pay More of Environmental Costs, Report Says."

– Dohring Co. Inc., Dec. 15/97, "Study Reveals Uneducated Car Buyers Could Mean Trouble for Global Warming Treaty."

– *Nature,* Jul. 31/97, "Clinton pulls out the stops in bid to win backing for carbon cuts."

– *New Yorker,* Jun. 9/86.

SELECTED
BIBLIOGRAPHY

- *Climate Change 1995, Second Assessment Report of the Intergovernmental Panel on Climate Change. 1996. Cambridge University Press.
- IPCC: Summaries for policymakers. 1996. UNEP
- U.N. Framework Convention on Climate Change, 1992. United Nations.
- FCCC Press Backgrounder, 1997. UNFCCC-Secretariat
- Kyoto Protocol to the UNFCCC. 1997. Un. Nations.
- *Canada Country Study, 1997. Environment Canada
- U.S. National Assessment of the Potential Consequences of Climate Variability and Change. USGCRP.
- *Workshop on Indices and Indicators for Climate Extremes, 1997. NOAA.
- *Workshop on the Social & Economic Impacts of Weather, 1997. NCAR/UCAR/NSF.
- *Climate Variability, Atmospheric Change & Human Health, Nov./96. Env. Can.
- *Adapting to Climatic Change and Variability in the Great Lakes– St. Lawrence Basin, Proceedings. 1997. Env. Can.
- *Climate Change, Cross-Canada Briefings Before Kyoto, 1997. Canadian Global Change Program.
- *Commission for Environmental Cooperation. 1999. "On Track? Sustainability and the State of the North American Environment." Part B. Montreal: CEC.
- *Harvey, D.L., 1996. Development of a Risk-Hedging CO_2-Emission Policy. *Climate Change* 34:1–71.

– *Karl, T. et. al., 1996. Indices of Climate Change for the United States. *Bulletin of the American Meteorological Society* 77–2:279–291.

– *Local Government Implementation of Climate Protection, 1997, ICLEI.

– *Mahlman, J.D., 1997. Uncertainties in Projections of Human-Caused Climate Warming. *Science* 278:1416–1417.

– *Mortsch, L. & B. Mills, Great Lakes–St. Lawrence Basin Project, Progress Report 1, 1996. Env. Can.

– *O'Meara, M., Nov–Dec 1997. "The Risks of Disrupting Climate" *Worldwatch.*

– Pielke, Jr., R. A., 1997. Rethinking the Role of Adaptation in Climate Policy. Global Environment Change, NCAR.

– *Pielke, Jr., R.A., 1997. Asking the Right Questions. *Bulletin of the American Meteorological Society* 78–2:255–264.

– *White, R. and D.A. Etkin, 1997. Climate Change, Extreme Events & the Canadian Insurance Industry. *Journal of Meteorological Hazards* 16:135–163.

– *World Climate Report,* The Greening Earth Society.

– Repetto, R., & D. Austin, 1997. The Costs of Climate Protection: A Guide for the Perplexed. WRI.

– National Action Plan on Climate Change. 1995. Government of Canada.

– Canada's Second National Report on Climate Change. 1997. Gov. of Canada.

– Primer on Climate Change, Env. Can.

– GCCIP Fact Series, Atmospheric Research and Information Centre, Manchester Metropolitan University.

INDEX